Colours, Commodities and the Birth of Globalization

Colours, Commodities and the Birth of Globalization

A History of the Natural Dyes of the Americas, 1500–2000

Edited by
Carlos Marichal and David Pretel

BLOOMSBURY ACADEMIC
LONDON • NEW YORK • OXFORD • NEW DELHI • SYDNEY

BLOOMSBURY ACADEMIC
Bloomsbury Publishing Plc
50 Bedford Square, London, WC1B 3DP, UK
1385 Broadway, New York, NY 10018, USA
29 Earlsfort Terrace, Dublin 2, Ireland

BLOOMSBURY, BLOOMSBURY ACADEMIC and the Diana logo are trademarks of Bloomsbury Publishing Plc

First published in Great Britain 2024

A catalogue record for this book is available from the British Library.

A catalog record for this book is available from the Library of Congress.

ISBN: HB: 978-1-3504-0811-1
ePDF: 978-1-3504-0812-8
eBook: 978-1-3504-0813-5

Typeset by Deanta Global Publishing Services, Chennai, India

To find out more about our authors and books visit www.bloomsbury.com and sign up for our newsletters.

Contents

Figures

Tables

Contributors

José Jobson de Andrade Arruda is Emeritus Professor of the University of São Paulo (USP) and long-time member of the Department of History and the Graduate Program in Economic History at the Faculdade de Filosofia, Letras e Ciências Humanas (FFLCH) at the USP, as well as professor at the Instituto de Economia at the UNICAMP. He is author of *O Brasil na Revolução Industrial (1780-1830)*, Bauru/EDUSC, São Paulo, 2009, and *Uma colônia entre dois impérios: a abertura dos portos brasileiros 1800-1808*, Bauru/EDUSC, 2008, as well as of a great number of educational texts in modern and contemporary history with emphasis on economic history and historiography.

Dominique Cardon is Emeritus Research Director of History at CNRS, Paris. Her research areas include the evolution of textile production in Europe and the Mediterranean world between Antiquity and the end of the Middle Ages, according to written and archaeological sources, as well as the history of the use of natural dyes over the very long term, with a view to comparisons between different regions of the world. She is the author of several books on natural dyes, in French and English, one of which received the 'Art and Science of Colour Prize', Bronze Award, of the L'Oréal Foundation in 2003. Her many publications include *The Dyer's Handbook – Memoirs of an 18th Century Master Colourist* (2016); *Le Monde des Teintures Naturelles* (2014); and *Natural Dyes: Sources, Tradition, Technology and Science* (2007).

Adrianna Catena has been a Leverhulme Early Career Fellow (2017–20) and affiliate of the Global History and Culture Centre at the University of Warwick. Her postdoctoral project examines the impact of New World dyes – cochineal, indigo, logwood and other lesser known colourants – on the Spanish and broader European dyeing sectors. Among her articles and papers are 'The Hatters' Blues: A Microglobal History of New World Dyes in Early Modern Spain'; 'The Colour of Saint Francis: Reflections on the Visual Culture of Poverty in Colonial New Spain' (2016); 'Indigo in the Atlantic World' (2018); and 'From Manila to Annobon: Indigo Production and Trade in the Spanish Pacific (XVI-XVIII)' (2022).

Rosa Torras Conangla obtained a BA in History from the Universidad de Barcelona (1987), a MA in Central American History from the Universidad de Costa Rica (2005); and a PhD in Mesoamerican Studies from the UNAM (2010). She currently serves as a full-time research associate at the Centro Peninsular en Humanidades y Ciencias Sociales (CEPHCIS) and teaches in the bachelor's degree programme 'Desarrollo y Gestión Interculturales' at the same centre at the UNAM. Some of her papers include 'Fronteras en conflicto. El caso de la explotación del palo de tinte en Palizada' (2014);

'Palo de tinte y territorialidad en la península de Yucatan, en las postrimerías del periodo colonial novohispano' (2019); and 'Haciendas y ordenamiento territorial en la región de Los Ríos campechana' (2021).

Jorge Enrique Elías-Caro is Professor and Researcher at the Faculty of Business and Economic Sciences of the University of Magdalena (Colombia), Vice-Rector and editor of the Publishing House of the same university. He is currently vice president of the Association of Latin American and Caribbean Historians (ADHILAC). He has published over twenty books, dozens of articles and book chapters on the economic, social and cultural history of the Caribbean and Latin America. His books include *Los mil y un Caribe . . . 16 textos para su (Des) entendimiento* (2009) and *Ciudades portuarias en la Gran Cuenca del Caribe* (2010).

Huemac Escalona has a PhD in History from the Universidad Pablo de Olavide of Sevilla, Spain. He is a member of the Instituto de Investigaciones Históricas (IIH) at the UNAM and works in the Oaxaca Unit, in the area of Colonial History. His lines of research focus on the social history of New Spain and address the processes of production and marketing of natural dyes as well as the impact of mining and educational policies on Indian towns. His publications include 'Basura por mostaza. La pugna por el acaparamiento de la semilla de añil en el partido de Zacatecoluca, Alcaldía Mayor de San Salvador, Reino de Guatemala, 1781-1784'; 'Aproximación a la actividad minera en Oaxaca, siglos XVI al XVIII' and 'Viaje a la tierra del "púrpura de los reyes": Thiery de Menonville y su tratado sobre la grana cochinilla'.

Andrea Feeser received a BA in 1984 from Williams College with a double major in History and Art History. In 1996, she received a PhD in Modern and Contemporary Art History from the City University of New York Graduate Center. Andrea is Professor of Art and Architectural History in the Art Department at Clemson University. Her research into the histories of place and community with respect to intertwined cultures and environments has had widespread impact, redefining perceptions from the tourism industry in Hawaii to colonial indigo production and use in the British Atlantic World. She explores the latter in her book *Red, White, and Black Make Blue: Indigo in the Fabric of Colonial South Carolina Life* (2013).

Ronny Viales Hurtado is Professor at the School of History and the Postgraduate Program in History at the Universidad de Costa Rica, of which he has served as director for several periods. Additionally, he is a researcher at the Centro de Investigaciones Históricas de América Central, where he has coordinated the Programa de Investigación 'Ambiente, Ciencia, Tecnología y Sociedad'. His research focuses on the economic, social and environmental history and historiography of Costa Rica and Central America. Notable among his publications are *Después del enclave. Un estudio de la región atlántica costarricense. 1927-1950* (1998) and *Concepciones y representaciones de la Naturaleza y la Ciencia en América Latina* (2009).

Frédérique Langue is *directeur de recherche* at Centre Nationale de la Recherche Scientifique (CNRS) within the Institut d'Histoire du Temps Présent (IHTP). She received her doctorate in history at the Sorbonne (Paris 1) and HDR from EHESS. Her interests lie in the field of 'l'histoire du temps présent' and the Iberian worlds, with a particular focus on Venezuela. She was the founder, webmaster and chief editor of the e-journal *Mundo Nuevo / Nuevos Mundos* from 2001 to 2012. Notable publications include, with Luc Capdevila, *Le passé des émotions d'une histoire à vif, Amérique Latine et Espagne* (2014); and, with Salvador Bernabeu, *Fronteras y sensibilidades en las Américas* (2011). She also authored the book *Los señores de Zacatecas. Una aristocracia minera en el siglo XVIII novohispano* (1999).

Carlos Marichal is Emeritus Professor of Economic History at El Colegio de México, where he has taught for more than three decades. He has a PhD in History from Harvard University (1977) and has been visiting professor in many universities in Europe, the United States and Latin America. He has published some twenty books, as author and editor, including the much cited *A Century of Debt Crises in Latin America*, 1820-1930 (1989) and his book *Bankruptcy of Empire: Mexican Silver and the Wars between Spain, Britain and France, 1760-1810* (2007), which received several prizes, including the Alice Hanson Jones Prize by the Economic History Association. In 2012, he received the National Prize of Mexico in Sciences and Arts. In 2019, he was invited to become a member of the American Academy of Arts and Sciences. Website: https://carlosmarichal.colmex.mx/en/index-en.html.

Anthony Goebel McDermott earned his PhD in History from the Universidad de Costa Rica. He is a Professor at the School of History and a researcher at the Centro de Investigaciones Históricas de América Central (CIHAC) at that university. He is part of the research team of the programme titled 'Ambiente, Ciencia, Tecnología y Sociedad' (ACTS). Currently, he serves as the director of the Central American Postgraduate Course in History at the Universidad de Costa Rica. His research focuses on the history of science and economic history and his publications include the book *Los bosques del 'progreso'. Explotación forestal y régimen ambiental en Costa Rica: 1883-1955* (2013) and the essay 'Environmental History of Commodities in Central America' (2021).

Karl Offen is Professor of Geography and the Environment at Syracuse University. He was previously Chair of the Environmental Studies Program at Oberlin College from 2019 to 2021. His research includes methodological approaches from historical geography, the history of cartography, political ecology and environmental history, and he explores themes in Afro-Amerindian, colonial and Atlantic world studies. Among his publications are Karl Offen and Terry Rugeley, eds., *The Awakening Coast: An Anthology of Moravian Writings from Mosquitia and Eastern Nicaragua, 1849-1899* (2014) and Jordana Dym and Karl Offen, eds., *Mapping Latin America: A Cartographic Reader* (2011).

David Pretel is Professor of Economic History at the Universidad Autónoma de Madrid. He has been visiting scholar at the universities of Harvard, Bristol, Nottingham, UCLA and MPIWG. His first book, *Institutionalising Patents in Nineteenth-Century Spain* (2018), examined the development of the Spanish patent system (1826–1902), providing a fundamental reassessment of its evolution in an international and imperial context. He is co-editor of the volumes *The Caribbean and the Atlantic World Economy* (2015) and *Technology and Globalisation: Networks of Experts in World History* (2018). His recent publications include articles in the journals *History of Science*, *Technology and Culture* and *Global Environment*, as well as chapters in the *Oxford Handbook of Commodity History* and the *Cambridge History of Technology*. His works, CV, and other details can be found on his website: www.davidpretel.com.

Georges Roque is a philosopher, art historian and honorary researcher of the Centre National de la Recherche Scientifique (CNRS-EHESS) and Centre de Recherches sur les Arts et le Langage (CRAL), Paris. He holds a PhD in Philosophy (Université de Montréal, 1981) and a PhD in art history (EHESS, Paris, 1996). His research focuses on the theory of colour and the history of art. Notable publications include *La cochenille, de la teinture à la peinture. Une histoire matérielle de la couleur* (2022); *Art et science de la couleur. Chevreul et les peintres de Delacroix à l'abstraction* (2009); and *Rojo mexicano. La grana cochinilla en el arte* (2017).

Pascale Villegas is Research Professor of the Centro de Investigaciones Históricas y Sociales (CIHyS) at the Universidad Autónoma de Campeche (Mexico) since 2010. She holds a PhD in Latin American Studies from the University of Toulouse, France. Additionally, she is an active member of the Asociación Mexicana de Estudio del Caribe (AMEC) and the Asociación de Historia Económica del Caribe (AHEC). Her publications include works such as 'Del puerto de Marsella a las casas de Campeche. El comercio de las tejas de barro, 1852-1932', 'La creación del estanco de palo de tinte y sus consecuencias para la provincia de Yucatan' and 'La extracción y exportación de palo de tinte a manos de colonos extranjeros. El caso de la B. Anizan y Cía'.

1

Introduction

The colours of globalization

Carlos Marichal and David Pretel

From early modern times, a relentless quest for vibrant colours led Europeans to create globalizing trading networks in the Atlantic World, with an enduring economic and cultural impact. Yet, in the vast narrative of globalization, a crucial chapter often remains unexplored: the production, trade and consumption of large quantities of natural dyes from the Americas between the sixteenth and twentieth centuries. Surprisingly, many historical accounts of global connections overlook these prized dyes, which were among the earliest products exported from Spanish America and Brazil. In the sixteenth century, their value was second only to silver and gold. As time passed, demand for them increased, fuelling textile industries in Europe and Asia and eventually other parts of the world as well. This demand had a far-reaching impact, shaping economies and cultures well beyond the shores of the Americas. But global history is not only about the relative quantities of commodity production and trade. Even in the new era of synthetic dyes and pigments of the late nineteenth and early twentieth centuries, certain natural dyes continued to enjoy widespread use and even found new applications, with demand surging notably during the World Wars.

This book explores the global history of the colourants of the Americas from the time of early globalization in the sixteenth century through the industrial revolutions in Europe and the United States, culminating in the golden age of artificial dyes of the so-called second technological revolution. It examines the extraction and processing of natural dyes in the Americas as well as the development of overseas trade in colourants and the rise of textile industries and consumption of dyestuffs in Europe, the United States and Latin America. While focusing on the production, trade and consumption of natural colourants since early modern times, this book more broadly sheds light on the emergence of economic globalization in the Atlantic World. The volume's sole concentration on the natural colourants of the Americas makes it a particularly novel contribution to global history. It takes a holistic approach by combining regional and national studies, exploring transoceanic connections, and conducting comparative analyses. This approach enriches the reader's understanding of the history of natural colourants in all its dimensions, highlighting the complex interplay among different regions and cultures.

The study of globalization has for the past several decades been a prolific line of historical enquiry.[1] Research into global commodity networks and trade in colonial America from early times to the present has likewise garnered extensive attention from scholars.[2] The export of metals and agricultural products from the New World, ranging from silver to sugar, has been widely recognized as a pivotal factor in the intensification of transoceanic interconnections. However, other commodities originating from North and South America, as well as Central America and the Caribbean, have received less attention in global histories, especially those aimed at a general audience. This is evident in the case of the production and trade of natural dyes , which is a vital, yet neglected, chapter in the history of economic globalization. The existing scholarship on these dyes, although valuable, has failed to consolidate in one volume a transnational overview that compares a variety of case studies, places and times.

While a central aim of the present volume is to explore how the production and trade of natural dyestuffs have shaped economic globalization, its primary emphasis lies in illuminating the diverse and contrasting global histories of these colourants of the Americas since the early modern era. The narrative unfolds not merely as an exploration of the mechanics of global integration through dyestuff commerce, but as a nuanced portrayal of the myriad historical trajectories, complexities and interconnected narratives that have characterized the production and use of these

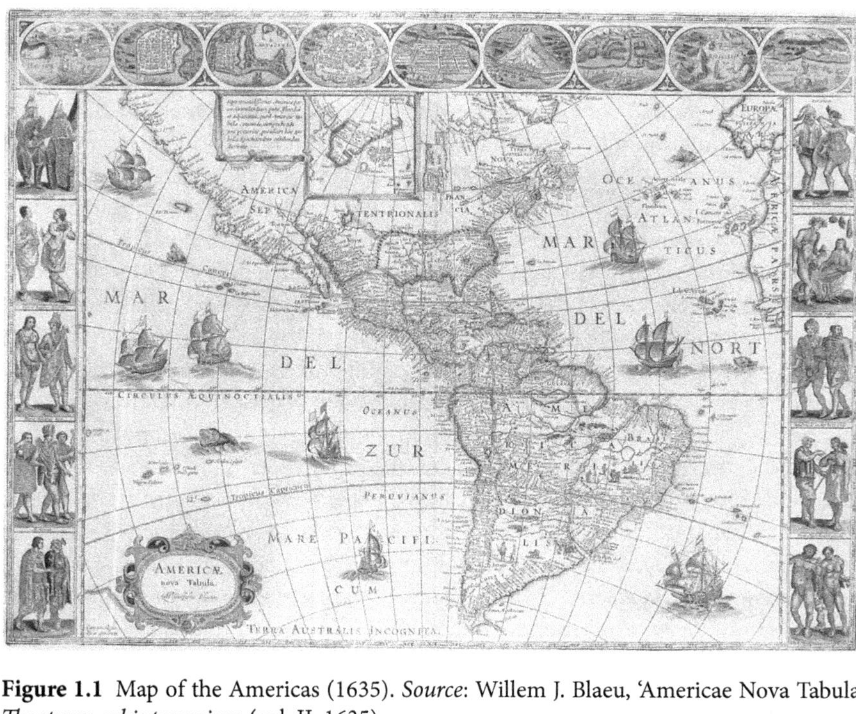

Figure 1.1 Map of the Americas (1635). *Source*: Willem J. Blaeu, 'Americae Nova Tabula', *Theatrum orbis terrarium* (vol. II, 1635).

vibrant natural dyes since ancient times. The volume spans several centuries – from the time of early globalization in the sixteenth century through the industrial revolutions in Europe and the United States, culminating in the twentieth century. Ranging from the indigo trade in the Atlantic to the secrets of the Indian production of cochineal consumed in Europe, the topical breadth of this collection will transport readers across continents and epochs, revealing the intricate threads that connect diverse cultures, economies and technological advancements.

Specifically, the book focuses on the extraction, manufacturing, trade and consumption of brazilwood, logwood, indigo, cochineal and fustic – all natural colourants from the Americas that helped shape the emergence of an increasingly global economy (see Figure 1.1). Borrowing a variety of analytical tools from commodity studies, economic history and political ecology, many of the chapters transcend nationally bounded historical narratives and explore transoceanic dynamics, specifically issues such as Atlantic trade routes, imperial ambitions and the cross-cultural exchange of knowledge and techniques in the world of textile production, fashion and art over centuries (Table 1.1).

A comprehensive understanding of the history of colourants of the Americas entails a nuanced analysis of both global and local dynamics. This multifaceted exploration operates at varying scales, delving into different stages of natural dye 'commodity chains' as well as local labour, techniques and market conditions. Within this volume, diverse approaches are employed. Some chapters implicitly utilize the analytical framework of 'commodity chains' to trace the global journeys (and transformations) of dyestuffs, while others explore the intricate production dynamics within 'commodity frontiers'.[3] This plurality of approaches elucidates the historical trade and production networks involving indigo, logwood, cochineal, fustic and brazilwood within their specific historical and socio-economic contexts (see Table 1.1). Collectively, these contributions shed light on the significant reliance of colourants manufactured in the Americas on extensive webs of intermediaries. Local traders and wholesale merchants played pivotal roles, serving as indispensable links in transporting dyes to the flourishing textile industries of both American and European origins (Figure 1.2).

Table 1.1 Natural Dyestuffs of the Americas

English	Spanish/Portuguese	Main Colours (Different Shades)	Main Historical Areas of Production
Cochineal	Grana cochinilla	Red, carmine	Oaxaca (Mexico)
Logwood	Palo de tinte, Palo de Campeche	Black, blue, red	Yucatan Peninsula and the Caribbean basin
Brazilwood, Paubrasilia	Palo de Brasil / Pau-Brasil, Pau-de Pernambuco	Red, purple	Atlantic forest of Brazil
Indigo	Añil	Blue	Central and North America and the Caribbean basin
Fustic, Mulberry	Palo de Mora, Mora	Yellow, khaki, green	Central America

Two recurring themes bring coherence to the book. One examines the history of production, including social, labour and technological histories; the other looks at the patterns of trade and consumption. The conjunction of both processes underscores the complex interplay between local practices and global exchanges in the realm of natural dyeing. The first crucial issue addressed by several contributions to this volume is the enduring significance of local dye-making techniques, modes of production and labour regimes in the extraction and processing of natural colourants and dyeing processes in the Americas, Europe and elsewhere. Obtaining colourants required an understanding of the natural conditions of various production methods (wild harvesting/foraging, agriculture and insect cultivation). Artisans and skilled workers possessed unique expertise deeply rooted in their local environments, making it challenging to transfer and disseminate these skills widely. The practical experiential knowledge and skills held by these craftsmen, which were integral to the production process, became a form of intangible capital. Their artisanal expertise not only influenced the production, distribution and trade of natural colourants in the Americas but also played a pivotal role in the birth of globalization.

Second, the book explores the local, regional and global trade and consumption patterns of natural dyes. A crucial aspect here is how the production of colourants in the Americas responded to changes in the textile industries since the sixteenth century. The bulk of exports of dyestuffs from the Americas went to Spain and Portugal but also made their way to a great variety of overseas markets, including reexports to the Italian ports of Genoa, Livorno and Venice. They were also soon rechannelled to the vast domains of the Ottoman Empire, from Constantinople, Smyrna, Aleppo and Alexandria to as far as Baghdad. Shipments from Lisbon and Seville simultaneously made their way to all leading French ports, to Antwerp and, hence, into the German heartland as well as the rest of Central and Eastern Europe. But the new and much sought after dyes also wended their ways to England and the Scandinavian port cities as well as those of Russia. By the seventeenth century, Mexican cochineal reached markets in Africa as well as East and South Asia. From the early eighteenth century, dyestuffs from Mexico, the Caribbean and South America would make their way to the younger British colonies in North America, while indigo was shipped out in large quantities from South Carolina to England.

To our knowledge, no recent monograph or edited volume in English has endeavoured to provide a comprehensive, long-term, comparative history of the dyes of the Americas within the broader context of economic globalization. While existing works, such as the 2012 book *The Materiality of Colour* by Feeser, Goggin and Tobin, delve into the history of colour during the early modern times across diverse continents, they do not specifically focus on the Americas, Atlantic networks or the context of economic globalization.[4] Similarly, resources such as *the Handbook of Natural Colorants*, edited by chemists, offer invaluable information on natural colourants but lack the expansive historical perspective that this volume aims to provide.[5]

Figure 1.2 An eighteenth-century dyers workshop in Paris. *Source*: Dyeing workshop at the Gobelins factory in Paris, from the *Encyclopédie ou dictionnaire raisonné des sciences, des arts et des métiers* (1772).

Several historians have recently produced excellent monographs that contribute to a burgeoning interest in the history of natural colourants, particularly within the framework of the material history of colour or the history of science. A notable example is Prakash Kumar's work, which examines the evolution of agricultural indigo from traditional peasant knowledge in early modern India to scientific methods.[6] In this country, the indigo industry thrived until the widespread adoption of synthetic colourants in the textile sector after the First World War. Another remarkable publication is Andrea Feeser's book on colonial South Carolina's indigo culture, which sheds light on the fortunes of plantation owners and African slaves.[7] For their part, Elena Phipps, Carmela Padilla, Amy Butler and Barbara Anderson, among others, have offered artistic and cultural analyses of cochineal red, examining its historical significance for the worlds of fashion and painting in the contexts of empire and globalization.[8] Similarly, Catherine McKinley has traced the captivating allure of indigo across different cultures and regions.[9]

The emergence and ongoing impact of artificial dyestuffs have been a subject of attention for quite some time.[10] For example, there is a growing body of literature exploring the consumption culture of synthetic dyes in Europe and North America from the late nineteenth century, as exemplified by the volume *Bright Modernity*, edited by Regina Blaszczyk and Uwe Spiekermann.[11] However, only a few of these works explore together the intricate industrial and scientific histories of artificial dyestuffs and natural colourants, with even less emphasis on the coexistence and complementarity of the two. Notable exceptions include Agustí Nieto-Galan's *Colouring Textiles* and the edited volume by Robert Fox and Nieto-Galan, *Natural Dyestuffs and Industrial Culture in Europe*.[12] These pioneering works provide invaluable insights into research, dyeing, and printing with natural dyestuffs from the perspective of the history of science and modern industry in eighteenth- and nineteenth-century industrial Europe. However, their scope does not encompass a detailed exploration of the diverse local techniques and indigenous knowledge of production in the Americas, nor the complexities of Atlantic trade networks and associated financial circuits.

Among the general histories of natural dyes, Dominique Cardon's *Natural Dyes* (2007) stands out as particularly valuable. In this monumental book, she explores the historical application of colourants from plants, animals, fungi and minerals in textiles, artefacts and artwork and their chemical composition.[13] An outstanding feature of Cardon's cross-disciplinary book is its comprehensive and in-depth exploration of particular dye origins and dyeing techniques throughout the world, including the extraction and production of red dye from cochineal insects, blue dye from indigo plants, and purple dye from marine molluscs. In this book and other works, Cardon also traces the influence of natural colourants in cultural practices and rituals across civilizations while emphasizing their resurgence in sustainable fashion and art.[14] In the words of Cardon, 'natural dyes and colorants are an essential part of the world's ecological and cultural heritage'.[15]

Natural dyes not only were produced and traded in large countries but were a source of capital accumulation and the development of credit systems as historian Jeremy Baskes has shown for the case of cochineal produced by peasants in colonial Oaxaca (Figure 1.3).[16] Leading Mexico City merchants advanced funds to Oaxaca merchants who, in turn, provided funds to local bureaucrats (*alcaldes mayores*) in the cochineal producing towns and villages. The functionaries would lend the monies to the peasants in order that they could plant nopal plants or pay for sustenance until the cochineals were harvested and sold. In exchange, the peasants agreed to return payment to the bureaucrats and allied merchants at a fixed price (lower than the current international price), generally 12 reales (one a half silver pesos) per pound of cochineal. This allowed merchants to be able to calculate profits with some accuracy and to make huge windfalls when prices in Europe rose abruptly, as they did with some regularity, depending on many factors. A rather similar system of dye cultivation, harvesting and processing was developed in the case of indigo (añil), both in southern Mexico (particularly Oaxaca and Chiapas) and in various regions of Spanish Central America (particularly Guatemala and valleys near the town of San Salvador). Historian David McGreery has described some of the key aspects of cultivation and production of the dye but also of the commercial and financial aspects which speak of features of early capitalism in the Americas from the late seventeenth century and onwards.[17]

Figure 1.3 Indian collecting cochineal (1777). *Source*: 'Indio que recoge la Cochinilla con una colita de Venado', in Jose Antonio de Alzate y Ramirez, *Memoria sobre la naturaleza, cultivo, y beneficio de la grana* (1777).

The present volume brings together scholars from various countries to establish a dialogue and provide comparative insights into intersecting strands of this subject based on their primary research and expertise. In so doing, this book offers a comprehensive historical overview spanning five centuries, delving into the most significant natural colourants of the Americas through rigorous primary research. The inclusion of chapters written by a diverse group of multinational and multilingual specialists, ranging from established experts to emerging scholars, guarantees a wide range of perspectives and in-depth analyses in the history of dyestuffs.

Chapter 2 by Carlos Marichal and the Epilogue by Dominique Cardon tease out the commonalities and contrasts among the case studies and construct a larger frame of analysis that will provide a basis for further studies. Marichal provides an overview of key problems in the geography, labour and trade trajectories of the natural dyes in the Americas between the sixteenth and late nineteenth centuries. After presenting a brief historical geography of the principal natural dyes of the early modern and modern worlds, he focuses on the production and labour history of brazilwood, Mexican cochineal, indigo and logwood. In particular, he explores the methods of cultivation, harvesting and subsequent primary processing to produce valuable dyestuffs. The chapter argues that the transformation of natural dyes into commodities by peasants, artisans and slaves – working in both towns and the countryside – provided essential material foundations for early as well as modern globalization.

In her Epilogue, Cardon reviews the case studies in this book by focusing on their contributions to the comparative history of natural dyes from the Ancient Régime down to the present. She points to the relevance of this subject in relation to the work done by dyers, weavers and artists in the contemporary era in dozens of countries. The appreciation of natural dyes and pigments as a key component of the creative work of people in many societies today speaks to the urgency of recreating a world more attuned with nature, in all its variety and richness. The study and knowledge of dyes today and yesterday therefore represents a cultural challenge that reveals the intersection of past and present in a vast and diverse coloured tapestry.

In their chapter, Catena and Escalona explore the transformation of indigo from an exotic product to a commonplace household item in colonial America. They delve into the early encounters and innovations of New World indigo production between 1560 and 1620. By examining the incorporation of indigenous practices and knowledge gathered through merchant networks, they demonstrate their participation in establishing transatlantic networks specialized in the dye trade. Their narrative highlights the impact on land, communities and production techniques, offering insights into the connections and transformations that occurred during this period, linking two distant worlds through indigo.

Brazilwood also played a pivotal role in early globalization, running parallel to the expansion of the sugar trade and the commerce in other commodities that fed European markets from the sixteenth century onwards. Jobson Arruda's chapter shows that brazilwood drove the first colonial export boom in Brazil. He emphasizes that brazilwood was a timber that produced a rich red dye and was the first natural resource export of all the Americas, as trade in the logs from these trees, harvested in the natural and ancient forests of Brazil began as early as 1501. Local production was dictated by

European demand but relied heavily on forced labour to cut, trim and transport tens of thousands and, later, millions of logs of brazilwood during centuries. The workers included indigenous forest workers hired by Portuguese merchants and colonists who subsequently began to also force African slaves to do this arduous work in various forested regions of Brazil as well as in the Amazon. The Portuguese monarchy's monopolistic policies tightly controlled brazilwood trade, shaping a narrative of exploitation and external dominance during colonial times.

Georges Roque, a leading specialist in the theory and role of colourants in art, demonstrates that studying natural dyes from the Americas is vital for understanding not only the textile manufactures of preindustrial and early industrial Europe but also for analysing the history of pigment supply for the palettes of artists and artisans. The natural dyestuffs that arrived to the ports of Europe from the early sixteenth century were bought by merchants who supplied the textiles workshops in many cities, but they also supplied artists who wanted the highest class of pigments as colourants. Roque focuses on the use of Mexican cochineal, the most expensive dye, which became a fashionable phenomenon, since it was used to dye the most precious fabrics in the royal courts and in the palaces of the nobles (clothes, curtains, chairs and armchairs, rugs, tapestries), as well as in clothing, religious, civil and military uniforms. Cochineal also quickly won a prominent place in markets serving artists, for example among Sevillian painters from as early as the 1560s, although also in use much later as has been verified by chemical analyses of the paintings of seventeenth artists such as Francisco Zurbarán, among many others. In contemporary Italy, cochineal was much appreciated by the great Venetian painters, including Titian, Veronese and Tintoretto, among others. And in Antwerp and Amsterdam cochineal became a favourite pigment among many Flemish and Dutch painters, including Rembrandt.

The geographical panorama of the history of American natural dyes forms a core concern of Karl Offen's chapter, which studies the political ecology of logwood harvested in the forests of Central America from the seventeenth through the early nineteenth centuries. He emphasizes the role of British traders and buccaneers who often entered into conflict with native communities that struggled to obtain or keep access to commons – including lands, maritime spaces, and natural resources. From the seventeenth century, the author describes the role of those 'masterless men' who routinely violated authority across transimperial spaces, seeking new opportunities for trade, plunder and the extraction of natural resources in the western Caribbean. Their activities influenced regional conflicts for centuries. Above all they sought control of the trade in dyewoods, especially logwood. This led to a long-lasting and lucrative contraband trade that fostered a British community along the Mosquito Shore that was part of the saga of the export of precious woods and natural colourants from native forests to the markets of northern Europe.

The following chapter by Jorge Elias-Caro describes the production of natural dyestuffs in the neighbouring Caribbean region of the viceroyalty of *Nueva Granada* (modern-day Colombia) during the eighteenth and early nineteenth centuries. There, the dry tropical forests hosted a great variety of tree species that allowed for extraction of many dyes. These included brazilwood, valued for the variety of red dyestuffs produced, *palo de Campeche* (logwood), for black dyes, *palo de Mora* that produced yellow pigments, and *dividivi* (tannin), which can be red, yellow or colourless, and used for dyeing and tanning leathers. Regional landowners and merchants exploited the local forests with a mix of labourers, free and slave, and spurred the creation of complex, mercantile and business networks and changing mentalities at local and regional levels. In some cases, beggars and young people who were held in juvenile correctional institutions also came to be employed to saw the logs. In all cases, muleteers played an essential role as they transported logs rich in dyes to the rivers and ports. From there, the logs were shipped to England, Holland, France and Germany, marking an era of early mercantile globalization.

Indigo production underwent profound transformations in several parts of the Americas during the eighteenth century, with far-reaching impacts on the Atlantic economy. Frédérique Langue offers a penetrating look into the production and trade of indigo in late colonial Venezuela, which passed from being a rather primitive ranching society to a more complex export economy that relied essentially on two products, indigo and cocoa, exported to a variety of markets in both Spanish America and Europe. The cultivation of indigo had previously been largely concentrated in Guatemala, Salvador and southern Mexico, but the demand for the precious and bright blue of this tropical plant outraced production there and as a result it was transposed to Venezuela. The author describes the harvesting and processing of the plant to produce the valuable blue dye, taking place mostly on mid-sized haciendas with free labour, although slaves were also employed in the larger plantations, just as in contemporary South Carolina. The complexity of the relations between planters, merchants and Spanish colonial functionaries provides a fascinating glimpse for the comprehension of power relations in what was in many ways a frontier economy of South America.

Andrea Feeser also concentrates on the story of indigo during the eighteenth century, in this case in South Carolina. She demonstrates that it was a staple that not only defined much of the colony's colonial politics and culture but also played a central role in transatlantic exchanges. The historiography has often recounted this story as a tale of planter achievements, notably those of the young Eliza Lucas Pinckney. While highlighting the work of a woman in the history of the Atlantic World, this chapter also emphasizes the process of expansion of the indigo plantations that were the product of slavery and Native American land dispossession. This chapter focuses on these two intertwined realities and details the lives of African and Indian slaves who furthered the production of indigo while carving out spheres of agency for themselves.

Torras and Villegas consider the history of the exploitation of logwood in the surroundings of Laguna de Términos (Gulf of Mexico) during the nineteenth century. According to the authors, the exploitation of *palo de Campeche* (logwood) laid the foundations for an extractive economy controlled by a multifaceted elite of merchants linked to Atlantic trade. Their chapter first explores the socio-economic structure of this extractive enclave and investigates the formation of commercial and landowner elites. This includes a focus on the arrival of European migrants and their integration into the regional elite of colonial origin. The chapter also examines the establishment of productive units for tree cutting, delving into questions of property, production, and labour structure. Additionally, the authors analyse the factors that contributed to the success in forest exploitation and discuss the constraints on local processing of

the heartwood of logwood. In so doing, Torras and Villegas provide a comprehensive summary of the logwood economy in the region and the major Atlantic commercial routes for logwood exported from the ports of El Carmen and Campeche.

The production of natural colourants had not only significant socio-economic impacts in the regions of extraction and processing but also environmental implications. In this regard, Anthony Goebel and Ronny Viales examine the economic and socio-environmental impacts of neotropical forest exploitation in independent Costa Rica. Their chapter delves into Costa Rica's timber exports, exploring global connections with foreign industries and the resulting socio-environmental consequences in peripheral regions like the Central Valley and the Atlantic-Caribbean area. The authors demonstrate that the plunder of tropical forests significantly reshaped Costa Rica's landscape. Amidst the global trend towards commercializing tropical forests and the surge in exporting 'cheap' woods, especially fustic, Costa Rica underwent profound changes in its forest cover. These transformations were propelled by local and foreign contractors meeting international market demands while facing ecological limitations.

Finally, David Pretel rounds out the book by focusing on the transition period from natural to synthetic dyes between the mid-nineteenth century and the early decades of the twentieth century. He argues that conventional histories of the rise of artificial dyestuffs, based on advances of chemical science, did not necessarily imply an immediate replacement of traditional colourant materials and techniques with modern ones. On the contrary, in the case of some Caribbean and Central American dyewoods, the substitution process was gradual and would last several decades, a long period during which tree extracts would continue to be of vital importance in the European and American textile industries, competing and coexisting with artificial dyes. For instance, logwood had won over the British market for dyes by the 1850s, displacing indigo and madder, and continued to be in high demand until the First World War, followed by decline between the 1920s and 1940s. The chapter thus proposes a less linear history that questions the temporality of the diffusion of synthetic dyestuffs. More importantly, it shows the continuity of traditional dyewoods and practical colouring skills well into the synthetic age.

In summary and in lieu of a final reflection, which is actually the object of the epilogue to this book by world renowned expert Dominique Cardon, it may be suggested that the historical survey of the varied origins, production and trade in American natural dyes can serve as a preliminary introduction to a fascinating subject that needs to be better recognized as part and parcel of the emergence of early globalization and subsequent trajectories down to the present day. In a panoramic overview of the history of natural dyestuffs in industrial Europe during the eighteenth and early nineteenth centuries, the historian of science Agustí Nieto-Galan argued that the history of dyes often was relegated to a secondary position in the works of cultural historians as well as leading social and economic historians.[18] While this is largely true, it is also a fact that specialists in various and diverse fields have recently advanced in major studies on several dyes for various time periods. There are, therefore, now ample opportunities to learn more and in greater depth on this subject that can provide useful insights and new knowledge for related fields of academic and artistic inquiry. As is well known, the study of dyes has proven to be very useful for archaeological studies carried out by experts on the application of ancient colours, for the history of art in the medieval era and Renaissance and later periods, as well as for specialist histories of textiles and dyes, the cloth trade and protoindustrialization. In addition, museum laboratories, as well as specialists of the natural sciences, especially in departments of chemistry in a number of European universities, have collaborated in incredibly sophisticated studies of the evolution and application of natural dyes in textiles, furniture and pigments for painting. From these varied perspectives, it becomes clear that there exist numerous new opportunities to insert studies of the history of natural dyestuffs within global history, serving as an attractive option for future studies and further reflection.

Notes

1 See, for instance, the following pioneering work A. G. Hopkins (ed.), *Globalization in World History* (London: Pimlico, 2002).
2 See, for example, Carlos Marichal, Steven Topik and Zephyr Frank, 'Commodity Chains and Globalization in Historical Perspective', in *From Silver to Cocaine: Latin American Commodity Chains and the Building of the World Economy, 1500–2000*, eds. Steven Topik, Carlos Marichal and Zephyr Frank (Durham and London: Duke University Press, 2006), 252–360; Amelia Peck, et al. (eds.), *Interwoven Globe: The Worldwide Textile Trade, 1500–1800* (New York: Metropolitan Museum of Art, 2013); Steven Topik and Alan Wells, *Global Markets Transformed, 1870–1945* (Cambridge, MA: Harvard University Press, 2014); Joan Bristol, 'A Trail of Precious Goods: Colonial Latin American Commodity History', *History Compass* 11, no. 11 (2013): 948–56.
3 Terence Hopkins and Immanuel Wallerstein, 'Commodity Chains in the World Economy prior to 1800', *Review* 10, no. 1 (1986): 57–70; Sven Beckert, Ulbe Bosma, Mindi Schneider and Eric Vanhaute, 'Commodity Frontiers and the Transformation of the Global Countryside: A Research Agenda', *Journal of Global History* 16, no. 3 (2021): 435–50; Jean Stubbs, William G. Clarence-Smith, Jonathan Curry-Machado and Jelmer Vos (eds.), *Oxford Handbook of Commodity History* (Oxford: Oxford University Press, 2024); Arjun Appadurai (ed.), *The Social Life of Things: Commodities in Cultural Perspective* (Cambridge: Cambridge University Press, 1986); Jonathan Curry-Machado (ed.), *Global Histories, Imperial Commodities, Local Interactions* (Basingstoke and New York: Palgrave-Macmillan, Cambridge Imperial and Post-Colonial Studies Series, 2013); Topik, Marichal and Frank, *From Silver to Cocaine*.
4 Andrea Feeser, Maureen D. Goggin and Beth K. Tobin (eds.), *The Materiality of Colour: The Production, Circulation, and Application of Dyes and Pigments, 1400–1800* (Farnham, UK: Ashgate, 2012).
5 Thomas Bechtold and Rita Mussak (eds.), *Handbook of Natural Colorants* (Chichester: Wiley, 2009).
6 Prakash Kumar, *Indigo Plantations and Science in Colonial India* (New York: Cambridge University Press, 2012). See also Ghulam Nadri, *The Political Economy of Indigo in India, 1580–1930: A Global Perspective* (Leiden: Brill, 2016).
7 Andrea Feeser, *Red, White, and Black Make Blue* (Athens: University of Georgia Press, 2013).
8 Amy Butler, *A Perfect Red: Empire, Espionage and the Quest for the Color of Desire* (New York: Harper Collins, 2006); Elena Phipps, *Cochineal Red: The Art History of a*

Color (New York: Metropolitan Museum of Art, 2010); Carmela Padilla and Barbara C. Anderson (eds), *A Red Like No Other One: How Cochineal Colored the World. An Epic Story of Art, Culture, Science and Trade* (New York: Skira Rizzoli, 2015); Other works by Elena Phipps are: Elena Phipps, 'Global Colors: Dyes and the Dye Trade from the Sixteenth to the Eighteenth Century', in Peck et al., *The Interwoven Globe*, 28–45; Elena Phipps, 'Textile Colors and Colorants in the Andes', in *Colors between Two Worlds: The Florentine Codex of Bernardino de Sahagún*, ed. Gerhard Wolf, Joseph Connors and Louis A. Waldman (Villa I Tatti, Florence: Harvard University Press, 2012), 256–80; On the production and trade of Mexican cochineal, see also Carlos Marichal, 'Mexican Cochineal and the European Demand for American Dyes, 1550–1850', in Topik, Marichal and Frank, *From Silver to Cocaine*, 76–92.

9 Catherine McKinley, *Indigo: In Search of the Colour that Seduced the World* (London: Bloomsbury, 2011).

10 Alexander Engel, 'Colouring Markets: The Industrial Transformation of the Dyestuff Business Revisited', *Business History* 54, no.1 (2012): 10–29; Simon Garfield, *Mauve, How One Man Invented a Colour that Changed the World* (New York: Canongate Books, 2016).

11 Regina Lee Blaszczyk and Uwe Spiekermann (eds.), *Bright Modernity. Color, Commerce, and Consumer Culture* (New York: Palgrave Macmillan, 2017).

12 Agustí Nieto-Galan, *Colouring Textiles. A History of Natural Dyestuffs in Industrial Europe* (Dordrecht, Boston and London: Boston Studies in the Philosophy of Science, Kluwer Academic Publishers, 2001); Robert Fox and Agustí Nieto-Galan (eds.), *Natural Dyestuffs and Industrial Culture in Europe, 1750–1880* (Canton: Science History Publications, 1999).

13 Dominique Cardon, *Natural Dyes: Sources, Tradition, Technology and Science* (London: Archetype, 2007).

14 Dominique Cardon and Gaëtan du Chatenet, *Guide des teintures naturelles: Plantes, lichens, champignons, mollusques et insectes* (Paris: Delachaux et Niestlé, 1990) and Dominique Cardon, *Le Monde des teintures naturelles* (Paris: Belin, 2003).

15 Dominique Cardon, 'Natural Dyes, Our Global Heritage of Colors', *Textile Society of America Symposium Proceedings* (2010). Presented at 'Textiles and Settlement: From Plains Space to Cyber Space,' Textile Society of America 12th Biennial Symposium, Lincoln, Nebraska, 6–9 October 2010. https://digitalcommons.unl.edu/tsaconf/12/.

16 Jeremy Baskes, *Indians, Merchants and Markets: A Reinterpretation of the Repartimiento and Spanish Indian Economic Relations in Colonial Oaxaca, 1750–1821* (Stanford: Stanford University Press, 2001).

17 David McCreery, 'Indigo Commodity Chains in Spanish and British Empires, 1560–1860', in Topik, Marichal and Frank, *From Silver to Cocaine*, 62–3.

18 Nieto-Galan, *Colouring Textiles.*

2

The natural dyes of the Americas

Geography, labour and trade (sixteenth to nineteenth centuries)

Carlos Marichal

From the time of the discovery, conquest and/or *encounter* by Europeans of the American continent and its native peoples, a number of transatlantic trades were born on a small scale but increasing quite rapidly in value and volume. The most important products shipped were gold and silver, accompanied subsequently by sugar, chocolate (*cacao*) and tobacco, but within this process of early mercantile globalization, the export of natural American dyes also played a significant role for almost four centuries. The most important natural dyes from the Americas were brazilwood (*pau-brasil*), Mexican cochineal, indigo (*añil*) from Central America and logwood (*palo de Campeche*), the latter from the regions bordering on the Yucatan, while additional natural dyes came from Nueva Granada (Colombia and Venezuela) and the greater Caribbean. These dyestuffs played a major role not only in stimulating international commerce but also in the production of textiles in Europe, the Middle East and Asia, since attractive colours were essential to the manufacture and sale of fine and expensive garments of many kinds. The different shades and relative brilliance of the expensive dyes – red, blacks, greys, blues – were indispensable to the making of luxury cloth for princes, aristocrats, clergy, merchants, military officers and high-level bureaucrats, although the natural pigments were also used by leading painters for hundreds of years.

The description and explanation of the history of such dyes can fill out important but, until recently, missing chapters on the trajectories of early modern trade and protoindustrialization in Europe, the Middle East, Asia and the Americas, as well as on the persistence and popularity of many natural dyes during the nineteenth century. Commerce and industry, however, are not the only dimensions which should be brought to the reconstruction of this vast subject, since it is also indispensable to focus on the social history of exploitation of the natural dyes, in particular, the role of peasants, forest peoples and slaves who carried out the laborious extraction and production of these international commodities over centuries. In this chapter, the objective is to place the case studies of the different American natural dyes in the context of historical geography, labour and trade, as a general framework that may stimulate interpretative comparisons in the long term.

The chemical properties of natural dyes allowed for the extraction of an amazing number of colours, which were manipulated by specialist dyers in many countries, using complex technical methods for the production of a great variety of textiles and artistic creations. The location of the natural colourant sources in the New World was related to botanical topographies, but their exploitation depended initially on the knowledge and techniques of the native populations who, in most cases, had cultivated the natural dyes long before the arrival of Spaniards, Portuguese and French explorers, navigators or colonists in the early sixteenth century. It was largely as a result of the information garnered from native Americans that the Europeans were able to identify the most valuable and abundant dyes and their sources. Subsequently there began exploitation of these resources, a second theme emphasized in this chapter, which briefly reviews key aspects of the production history of brazilwood, Mexican cochineal, indigo and logwood, emphasizing the methods used to harvest and process them in diverse rural societies and economies; in fact, it can be argued that the different kinds of labourers who carried out the work actually provided the material foundations of early globalization. A third subject is the nature of the global trade of the dyes, from the point of view of both supply of these commodities and demand from textile and art markets in Europe and much of the rest of the world, from the sixteenth to the early twentieth centuries.

Origins and historical geography of American dyestuffs

The first American natural dye to be found after the arrival of the Portuguese to Brazil was derived from a tree that was soon baptized as *pau-brasil* (brazilwood). Information on the rich red dyes produced by these trees was picked up almost immediately by Portuguese adventurers and merchants travelling with the earliest explorations in 1500, when they encountered native populations living in forests on the Atlantic coast. Almost simultaneously, French ships – mostly from Rouen – began arriving in considerable numbers to the northeastern coasts of Brazil, continuing to do so for many decades despite the protests of the Portuguese Crown. In 1546, twenty-eight ships left Le Havre for *Terra de Santa Cruz*, as Brazil was originally known, and in 1549, nine ships were registered in Rouen for departure to Brazil: the shipments brought exports to France of some 10,000–20,000 tons of brazilwood per year for several decades. The French were very skilful in establishing alliances with local Indian tribes particularly the Tupinambâ, Tamoio and Potiguar peoples; on the other hand, their enemies, the Tobajara and Tupinikin, were closer to the Portuguese.[1]

From the 1530s the Portuguese Crown declared a royal monopoly over the extraction and trade of brazilwood but negotiated directly with navigators and contractors from Portugal to assure at least partial control of the trade. At the same time, constant attempts were sustained to curtail the activities of the dozens of French ships that arrived at different points of Northeastern Brazil – particularly in the region of Pernambuco – during most of the century to take on logwood, obtained by trading with the local

Indians in exchange for European trinkets, knives and alcohol. The geography of dye exploitation also depended on a varying degree of official regulations, as the Spanish and Portuguese crowns and their overseas administrations soon identified the dyes as valuable commodities to be taxed or monopolized by the State. As a result, from early on, officials encouraged production in certain territorial zones, but could also prohibit exploitation in others so as to maintain some control over production and local trade. Despite the French competition, the Portuguese monarchy obtained substantial profits from its monopoly of dyewood that became one of the major sources of royal income.

Some twenty years after the first shipments to Europe of brazilwood, another commercially valuable, but very different, natural dye was found in Mexico. The initial discovery of cochineal by Europeans came with the conquest of the Aztec Empire by the Spanish conquistador Hernán Cortes, who reported to Emperor Charles V as early as 1523 that he had identified a local dyestuff that could be used by Spanish and European textile manufacturers of the age. The Spanish quickly appreciated the extraordinary qualities of *Grana cochinilla* which was an essential component of the tribute paid by many peoples subject to the Aztecs and regular shipments of exports to Spain began as early as 1526.

The administration of Cortes adopted the practice of collecting tribute – including much cochineal – on the basis of the *Matrícula de tributos*, an early-sixteenth-century Mexican document which included extraordinarily detailed lists of products paid by thousands of peasant communities to Emperor Moctecuhzoma II, before his death in the year 1520. The red dye was not only an important component of the tribute paid to the Aztec emperors but was also sold in urban markets, particularly in Tenochítitlan (Mexico city) as well as some other regional markets. Cochineal was used extensively in the prehispanic societies to dye fabrics and leather, as face paint or cosmetics by women, as well as in the codices and in some buildings, particularly on wood.

Cochineal was obtained from a very small insect, *grana cochinilla* (*Dactylopius coccus*) cultivated by Mexican peasants on the nopal plants, a type of cactus (prickly pear), especially in the regions of Tlaxcala, Puebla and Oaxaca. A distinction must be made between the cultivated cochineal (*grana fina*) developed by Mexican peasants over at least a thousand years, and the wild cochineal (*grana silvestre*) that can be found in other parts of both Central and South America. When dried, and later ground and soaked in a liquid, cochineal provides an intense colouring material that can be used to dye a number of bright or deep shades of red on textiles. Cochineal can also be used for pigments and was adopted by the leading painters of Spain and Italy from the 1540s onwards, as can be attested by the laboratory analysis practiced in various European museums of paintings by Titian, Tintoretto and many other artists.[2]

In the case of Mexican cochineal, extensive production in different places took place in the sixteenth century, but subsequently the Spanish Crown adopted policies to regulate cultivation and eventually concentrated it almost entirely in the Oaxaca region. This became an almost structural feature during the rest of the colonial era. For the Spanish Crown, there were clear fiscal advantages to the production of this dyestuff, which was highly valued because of the long-lasting and brilliant red colours when applied to cloth and leather. Since the Indian communities (called 'repúblicas de indios') were obliged from the sixteenth century to pay tribute to tax collectors of the colonial administration, it was soon stipulated that in Oaxaca they should do so preferentially in cochineal. The complex structure of fiscal and mercantile control of the cochineal production and trade operated on the basis of a close alliance between merchants and local bureaucrats, who exploited the Indian communities as much as they could, although they also had to negotiate with them. Researchers Brian Hamnett and Carlos Sánchez Silva have underlined the alliances between mercantile and administrative elites in major towns near the areas of cochineal cultivation from the sixteenth century through to the end of the colonial regime, and they have argued that the merchants and official functionaries utilized exploitative debt mechanisms to force peasants to produce the cochineal and sell it to them at low prices.[3]

However, coaction was not the only factor involved in production by peasants of cochineal. Jeremy Baskes has convincingly argued that incentives (provided by both merchants and the viceregal administration) help explain the continued specialization of Oaxaca peasants in the cultivation of the cochineal insects and the production of the dyestuff. Certainly, it would appear that the *repartimiento* system (which lasted until 1787) proved quite successful in assuring a consistently large cochineal harvest each year. In very basic terms, this system functioned as follows: leading Mexico City merchants advanced funds to Oaxaca merchants who, in turn, provided funds to local bureaucrats (*alcaldes mayores*) in the cochineal producing towns and villages. The functionaries would lend the monies to the peasants in order that they could plant nopal plants or pay for sustenance until the cochineals were harvested and sold. In exchange for the funds advanced the peasants, the latter agreed to return payment to the *alcaldes mayores* with cochineal at a fixed price (lower than the current international price), generally 12 *reales* (one a half silver pesos) per pound of cochineal.[4] Price levels, locally, therefore remained remarkably stable, a fact which allowed merchants to be able to calculate profits with some accuracy and to make huge windfalls when prices in Europe rose abruptly, as they did with some regularity, depending on many factors.

Apart from cochineal, when the Spaniards began establishing their dominion over the different regions of Mexico and Central America, they found indigo, another natural colourant growing wild that attracted the attention of merchants and functionaries because it provided a dye of the same or superior quality as that from India, which was extremely expensive.[5] The kinds of indigo producing plants in the world is very great, as may be noted by the fact that as early as 1762 Linnaeus had classified ninety varieties.[6] In colonial Spanish America, however, the principal ones cultivated were *Indigofera suffruticosa, Indigofera bastarda, Indigofera añil* and *Indigofera tinctoria.*

According to historian Dauril Alden in a classic study on indigo in the Americas, production first began in central Mexico in the mid-sixteenth century. Historian Alicia Contreras has demonstrated that commercial production of indigo also began in the Yucatan from the 1570s.[7] Many *encomiendas* (large properties of the Spanish ruling elite) used Indians to cultivate indigo as a cash crop, making considerable profits. Nonetheless, this did not last long in Yucatan, as restrictions were placed on the use of Indian labour in the indigo haciendas and later, during the seventeenth century, indigo production in Yucatan declined notably.

As early as the 1560s, another indigo variety was discovered in Guatemala and soon was traded on international markets. This plant was cultivated in the interior valleys of Honduras and along the Pacific slopes of Central America from northwestern Guatemala to Lake Nicaragua. Peasant labour was used both to harvest these plants and to dry them and process the dye, which was transported in leather bags. This colourant eventually became known as 'Guatemalan indigo', after the administrative district from whence it came, but actually three-quarters of the Central American indigo eventually were harvested in the coastal valleys and plains of El Salvador, where the best growing conditions were found and where there existed an abundant labour force composed mainly of the sedentary Pipil Indians.

The last of the four most important natural American dyes gained acceptance in European markets more slowly: this was logwood or *palo de Campeche* (*Haematoxylum campechianum*), known by the Mayan Indians as *ek*; alternate names long in use have been India wood, *palo de tinte*, *palo de Campeche*, *Blauholz* (German) and *Bois de Campéche* (French). The dye is produced from the core of a spiny tree that can grow as much as 15 metres, with a trunk that has a high content of hematoxylin, of great use for colouring textiles and leather. According to scientific descriptions logwood is native to the riparian lowlands of the Yucatan Peninsula and northern Belize, where it grows gregariously in swampy and calcareous ponds. It was actually in Yucatan that the Spanish discovered forests with an abundance of these tree, and by the mid-sixteenth century, they were exporting large cargoes of the valuable heartwood.

According to historians Alicia Contreras, as well as Rosa Torras and Pascale Villegas, the exploitation and trade in this tropical wood as an export to Europe for the fabrication of a valuable dye began relatively soon after the conquest of Mexico.[8] From 1575 it was reported that ships of the Royal Navy returned to Seville with 1,350 *quintales* (one quintal was equivalent to one hundred pounds or 46 kilograms) of logwood. Historian Manuela García Bernal has found evidence that some 19,714 *quintales* were embarked in the port of Campeche between 1575 and 1598.[9] Subsequently, this trade stagnated, but after the mid-seventeenth century, British buccaneers, interlopers and baymen developed a booming business in logwood on the coasts of Yucatan and modern-day Belize. The trade in *palo de Campeche* later expanded due to increasing demand for the dye, supplied by local as well as international merchants, and remained important through the end of the nineteenth century.

Dynamics of labour in the production of natural dyes in the Americas

While studies on commodities in the age of early globalization emphasize the role of merchants as critical agents of international trade, equally important were rural workers who produced the valuable goods that fed a growing world demand. The merchants were key intermediaries, but peasants, artisans and slaves, working both in towns and in country, actually provided the material foundations of the birth of globalization, as can be seen in the study of labour practices in the production of the natural dyes of the Americas. A general review of the types of workers engaged indicates that there were four broad categories: (1) forest workers – most of them native Indians – who were experts in extracting dyewoods and negotiated with the European merchants the provision of trunks of brazilwood and logwood that they cut down in exchange for European goods and trinkets; (2) slave labourers – both Indians and Africans – which became common in the seventeenth century to obtain brazilwood and logwood. Slave labour was also the most common form of labour exploitation in the indigo plantations that multiplied throughout the Caribbean and in South Carolina during the eighteenth century; (3) Mexican peasants who were engaged in the cultivation and harvest of the cochineal that lived on the nopal plants in small plots at their homes; (4) Central American peasants who were the main actors in cultivation, harvest and early stages of production of indigo, although some became day labourers in local plantations.

In the case of brazilwood from the Atlantic forests of Brazil and the logwoods that grew in different parts of the Gulf of Mexico, the principal workers were native Indians who lived in forests or in rural agriculture areas. They were engaged to cut trees, strip clean the trunks, carving out the rich red cores and transporting them on rivers or directly to the beaches, where they were loaded on ships bound across the Atlantic Ocean. After being transported to Europe, the dyewoods were sold to textile manufactures, some of whom operated on a small scale but others on a large and almost industrial scale. The Indians who harvested the trees in Brazil had to be attracted to the harsh work of cutting logs and hauling them to the coasts by a series of material incentives, including trade in a variety of cheap goods, but tended to die rapidly because of the introduction of diseases from the Old World.[10]

As a result of the rapidly declining coastal population, merchants and colonists involved in the brazilwood trade financed campaigns to find new reservoirs of indigenous labour from other regions, including, São Paulo, Goiás, Mato Grosso, Maranhão, Pará and Rio Negro. Royal functionaries justified this process of Indian enslavement, as did later the Jesuits, who also became involved, on the basis of the argument of a *just war* against heathens.[11] Most telling in this regard are the works of the historian and geographer John Hemming, who has demonstrated in a detailed history how the Jesuits operating their missions on the riverbanks of the vast regions of the Amazon during the late sixteenth century participated actively in the submission and, often enough, of hunting down and enslavement of Indians.[12] In the 1590s, and for a for a brief time, the Portuguese Crown granted the Jesuits a monopoly on the cutting, collection and transportation of timber to the ports of embarkation, although it must also be noted that mortality rates in Jesuit-run Indian villages ranged from 30 per cent to 75 per cent.[13] Later, the exploitation of the forests of brazilwood took place using more varied forms of labour, both contracting with native dwellers of the Amazon forests and exploiting slaves to cut the valuable trees and drag the trunks to the rivers, whence they were transported to the multiple Atlantic ports of Brazil such as Fortaleza, Belem, Natal and Recife.

The cutting down of brazilwood trees in the initially dense forests as well as stripping and transportation of the logs proved arduous work. Historians who have specialized in the history of forests in Brazil, such as Warren Dean and Shawn Miller, indicate that the use of fire was increasingly applied at the base of the trees, as well as for thinning of clearings, where both Indians and the first colonists could plant other crops such as manioc. But the results of such overexploitation were that vast tracts of the youngest trees of brazilwood were destroyed, breaking the cycle of natural reproduction of the forests which were heavily devastated during most of the colonial era. [14]

The Indian labourers were initially free and carried on the strenuous work, which required much effort both to cut the trunks and to transport them, being hauled by at least three men each. But before hauling and placing in canoes on rivers or on boats at the coast, it was needed to strip the trunks, and cut down to the sapwood so as to retain just the core, the red part that was segmented into logs of 1 to 2 metres, each with a weight of about 30 kilos. This allowed for transport on the shoulders of the Indians, as they needed to travel distances of up to 25 miles to deposit them in a series of precarious warehouses, where all the logs destined for Europe were amassed.

Although there was extensive resistance from workers extracting and mobilizing brazilwood, eventually, the Indian tribes were decimated, as much by disease as by other factors. As a result those that were able to remain deep in the Amazon had a better chance of survival, while on the Brazilian northeastern coasts, the composition of the population changed quite quickly, both because of colonization by Portuguese and even more as a result of the introduction of large numbers of Black slaves transported forcibly from Africa to work on the burgeoning sugar plantations which expanded rapidly from the late sixteenth and early seventeenth centuries.[15] Inevitably, this new type of agricultural production also impelled the destruction of the forests bordering the Atlantic coasts, and many slaves were obliged to participate in the woodcutting business. According to various calculations by specialists in environmental history, over the first two centuries of colonial rule, and to a lesser extent in the following centuries, a huge number of tons of brazilwood were extracted as a result of the felling of millions of trees.

The sources and labour requirements for the production of Mexican cochineal were very different from that of brazilwood. The cultivation of the tiny cochineal insects was quite complex, as can be seen in the labour-intensive steps required to produce commercial dyes, and actually evokes many similarities with that used by Chinese peasants who cultivated the silkworms on mulberry plants, before export to local and foreign markets. In Mexico, and more particularly in Oaxaca, the high population density of peasant communities in this mountainous territory was an important precondition for the highly labour-intensive cultivation of the cochineal on nopal plants.

The expert historian of dyes, R. A. Donkin observed that generally Indian holdings where cochineal was cultivated were family plots, although towards the end of the colonial era, some exceptional units of production were quite large.[16] On the whole, nevertheless, most cochineal production was carried out by peasant families in small terrains, the main reason being that the intensive labour required could be divided among family members, including women, children and elder men, while most men were involved in cultivating maize and tending to cattle and sheep.[17] Careful and delicate labour was needed to select the insects and move them from one nopal leaf to another, as well as to eliminate the many pests that fed on the cochineal insects. Furthermore, in the case of Mexican cochineal, tens of thousands of the insects had to be carefully selected by hand, and then boiled or baked and dried in the sun to produce just one pound of the finished dye, which was later cut into small cakes and placed in leather bags in order to be able to transport them by land to local markets and thence by sea on to overseas trade fairs.

A rather similar system of dye cultivation, harvesting and processing was developed in the case of indigo, both in southern Mexico (particularly Oaxaca and Chiapas) as well as in various regions of Spanish Central America (particularly Guatemala and valleys near the town of San Salvador). David McGreery has described some of the key aspects of cultivation and production of the dye. The indigo plants were harvested from two to four times a year. Once cut, they were processed by submersion in water and put into vats where they were mixed with other ingredients. Once dry, the indigo was cut into small cakes and put in leather bags for transport to markets.[18]

The historian Alicia Contreras discovered a remarkably early document from 1577 that described these same basic tasks required for the cultivation of indigo and production of the indigo dye in considerable detail. The methods used were quite similar whether in El Salvador, Guatemala, Chiapas, Oaxaca or the Yucatan. The document noted that the indigo farms and mills had a high water tank and ferris ('noria') made of wood which allowed for the transport of the liquid by little canals to large heating vessels from which the steaming water was sent on to vats in which the indigo plant had been placed. The vats were covered with wooden planks, and at certain moments the mixture was stirred with wooden poles. The water was drained and then began the process of beating and shaking of the indigo in another vat until it was thoroughly mixed, after which the last water was drained. Following a sequence of additional steps, the now pure indigo colourant was placed on cloths on tables and left to dry in the sun, after which it could be cut into small cakes for sale.[19]

Most of the labour engaged in cultivation of indigo during the colonial era in southern Mexico, Guatemala and El Salvador was based on relatively small family farms. But this plant also could be exploited in large plantations as described in 1770 by Paul Beauvais-Raseau, who published a most informative work titled *L'art de l'indigotier*, which provides a general view of the history of indigo in various regions of the world where it was grown and then described in detail indigo cultivation and production in the Caribbean; there small peasant cultivation was soon surpassed by slave plantations. Beauvais-Raseau had been a captain of the militia on Saint Domingo in the 1760s and had carefully observed rural production methods. He noted that indigo cultivation could be combined with other agricultural activities since the indigo plantations required most labour during the months of July, August and September when the plants were harvested, fermented and the dye extracted.[20] This production calendar allowed for a balance with other types of rural tropical production such as sugar and coffee harvests that took place basically between November and March and experienced a huge boom in Saint-Domingue and the other French islands in the Caribbean.

During the third quarter of the eighteenth century, annual production of indigo in the French islands of the Caribbean approached two million pounds, which was somewhat more than the Central American figures of that time. On the eve of the great rebellion in 1791 by the Black slaves of Saint-Domingue (soon to be named Haiti), it was calculated that there were more than 3,160 indigo establishments (mostly small plantations) on the island, and indigo ranked fourth in exports-behind sugar, coffee and cotton.[21]

The exploitation of natural forests in order to obtain logwood was promoted by a varied cast of merchants and adventurers who negotiated contracts with the peasants and forest peoples of eastern Yucatan, Honduras and Campeche, as well as those of Nueva Granada (Colombia). Initially, the rather limited trade in logwood was based on links between traders and local Indians who were familiar with the forest and were hired to cut the trees, scrape the logs until the red interior core was obtained and later carry the redwood long distances on their shoulders to rivers or to the coast. Eventually, however, groups of Spaniards began to take over large swaths of forest and establish the basis for large establishments. Nonetheless, the trees were also located in smaller properties and in royal lands. The so-called cutters (*cortadores*) were in charge of hiring the work force that would go into the forces to cut down the trees. Some of these men had sufficient capital to carry on this activity regularly, while others were seasonal as they had less resources or simply were interested in a quick sale of the dyewood. The cutters got their work force from local Indian communities which were eventually submitted to the familiar *repartimiento* system, although mestizos, mulattos and free Blacks also frequently participated in the work for a daily wage. The cutters supervising the process provided the work squads of from ten to twenty men with axes and knives and some other simple instruments, as well as supplying them with daily food rations, mainly corn, beans, dried meat, chilli, salt and also the vessels to cook their food in the woods.

As in the case of brazilwood, a key problem was the relative lack of Indian workers, due to the harsh working conditions but, above all, to the decline of the Maya population as a result of disease, which left large zones of forests with few inhabitants. As a result the cutters attempted to force Indian workers, as well as mestizos and mulattos and free Blacks, to sign contracts in exchange for advances of money. These labourers thus became, in effect, indentured servants. Nonetheless, the local geography allowed for frequent escapes and for a very unstable labour force.[22]

In the seventeenth century a special and parallel aspect of the logwood business was the role of English buccaneers and other traders who established themselves, at least temporarily, on the coasts of British Honduras, later known as Belize, but also moving around the coasts of the entire logwood-producing areas on their ships. The first of their logging camps were established between 1640 and 1660, and these early wood-cutters were soon called baymen. According to Karl Offen, the baymen moved the thousands of felled trees mostly by rafting them down the rivers. Before loading onto the ships bound for Europe or New England, however, a great amount of labour was required, as the bark and sapwood had to be stripped before obtaining the logwood core that was the valuable commodity. They brought in slaves from Jamaica as well as directly from Africa to do the hardest work in the cutting, hauling and cleaning of the logwood. Nonetheless, mortality was very high and there are many testimonies of their sufferings from disease, malnutrition and inhumane atrocities at the hands of their owners.[23]

In summary, the extraction and production of the diverse natural American dyes illustrate the complexity of the first stages in the transformation of basic commodities from the Americas that contributed to the subsequent processes of textile protoindustrialization in the *Ancien Régime*, both in Europe and in other world regions. From the mid-sixteenth century, the success of cochineal, brazilwood, indigo and logwood dyes in winning markets among manufacturers of luxury textiles obliged the numerous workshops of traditional dyestuffs in Spain, Italy, France, Flanders, Germany and England to adapt to the new products and explore a variety of mordants to determine which were more effective in fixing the dyes to woollens, linens cotton and silk fabrics in order to obtain the brightest, deepest and most durable results. The European textile entrepreneurs hired dyers who were generally small masters not tied to the one employer and were paid on a piece-rate basis – in order to apply the dies and appropriate mordants to the different types of fine cloth.[24] Similar methods were applied in textile centres in the Middle East and central Asia, as well as in Spanish America itself. In summary, the labour and technical prowess required in both the Old and New World to produce both the inputs and the finished manufactured articles were considerable, and this also helps explain the high prices of those dyes which were essential to the production of the finest and most costly luxury textiles of the age.

The global trade in American dyes

A subject complementary to the geography and the labour methods used to produce the different American natural dyes is that of international commerce. The subject is complex because statistical information on the specific trades is quite scarce for the sixteenth and seventeenth centuries. On the other hand, as may be observed from bibliographical references cited in the chapters of the present collective volume, there are more quantitative data on the eighteenth century, a time of rapid expansion for the export of all the natural dyes from the Americas, as well as for the nineteenth century, when commerce in some of these commodities continued to run strong for decades.

From the early arrival of the Spanish and Portuguese to the New World, the bulk of exports of natural American dyes went to Spain and Portugal, but reexports to other European regions began from the mid-sixteenth century, including the Italian ports of Genoa, Livorno and Venice, from where the dyes were rechannelled to the vast domains of the Ottoman Empire, from Constantinople, Smyrna, Aleppo and Alexandria, but also reaching as far as Baghdad. Shipments of American dyestuffs from Lisbon and Seville also made their way to all the leading French ports, as well as to Flemish and Dutch textile centres and, thence, into the German heartland and the rest of Central and Eastern Europe. These commodities also travelled to markets in cloth manufacturing towns in England, the Scandinavian port cities and to those of Russia. Subsequently, some of the dye trade was directed to Africa as well as East and South Asia, although these are subjects that require further research. From the mid-eighteenth century, dyestuffs from Mexico, the Caribbean and South America began to be sold in the British colonies in North America, and after 1783 to the newly independent United States.

The introduction of American natural dyes to Europe from the early sixteenth century had a significant impact in most countries and regions because textile markets were among the most important in late medieval economies. In medieval Europe there were two different cloth and dye markets that corresponded largely to class differences, since social hierarchies were characterized by very different levels of status, income and wealth. In the first place, it should be recalled that the bulk of clothing in peasant communities, where more than 80 per cent of the population lived, was locally woven and sewn by local artisans, but mainly by peasant women. Most wardrobes were mainly made up of woollen cloth and were rather drab, being generally characterized by grey colours as well as browns, pale blues, yellow-browns and blacks. This was due to the fact that local vegetable dyes tended to be used by peasants and artisans: the colours thus derived were not bright and tended to fade quite quickly as generally they were not colourfast.[25] It is true that there were several vegetable dyes that produced brighter dyes, for example madder for reds and woads for bright blues, but it is also a fact that as dyestuffs, they tended to be quite expensive for peasants, who would occasionally spend more money to buy better quality and brighter cloth but only for ceremonial purposes and for certain festivals.

On the other hand, a considerable number of luxury textile centres existed to supply aristocrats, high-level church and state functionaries, rich merchants as well as the royal and princely courts.[26] The level of consumption of these sectors was very high for luxury cloth of a great many types, special garments, blankets and bed materials, curtains and upholstery in wealthy homes and palaces. In these cases, bright and fast colours were considered essential for most kinds of cloth, whether woollens, silks or linens, as well as cottons after the mid-seventeenth century. Among the most sought after colours were reds and purples, although bright and deep blues were also much prized. Nonetheless, the supply of fast dyes in these colours in Europe and the Middle East was limited, and the best were mostly used for royalty and for ecclesiastical garments. The dyes imported from Southeast Asia and India by Arab and later Portuguese merchants were extremely expensive, and, as a result, American products soon came to compete most favourably in terms of both quality and durability.[27]

Despite the early French competition for brazilwood, most trade in this commodity was controlled from the early sixteenth century by Portuguese merchants, and according to the calculations of historian Jobson Arruda, each shipment allowed on average for a gross profit 300 per cent on original capital invested. At the same time the monopoly control that the Portuguese Crown exercised over brazilwood trades became one of the major sources of royal income. According to Arruda although fiscal receipts did not surpass 5 per cent of the total official income of the Portuguese Empire, it was sufficient to cover a substantial part of the costs of maintenance of the overseas territory (Brazil) during the late sixteenth century.[28]

Later, in the early seventeenth century, when Portugal was under Spanish dominion, the royal administration returned to the policy of contracts, by which, the private entrepreneur were to pay 40,000 cruzados annually for the right to transport as much as ten thousand quintals of wood. After the Dutch invasion of the northeastern sugar coast of Brazil in 1628, where there was the best-quality redwood, shipments began to be sent directly to Amsterdam and other textile centres in Holland. Finally, after the expulsion in 1649 of the Dutch from Pernambuco, the Portuguese government created the Brazilian Trade Company, which was given a monopoly on the business of cutting and shipping brazilwood. This mercantile monopoly had agents all over

Europe and registered the largest sales of brazilwood in the ports of France, northern Germany, Holland, Genoa and England, in that order. [29]

The trade in brazilwood continued to be, in principle, a royal monopoly as it had been for two centuries, although its relative significance had declined markedly during the great age of gold exports from Brazil (1670–1770). During the middle decades of the eighteenth century, the brazilwood trade continued, although it was not particularly buoyant, yet subsequently international wars led both the Portuguese Crown and the contractors to reap extraordinary profits from the sale of the dyewood through many European markets. The European wars of this era impacted market prices in London, Amsterdam and other ports and they increased enormously. The profits for the royal monopoly on brazilwood of Portugal were truly extraordinary at the end of the *Ancien Régime*. Nevertheless, local Brazilian contractors, and presumably workers, did not benefit correspondingly.

The history of the trades in Mexican cochineal ran parallel to that of brazilwood but reflected numerous differences. From the late 1520s, Mexican cochineal began to appear on European markets in small quantities and soon gained wide acceptance as the finest and most durable crimson dyestuff for textiles, particularly woollens and silks. According to a classic historical study: 'Cochineal possessed from ten to twelve times the dyeing properties of kermes; it also produced colours far superior in brilliancy and fastness'.[30] This dyestuff thus quickly won growing markets in the leading luxury textile manufacturing centres of Europe, including Segovia in Spain, Suffolk in England, Rouen and Lyon in France, Malines and various centres in Flanders, as well as the largest, traditional producers and markets for luxury textiles: Florence, Milan and Venice in Italy. According to art historian, George Roque, cochineal dyestuffs were key for colouring luxury cloth but also as pigments for the leading painters of the sixteenth century.

Interdisciplinary studies provide historical evidence on the rapid expansion of European demand for cochineal from early times. A laborious chemical research programme on hundreds of samples of medieval and early modern dyed textiles has provided 'concrete evidence to substantiate the historical assertion that Mexican cochineal within fifty years of its introduction into Europe (c. 1520-1530) fully displaced kermes in scarlet textile dyeing'.[31] Nevertheless, it should also be noted that for decades there was considerable opposition to the cochineal imports on the part of traditional producers of madder (*rubia tinctorum*) dyes throughout Europe, including low-level textile guilds that could hardly afford to buy cochineal. It should be noted, in this regard, that madder had been used since ancient times as a red vegetable dye for leather, wool, cotton and silk.

In any case, cochineal soon won out among all reds of other dyes in the luxury textile industries of Italy, which remained among the most important of sixteenth-century Europe and therefore were among the major markets for expensive dyestuffs. Substantial quantities of the 'grana cochinilla' sent from Veracruz to Seville and Cádiz made their way to the port of Livorno, the entrance to the great manufacturing and financial centre of Florence.

The Spanish economic historian Felipe Ruiz Martín analysed the correspondence of contemporary Spanish merchant bankers to trace the exports to Florence, where a booming luxury textile industry consumed large quantities of dyes.[32] Cochineal soon replaced kermes as the red dye of preference. But he also noted that a substantial volume of cochineal was transhipped from Livorno to Venice where it was used to dye the cheaper textiles – 'pannina' – sent to Constantinople. According to both Spanish and Genoese merchants involved in this trade, the crimson dyestuff was always profitable, and in fact its price quadrupled over the sixteenth century even as the volume of trade rose rapidly.

From the mid-sixteenth century onwards, leading European merchants and merchant bankers became particularly interested in cochineal because it was a high-value commodity with low weight; in this regard it was similar to other leading global commodities of the era such as silks, pepper and spices as well as precious metals which were easily transportable by sea or land and allowed for great profits because of their high but fluctuating prices. As a result, it was not strange that cochineal should have become the object of much financial speculation. In 1788, for example, two of the leading merchant banks of Europe, Hope and Company of Amsterdam and Baring Brothers of London, were hard at work trying to attempt corner all the cochineal markets in Europe. Marten Buist, the historian of the merchant-banking firm of Hope and Company, has described the great cochineal speculation of 1788 in considerable detail. The operation involved buying up most of the stock of cochineal in all the principal European ports: Cádiz, Marseilles, Rouen, Genoa, Amsterdam, London and even Saint Petersburg, with the object of obtaining a virtual monopoly. The transactions required particular attention to acquisition of practically all the dyestuffs received from Mexico in Cádiz, before they were reexported. As a result, the monopoly was nowhere complete and attempts to rig prices failed, causing substantial financial losses to the main partners in the speculation.[33]

Yet European merchants were not alone in the international cochineal business and speculations. Some of the great eighteenth-century mercantile firms of Mexico City and Veracruz were also heavily involved in the management of this complex commodity chain on the American side and in its connections to both Europe and Asia. Similarly, Brian Hamnett's pioneering work describes the complex transactions of other Mexican merchant firms which were heavily involved in the cochineal trade in the late eighteenth century.[34] Further studies by various historians on the operations of the wealthy house of the Iraeta merchant family of Mexico City reveal the complexity of the control of trade inside New Spain and of the connections to Cádiz merchants, on the one hand, and to Asian markets for cochineal via the Manila Galleon, on the other (Figure 2.1).[35]

The trade in Mexican cochineal continued to increase in the mid-to-late decades of the eighteenth century but there were limits to this expansion. The Spanish monopoly on cochineal, as produced in Oaxaca, certainly put a brake on the development of alternative centres of production. Additionally, it should be recalled that the principal markets for cochineal remained those of high-quality and expensive textiles, and therefore were not as susceptible to rapid expansion, as were the cheap cottons produced by contemporary British factories – when the Industrial Revolution took off. Analysis of the production and trade in Mexican cochineal during the quarter century spanning the years 1758 to 1783 demonstrates that this was clearly an age

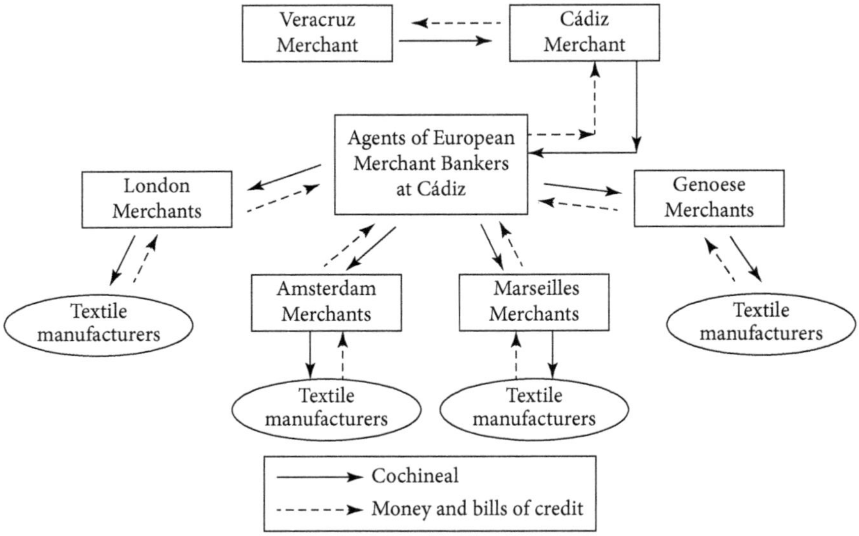

Figure 2.1 The commodity chain of cochineal from Veracruz to Europe (*c.* 1780)

of prosperity as annual production in Oaxaca averaged 922,600 pounds, worth over 2 million pesos per year, the benefits of which were reaped by local merchants and Oaxaca peasants. However, a marked drop in production levels took place after 1784, as a result of the terrible impact of the plagues and demographic crisis of 1784–5 (during which over 300,000 people died in New Spain) that is believed to have deeply affected the Oaxaca peasant communities and disrupted production. By 1803 harvest reached a nadir of less than half a million pounds per year. That production should have fallen so abruptly after 1784 and continued to remain depressed despite the continuing Oaxaca monopoly of cochineal would seem to suggest that it was the disruption of the complex credit-mercantile mechanism, which has been described by historian Jeremy Baskes, that probably contributed to regional decline.[36] But other authors have also insisted that additional factors were involved, such as increasing taxation on this commodity as historian Carlos Contreras argued on the basis of archival research.

In medieval Europe, woad was the most widely used blue dye, but it has been well established by laboratory studies that indigo was also present in many luxury textiles of the time, and appreciated because of its deep shades and excellent lightfastness on wool. However, indigo use was not overly extensive because it was imported from India and hence remained one of the most expensive dyestuffs available to European cloth manufacturers. In the medieval era it was brought to the Mediterranean by Arab merchants who shipped their commodities across the Indian Ocean and then by caravans up to Alexandria.[37] After the 1490s, as a result of the circumnavigation of Africa by navigators from Portugal, it was also imported to Europe by many of the same traders who brought a red dye known as 'brasil', from South Asia, a term later used from 1500 the Portuguese to describe brazilwood. As is evident, these late medieval antecedents are important to consider and explain why, from the first moment of the discovery of the Americas, there should have been a marked interest in finding new natural dyes which could subsequently be traded.

Indigo was sold in sixteenth-century Europe to dye wool cloth in black or blue, as well as silks and cottons in black, for religious reasons in both Catholic and Protestant countries but also because of fashions adopted by monarchs as can be seen in the contemporary paintings of the court of Burgundy from the beginning of the sixteenth century, and later in the Spanish courts of Charles V and that of his son, Philip II.

As the supplies of indigo from the Americas became more plentiful, the woad interests in Europe began to campaign against this new product which was denounced as poisonous and as 'food for the devil'. The woad plants were cultivated in northern Italy as well as several regions of France, England and Germany, yielding blue-coloured dye from its leaves, but inferior to that obtained from the indigo plants. By the end of the sixteenth century the woad producers had succeeded in persuading governments in various German states, France, and England to prohibit use of the dye, whether brought from India by Portuguese merchants or from America by Spanish merchants. Nonetheless, in most of the rest of Europe, particularly Italy and Flanders, imported indigoes were favoured by textile manufacturers and their associated dyestuff suppliers.

While the cheaper American supplies of indigo gradually increased, the trade in the more expensive indigo brought by the Portuguese from India declined, a fact also accentuated by loss of control of many of their trading stations in Asia such as Ormuz, Malacca, Kochi, and the Maluku Islands. As a result, indigo (from India) became one of the chief articles of trade of both the Dutch and the British East India Companies during the seventeenth century.[38] Nonetheless, despite efforts to increase indigo production in Asia, this commodity was surpassed in the mid-seventeenth century by American varieties of indigo, both 'Guatemala indigo' and, subsequently, by 'French blue' from Saint-Domingue. The competition in Europe among merchants of different countries therefore intensified during the seventeenth century. In the peak years of the seventeenth century, the captaincy general of Guatemala sent as much as half a million pounds of the dyestuff to Europe. That the Guatemala indigo was then of the highest quality was confirmed by the principal druggist of Louis XIV, king of France, in his famous history of natural drugs, in which he lauded the Central American product as the best, followed by those of Saint-Domingue and Jamaica.[39]

The boom in the indigo trade encouraged the leading merchant houses of Guatemala to participate and soon to dominate the export trade of this colourant in the Americas.[40] The royal administrators also favoured monopoly because it facilitated tax collection, particularly the trade taxes known as *alcabalas*. As José Antonio Molina has demonstrated, however, it is important to consider the control of production of indigo by local landowners and related small producers who sold their dyes at local and regional trade fairs in numerous rural towns of San Salvador and Guatemala in the eighteenth century.[41] It was there that the agents of the big Guatemala merchant houses arrived to buy the dyestuffs and, often enough, also advanced loans to guarantee future business. The process of indebtedness became engrained, and though a few planters were able to cover their debts and sell on the spot market, the tendency was for maintenance of a dependent relation with the big merchants. Indigo was transported by mule to the capital of Guatemala, and later a portion was sent onwards to various ports for overseas shipment. However, since there was a scarcity of ships arriving at the Central American ports, during the seventeenth and early eighteenth centuries, the bulk of the indigo dyes were generally sent by mule as far as Veracruz, where the great Spanish fleets arrived once a year. After the establishment of free trade within the Spanish Empire in the 1780s, shipments began to increase from local Caribbean ports such as Omoa and Santo Tomás, but on a relatively small scale. It should also be noted that from 1780 indigo production also expanded rapidly in Yucatan, as demonstrated by Alicia Contreras.[42] During this period, the bulk of dyewoods as well as cochineal from Yucatan, Mexico and Central America were sent on royal ships and a number of private merchant vessels from Veracruz to Havana and from there on to Cádiz, in convoys that were well defended in this era of constant naval warfare.

According to the commercial statistics analysed by John Fisher for the early 1780s, the value exports of indigo sent to Spain was less than a half of that of the cochineal shipments, but by the 1790s the figures for indigo tripled and came to surpass cochineal.[43] Much of the dye was consumed in Spain and substantial amounts went to Barcelona, where it was resold to the textile industries that were expanding rapidly throughout Catalonia. The indigo boom, however, was not limited to Central America. Other regions of the Americas had already begun to gain strength from the 1740s onwards. Among New World producers, the French-controlled island of Saint-Domingue became the most formidable, and 'French blue' was the only serious competitor of 'Guatemala' indigo in European markets in terms of quality and price. The eighteenth-century indigo boom of the French Caribbean islands was related to the decline in woad production in France and hence led to increased demand for this dye. According to Dauril Alden, from the late seventeenth century the French government desired that the country become self-sufficient in the blue dye, in great demand among its luxury textile manufacturers and artisans.[44]

From the early eighteenth century, much of the indigo made its way from the Caribbean to Bordeaux and Nantes which became major importers of this dye from Saint-Domingue and the other French Antilles, and as a result, dye manufacturing workshops were set up near the ports and their products sold to domestic textile centres and much also abroad. At the same time, Marseilles became a leading shipping hub for exports of all kinds of primary goods from the Americas. In the case of indigo, local merchants began supplying textile workshops in Italy and Switzerland but also shipped to the Ottoman Empire, in particular to textile merchants and producers in Smyrna, Aleppo, Constantinople, Salonica and Egypt.[45]

Curiously enough, the British were less successful than the French with indigo cultivation in their Caribbean island possessions, despite having started earlier and promoted indigo in Jamaica after taking control of the island in 1655. By 1672, there were some sixty indigo works localized in the parish of Vere (Middlesex County) producing some 50,000 pounds a year. But according to the contemporary observer, Beauvais-Raseau, the indigo species cultivated on the island were quite different and of lesser quality than those of Central America and Saint-Domingue.[46] Prices of Jamaica indigo remained lower than its competitors, and as a result from the early-eighteenth-century cultivation of the dye plant did not expand rapidly as sugar, which became the great cash crop in the first half of the eighteenth century when Jamaica became the largest producer of sugar in the world.

An additional geographical zone of intense production of indigo began in South Carolina from the 1740s and provided England with considerable supplies for burgeoning textile industries. The planters of lowland South Carolina had experimented with indigo in the late seventeenth century but found rice more profitable and largely abandoned the dye for the next half century. However, by the mid-eighteenth century demand for indigo in England, together with a crisis in rice markets, spurred South Carolina landowners to increase the cultivation of indigo. According to the manual of 1770 published by Beauvais, the planters of South Carolina experimented with three types of indigo, depending on the nature and quality of the soils where they were planting. From the best terrain, the harvest of one acre could produce up to seventy pounds of indigo, which could attain a price of fifty pounds sterling. According to McCreery, although South Carolina exports of indigo were qualitatively inferior to those of Saint-Domingue or Guatemala, the volume of production increased until reaching over one million pounds per year. But with the war of independence of the thirteen colonies between 1776 and 1783, indigo production dropped rapidly and later was replaced mainly with cotton as the main local crop.

Another region in the Americas where indigo plantations expanded was the captaincy general of Venezuela. Spanish authorities promoted this agricultural innovation and helped transfer seed from Guatemala in the 1760s.[47] An equally new and somewhat surprising indigo boom took place in colonial Brazil between 1770 and 1800. The systematic cultivation of indigo and its transformation into a dyestuff was the result of a campaign by Portuguese authorities to develop new export commodities that could help alleviate the decline in trade caused by the decline of the century-long Brazilian gold boom, which had begun to peter out after 1770. Although some of the indigo sent from Brazil to Portugal was destined for the small textile manufacturing sector in the mother country, most was reexported, some to Great Britain, France and Italy, but mostly to the Netherlands which had become the largest customer. Nonetheless, it soon became evident that this boom was losing steam, and by the beginning of the nineteenth century, it faded due to the impact of the Napoleonic Wars.

The early history of the global trade in logwood is as fascinating as that of cochineal, brazilwood and indigo, but it has received somewhat less attention in the literature. As we have already suggested, while logwood soon was recognized a superior and permanent dye, many European and particularly English dyers opposed the import of the cheaper, imported hardwood dyes from Mexico and Central America. In 1581 an Act of the English Parliament strictly forbade the use of logwood for dyeing and authorized the burning of confiscated logwood. After numerous protests by merchants and textile manufactures, these statutes were finally repealed by Charles II in 1662 who declared:

> the ingenious industry of modern times hath taught the dyers of England the art of fixing colours made of logwood [. . .] so as that, by experience, they are found as lasting as the colours made with any other sort of dyeing-wood whatever, and

> on this ground it repeals so much of the Statute of Elizabeth as related to logwood, and gives permission to import it and use it for dyeing.[48]

Subsequently, British merchants interested in the logwood trade began an aggressive expansion into the Central American territories where there were the richest forests of *campeche*, the trees that were the source of the dyewood in question. They did so in alliance with hundreds of buccaneers who had already begun to profit from this line of business, mostly contraband, as is confirmed by the studies of Karl Offen, among other historians. The trajectories of trade in logwood in the mid- and late eighteenth century reveal a notable expansion as a result of the great increase of demand by European, and particularly British textile producers.

The commercial circuits of the dyewood in the Spanish domains between 1750 and 1807 have been studied by Contreras Sánchez, who has highlighted the official routes of the logwood shipments between Campeche and numerous ports – Cádiz, Santa Cruz de Tenerife (Canary Islands), Santo Domingo, Havana and Veracruz – as well as smuggling routes. Before the 1750s, the Spanish Crown had paid inadequate attention to the trades from Campeche and the Yucatan, and actually dissuaded them because the dominance of the convoy system known as the regime of the *Flotas* had favoured trade and shipments from Veracruz and Havana, but this system did not facilitate local navigation and did not provide adequate protection from British privateers. As a result, a substantial part of the logwood trade depended on contraband with British traders and shippers who paid local merchants and producers and picked up the logs at any point available on the coast of the forests.

In order to counter the effects of the contraband trade, in 1751 the Spanish authorities created in Campeche a new body known as the *Companía de negociacion y giro del palo de tinte*.[49] Local producers soon flocked to the port and brought almost 200,000 quintales of the commodity for sale. However, there were not sufficient funds to pay all sellers and insufficient ships to send it out, and as a result most of the logwood was warehoused and soon enough began to rot. The failure of this royal initiative caused official functionaries to lose interest in the Campeche trade until the 1770s. Subsequently, taxes were reduced notably and free trade with all Spanish ports was ratified, with positive effects on the regional logwood trade. Between 1784 and 1794, ships with almost half a million quintals of logwood left for Spanish ports, while another 300,000 quintals were sent to Havana and 230,000 tons to Veracruz.[50]

Nineteenth-century trends

Wars deeply affected the trade of all the American dyes. This became manifest after the outbreak of naval war between Spain and Great Britain in 1797, as Spanish ships could no longer move freely in the Caribbean with the result that contraband trade began to boom. Similar circumstances affected the brief indigo booms of the late eighteenth century in Venezuela and Brazil. The reasons for the collapse of the South Carolina indigo industry were also directly related to the effects of the war of independence of the thirteen colonies (1776–83), which caused a drop in trade with Great Britain, the main customer for the dyestuff. On the other hand, the virtual demise of the indigo export business in Saint-Domingue can be explained by the impact of the revolutionary wars in the island during the 1790s and finally by the independence of Haiti in 1804, an event that radically reduced trade with France.

In the case of the most important consumer of indigo, Great Britain, all these events implied that there was urgent need to find new supplies for its dynamic and ever-expanding textile industries. The response was swift, especially by the East India Company which recruited West Indies planters familiar with indigo cultivation and sent them to the Presidency of Bengal, a region that had not been an especially significant indigo producer but soon became the leader in production in India. The Company provided credit abundantly to local landlords and to British merchants who in turn financed the ryots, small farmers and shareholders, as they began to expand production of the dyestuff. Also active were merchant houses of Calcutta that financed cultivation, harvesting and production of the valuable dyestuff. Between 1805 and 1814 exports from India to England averaged 5,600,000 pounds per year and increased subsequently. Historian Dauril Alden explained how these changes affected the production of Central American indigo: 'In Guatemala, production (which stood at nearly 1,150,000 pounds in 1793) fell by as much as two thirds over the next twenty-five years, and remedial efforts failed there, as elsewhere in the New World'. [51]

From the early nineteenth century, each of the natural American dyes followed different trajectories, but all were affected by the wars of independence which, generally speaking, provoked a sharp rise in prices because of the scarcity of supplies but, even so, production tended to decline. Dyewoods also experienced varying trajectories. In the case of brazilwood the story was quite different. Just before the invasion of Portugal by Napoleon and his armies, the Portuguese court transferred to Río de Janeiro and in principle maintained its old royal monopoly. But certain changes were soon applied, and according to Jobson Arruda, brazilwood continued to be exported but tended to fall after independence. In the by-laws of the Bank of Brazil, created precisely in 1808, and with the purpose of capitalizing it, rights were granted to commissions in all areas of brazilwood on the Brazilian coast. After the proclamation of independence in 1822, this income was, in principle, intended to pay debts owed to Portugal, a kind of compensation for the recognition of political emancipation, as well as to cover the expenses necessary for the installation of the first diplomatic representations of the new Brazilian State abroad. Nonetheless, there is not much evidence on volume of sales.

In the case of Mexican cochineal, the story of decline in the nineteenth century was complex and prolonged. During the years 1804–19, production of Oaxaca cochineal continued to decline (stabilizing at a plateau of 328,000 pounds per year) but was compensated in good measure by the rise in the international price of the dyestuff which rose to an average of 26 silver reales per pound during these years of war and intermittent interruption of navigation between Mexico and Europe. Oaxaca peasants and merchants benefitted as international conflict pushed the prices of this relatively scarce commodity steeply upwards, but, paradoxically, local production still continued to fall in these difficult years. However, after Mexican independence in the 1820s, cochineal began to be cultivated successfully and on a large scale in Guatemala and subsequently in the Canary Islands. The old colonial monopoly by Oaxaca disappeared. Indeed, cochineal became the leading export from both Guatemala and the Canary Islands between the 1820s and the 1860s.[52] The results of the increase in cultivation and production of the dyestuff were dramatic, causing a steady price decline per pound. Despite this turn of events, Oaxaca peasants responded by increasing production after 1824, although profitability was falling year by year. In 1880, according to Mulhall, the Canary Islands were still exporting over three million pounds of cochineal but soon collapsed; according to the same source, Great Britain imported cochineal to the tune of 430,000 pounds sterling in 1880, but only 50,000 pounds sterling in 1888.[53]

From mid-century, however, came advances in the chemical dye industries in Germany, and progressively natural dyes were substituted by synthetic ones until cochineal became something of a curiosity in the early twentieth century. As a result, the secular commodity chain based on cochineal that originated in Oaxaca and spanned a great number of markets in Europe and Asia disintegrated. Cochineal continued to be used for a very limited number of textile dyes, although as a pigment it continued to be much used by European artists.[54]

During and after the Napoleonic Wars, indigo cultivation dropped precipitously in Central America, but also throughout the Caribbean, South Carolina and Brazil. Quite different was the trajectory of indigo in British India, which continued to be an important commodity and exported on a rising scale until near the end of the nineteenth century, being controlled by English merchants who managed to run a virtual world monopoly. It is true that that with the decline of woad in Europe, no natural dye product could compete with the blue of indigo, but beginning in the 1850s an entirely new source of colours, aniline dyes made from coal tar, seemed to herald the end of natural colours. Still, indigo held on for almost fifty years more, because of initial failures to develop a chemical substitute specifically for the blue dye. After the turn of the century, however, synthetic anilines ran this natural dye industry to the ground. Soon the synthetic anilines overran markets, and by 1914 natural indigo production in India had dropped to less than 1,000 tons.[55]

In sharp contrast to other American natural dyes, Central American logwood exports began to recover after the wars of independence as logwood experienced a relative boom in Central America during the mid- and late nineteenth century, particularly in the regions of the Gulf of Mexico near Campeche, whence a large export trade was carried.[56] The logwood was in demand to dye colours in black, much in fashion in Europe and the United States as a result of the new fashions of bourgeois society: clerics, teachers, members of the legal and medical profession and domestics all wore black through most of the nineteenth century. As a result, demand for logwoods rose correspondingly. Torras and Villegas provide estimates of shipments on the order of 400,000 *quinales* per year between 1845 and 1900. At the same time, logwood exports from Belize (mainly exported to England and New England) reached an average of 15,000 tons a year between 1870 and 1890, although later declining precipitously.

In the case of Yucatan, both David Pretel and Martha Villalobos González have explored the theme of the continuing trade in logwood at mid-century, but in the midst of a terrible, regional war. This is a particularly complex story which requires some understanding of the great rebellion by the Mayan Indian peasants against the white elites which took place in 1847 in the eastern regions of the great peninsula and continued for years. Gradually, the Mexican troops and the white elites were able to defeat the well-armed peasants, killing thousands, and assassinating their leaders. Nonetheless, it is worth noting that during the prolonged conflicts, the Mayan rebels used the sale of the dyewoods (that they cut in the eastern forests of the peninsula) to obtain arms from British traders.[57] As a result of this war Yucatan lost its importance as a logwood producer and exporter, a fact which was compensated, in considerable measure, by the great henequen export boom that began in the 1860s and continued down to the twentieth century. Nonetheless the bulk of the Mayan peasants did not prosper despite this new boom but rather were submitted to prolonged, servile and near-feudal conditions until the outbreak of the Mexican revolution in 1910.

A final, but important, aspect of the logwood trade in the nineteenth century was the fact that markets and consumption for this commodity changed. The demand for this commodity was impelled not only because of its extensive use for dyes in the textile and leather industries of northern Europe and the United States but also as a result of a new and large demand for inks, particularly in schools, where the rather low-quality inks derived from logwood encountered a great and expanding demand over the century. In this regard, it may be recalled that in 1847, a German scientist, F. J. Runge, introduced *logwood* (or *chrome*) ink which has subsequently been considered as the first modern ink.[58] But additionally, these inks were also much appreciated by artists in the late nineteenth century, as has been shown by the analysis of 48 letters and 100 drawings by Van Gogh which demonstrates that he used logwood inks.[59]

Notes

1 Olive Patricia Dickason, 'The Brazilian Connection. A Look at the Origin of French Techniques for Trading with Amerindians', *Revue française d'histoire d'outre-mer*, 71, no. 264–265 (1984): 129–46.

2 See Georges Roque, 'Introduction' to Book/Catalogue of a great exposition titled *Rojo Mexicano, la grana cochinilla en el arte* (Mexico: Palacio de Bellas Artes, 2017), 18–61.

3 Brian Hamnett, *Politics and Trade in Southern Mexico, 1750–1821* (Cambridge: Cambridge University Press, 1971); and Carlos Sánchez Silva, *Indios, comerciantes y burocracia en la Oaxaca poscolonial, 1786–1860* (Oaxaca: Instituto Oaxaqueño de las Cultura/Universidad Autónoma Benito Juárez de Oaxaca, Oaxaca, 1998).

4 Jeremy Baskes, *Indians, Merchants and Markets: A Reinterpretation of the Repartimiento and Spanish Indian Economic Relations in Colonial Oaxaca, 1750–1821* (Stanford: Stanford University Press, 2001).

5 On the early history of indigo exports from Central America see Murdo MacLeod, *Spanish Central America* (Berkeley: University of California Press, 1973), 176–87.

6 Carl Linnaeus, *Mantissa plantarum: Generum editionis VI. Et specierum editionis II* (Holmiæ: Impensis Direct. Laurentii Salvii, 1767), 271–3, see online at https://www.biodiversitylibrary.org/item/151919#page/5/mode/1up; also Carl Linnaeus, *Species plantarum: exhibentes plantas rite cognitas ad genera relatas, cum diferentiis specificis, nominibus trivialibus, synonymis selectis, locis natalibus, secundum systema sexuale digestas Tomus II* (Holmiæ: Impensis Direct. Laurentii Salvii, 1753), 751,

online copy at https://www.biodiversitylibrary.org/item/84236#page/5/mode/1up; and Carl Linnaeus, *Systema vegetabilivm: secvndvm classes ordines genera species cvm characteribvs et differentiis* (Gottingae: Typis et impensis Jo. Christ. Dieterich, 1784), 677–9, see online edition at https://www.biodiversitylibrary.org/item/100722#page/5/mode/1up.

7 Alicia Contreras, *Capital comercial y colorantes en la Nueva España, segunda mitad del siglo XVIII* (Mérida: Universidad Autónoma de Yucatan and El Colegio de Michoacán, 1996).

8 Rosa Torras y Pascale Villegas, 'La extracción y exportación de palo de tinte a manos de colonos extranjeros. El caso de la B. Anizan y Cía', *Secuencia* 90 (2014): 79; and Contreras, *Capital comercial y colorantes en la Nueva España.*

9 Manuela Cristina García Bernal, *Campeche y el comercio Atlántico yucateco (1561–1625)* (Campeche: INAH/Gobierno del Estado de Campeche/Consejo Nacional para la Cultura y las Artes), 150.

10 See Jobson Arruda's chapter in this volume.

11 Carlos Alberto Zeron, *Ligne de foi: La Compagnie de Jésus et l'esclavage dans le processus de formation de la sociéte coloniale en Amérique portugaise (XVe-XVIIe siècles)* (Paris: Honoré Champion, 2009).

12 John Hemming, *Ouro vermelho. A conquista dos índios brasileiros* (São Paulo: Edusp, 1995), which is the translation of the first volume (Red Gold) of a trilogy on the history of the Amazon.

13 John Hemming, *Red Gold: The Conquest of the Brazilian Indians* (Cambridge, MA: Harvard University Press, 1977), 139–41.

14 Warrren Dean, *A ferro e fogo. A história e a devastação da Mata Atlântica brasileira* (São Paulo: Cia das Letras, 2004), and Shawn W. Miller, *Fruitless Trees. Portuguese Conservation and Brazil's Colonial Timber* (Stanford: Stanford University Press, 2000).

15 Stuart Schwartz, *Sugar Plantations in the Formation of Brazilian Society: Bahia, 1550–1835* (Cambridge, MA: Harvard University Press, 1985).

16 Robert Arthur Donkin, 'Spanish Red: An Ethnogeographical Study of Cochineal and the Opuntia Cactus', *Transactions of the American Philosophical Society* 67, no. 5 (1977): 1–83.

17 Excellent eighteenth-century images of the work to be found in Alejandro Avíla Blomberg and Carlos Sánchez Silva (eds), *La grana y el nopal en los textos de Alzate* (Mexico: Archivo General de la Nación/Conaculta, 2005).

18 David McCreery, 'Indigo Commodity Chains in Spanish and British Empires, 1560–1860', in *From Silver to Cocaine: Latin American commodity Chains and the Building of a World Economy, 1500–2000*, ed. Zephyr Frank, Carlos Marichal, and Steven Topik (London: Duke University Press, 2006), 62–3.

19 Contreras, *Capital comercial y colorantes en la Nueva España*, 49–50.

20 Paul Beauvais-Raseau, *L'art de l'indigotier* (Paris, 1770): 1–5, 29–34. See digital edition at https://gallica.bnf.fr/ark:/12148/bpt6k1067616x/f153.item.

21 See statistical broadside produced by the French government titled *Tableau des Finances et Commerce de la partie Françoise de St. Domingue* (Paris, 1791). This broadside shows economic statistics for Saint-Domingue between 1785 and 1792. A facsimile edition is to be found at: https://www.brown.edu/Facilities/John_Carter_Brown_Library/exhibitions/remember_haiti/econ_tableau-finances-commerce.php.

22 The description of labour practices can be found in detail in Alicia Contreras, *Historia de una tinórea olvidada: el proceso de explotación y circulación del palo de tinte, 1750–1807* (Mérida: Universidad Autónoma de Yucatán, 1990), 37–45.
23 Karl H. Offen, 'British Logwood Extraction from the Mosquitia: The Origin of a Myth', *Hispanic American Historical Review* 80, no. 1 (2000): 113–35.
24 For an excellent resumé on late medieval workshops see Raymond de Roover, *The Rise and Decline of the Medici, 1397–1494* (New York: Norton, 1966). The Medici merchant bankers ran great textile establishments in Florence and were among the most important suppliers and traders in luxury textiles sold throughout Europe in the early Renaissance.
25 On the long history of dyes see Franco Brunello and Manlio Brusatin, *The Art of Dyeing in the History of Mankind, Histoire des Couleurs* (Paris: Flammarion, 1986).
26 For details on Florentine dyeing techniques see Dominique Cardon, 'À la découverte d'un métier médiéval. La teinture, l'impression et la peinture des tentures et des tissus d'ameublement dans l'Arte della lana (Florence, Bibl. Riccardiana, ms. 2580)', *Mélanges de l'Ecole française de Rome. Moyen-Age* 111, no. 1 (1999): 323–56.
27 François Delamare and Bernard Guineau, *Colour: Making and Using Dyes and Pigments* (London: Thames and Hudson, 2000), 47.
28 See Jobson Arruda's chapter in this volume.
29 Francisco Zamora Rodriguez, '"Nervo deste Reyno" La red europea de agentes de la Companhia Geral de Comércio do Brasil', *Topoi Revista de História* 18, no. 36 (2017): 666–89.
30 Raymond Lee, 'American Cochineal in European Commerce, 1526–1625', *Journal of Modern History* 23 (1951): 205–24.
31 Judith H. Hofenk-De Graaff, 'The Chemistry of Red Dyestuffs in Medieval and early Modern Europe', in *Cloth and Clothing in Medieval Europe*, ed. N. B. Harte and K. G. Pointing (London: Heinemann, 1983), 75.
32 Felipe Ruiz Martín, *Lettres marchandes échangées entre Florence et Medina del Campo* (Paris: Ecole Practique des Hautes Etudes, 1965), in particular, chapter 6.
33 See the description of the participation of Hope in these transactions in Marten G. Buist, *At Spes Non Fracta: Hope and Company, 1770–1815, Merchant Bankers and Diplomats at Work* (The Hague: Martinus Nijhoff, 1974).
34 Hamnett, *Politics and Trade in Southern Mexico.*
35 Cristina Torales (ed.), *La compañía de comercio Francisco Ignacio de Yraeta (1767–1797)*, 2 vols. (Mexico: Instituto Mexicano de Comercio Exterior, 1985).
36 Baskes, *Indians, Merchants and Markets.*
37 Jenny Balfour-Paul, *Indigo in the Arab World* (Richmond, Surrey: Curzon, 1997).
38 For useful perspectives on the Asian trades see Jenny Balfour, *Indigo* (London: British Museum Press, 1998).
39 Pierre Pomet, *Histoire générale des Drogues* (Paris, 1694), 154. Online version at http://gallica.bnf.fr/ark:/12148/btv1b8626561c.image.
40 McCreery, 'Indigo Commodity Chains', 65.
41 Jose Antonio Fernández Molina, *Pintando el mundo de azul: El auge ahilero y el mercado centroamericano, 1750–1810* (El Salvador: Biblioteca de Historia Salvadoreña, 2003).
42 Contreras, *Capital comercial y colorantes en la Nueva España*, 45–6.
43 John Fisher, *Commercial Relations between Spain and Spanish America in the Era of Free Trade, 1778–1796* (Liverpool: University of Liverpool, 1985), 68.

44 Dauril Alden, 'The Growth and Decline of Indigo Production in Colonial Brazil: A Study in Comparative Economic History', *The Journal of Economic History* 25, no. 1 (1965): 42.

45 Marguerite Martin, *Les marchés de l'indigo en France: Flux, acteurs, produits (XVII*[e] *– XVIII*[e] *siècles)*, doctoral thesis (Université Paris I Panthéon Sorbonne, 2016).

46 Paul Beauvais-Raseau, *L'art de l'indigotier* (Paris, 1770), 56. Online version at https://archive.org/details/lartdelindigotie00beau.

47 Manuel Lucena Salmoral, *Vísperas de la independencia americana: Caracas* (Madrid: Editorial Alhambra, 1986), 124–5.

48 Robert J. Lancashire, 'The West Indies Chemical Works Ltd. and Logwood in Jamaica', University of West Indies, Working paper on line: http://wwwchem.uwimona.edu.jm/courses/CHEM2402/Textiles/Logwood.html.

49 Contreras, *Historia de una tinórea olvidada*, 54.

50 The trade data and maps of trade routes can be found in Contreras, *Historia de una tinórea olvidada*, 64–92.

51 Alden, 'The Growth and Decline of Indigo Production in Colonial Brazil', 59–60.

52 See Jacques Heers, 'La búsqueda de colorantes', *Historia Mexicana* 11, no. 1 (1961): 1–27, and Manuel Rubio Sánchez, *Historia del cultivo de la grana o cochinilla en Guatemala* (Ciudad de Guatemala: Tipografía Nacional, 1994).

53 M. Mulhall, *Dictionary of Historical Statistics* (London, 1892), 226–7.

54 See Roque, 'Introducción', 18–61.

55 Teresinha Roberts, 'History of Indigo & Indigo Dyeing', *Wild Colours natural dyes* (2023). Online version at http://www.wildcolours.co.uk/html/indigo_history.html.

56 Torras and Villegas, 'La extracción', 80.

57 See David Pretel, 'The Maya Forest and Indigenous Resistance during the Caste War', *Global Environment: A Journal of Transdisciplinary History* 14, no. 1 (2021): 120–45; and Martha Herminia Villalobos González, *El bosque sitiado. Asaltos armados, concesiones forestales y estrategias de resistencia Durante la Guerra de Castas* (México: CIESAS/Porrúa, 2006).

58 Agata Kłos, 'Non-Invasive Methods in the Identification of Selected Writing Fluids from Late 19th and Early 20th Century', *CeROArt* (2014). See digital edition at https://doi.org/10.4000/ceroart.3950.

59 Birgit Reissland, Art Ness Proaño Gaibor, et al., 'Exploring the Late 19th-century Landscape of Ink Manufacturing Via a Collection of 90 Bottles', *ICOM-CC 18th Triennial Conference* (Paris: International Council of Museums, 2017), 1–11. Online version at https://openheritagescienceblog.files.wordpress.com/2017/09/0105_385_reissland_icomcc_2017.pdf.

3

The making of colonial blue

Mesoamerican indigo in the Iberian Atlantic, 1560–1620

Adrianna Catena and Huemac Escalona

As Europe entered the age of exploration, new, exotic dyes remained high on the list of goods to discover – and here the Indies surpassed all expectations. It was with American dyestuffs, as Ana Roquero reminds us, that European master dyers finally resolved 'the greatest challenges in their field: how to obtain the perfect black, and a fiery red'.[1] Around 1560, they would also find a native source of indigo, and with it the ideal blue – a stronger, more versatile dye than Mediterranean woad. Production boomed over the following decades in parts of New Spain and along the Pacific Coast, from present-day Guatemala to Nicaragua; Crown and settlers were enthusiastic. By the turn of the century, indigo ranked third (with hides) among the most valuable of Spain's New World imports.[2]

The introduction of New World dyes had a large and lasting impact on the early modern European textile industries, stimulating technical innovation and contributing to broader transformations within the dyeing sector. Yet all this change was not without its detractors. Over the course of the sixteenth century, increasing access to New World dyes provoked enthusiasm, but also conflict and resistance from artisans and merchants.[3] It may be because these were ultimately stories of success, that the initial moment of encounter with these dyes, their reception and accommodation within the dyeing sector, has received limited attention from historians. This is especially the case for Spain, a neglect it hardly merits, when it lies at the centre of this experience – it was in the textile centres of Andalucía and Castile that samples of American dyes were first received, examined and tested.

When Spaniards turned to the large-scale production of indigo in the Americas in the late sixteenth century, they found themselves on new and unfamiliar ground, drawing on local knowledge and expert labour, and what information they might gather from networks of merchants that reached the Asian indigo trade.[4] So it was also here that indigo began the transition from an unfamiliar, exotic product, to a commodity that, by the eighteenth century, had become an ordinary household item on either side of the Atlantic.[5] This chapter considers some of the earliest episodes in the history of colonial New World indigo: an initial moment of encounter with the dye, and the trials and tests that followed, considering the role of expert indigenous labour and skills in the development of production technologies.[6] A second section turns to the emerging merchant networks, linked across the Atlantic and specializing in the dye trade, that would enable a first commercial boom; while a third considers the reception of this new, Spanish blue, among Old World dyers. Focusing on the initial development, or birthing pains, of a new, Atlantic commodity – the production, circulation and consumption of Mesoamerican indigo between 1560 and 1620, this chapter highlights the connections and transformations arising in this period across two distant and different spheres, intimately connected by indigo.

'Consulting with the Indians of this land'

Unlike cochineal, spotted almost immediately by Cortés and commercialized across the globe before the mid-sixteenth century, indigo is markedly absent from many of the earlier chronicles, seemingly unremarkable and unobserved. Doubtless, however, the new residents would have been witness to its various indigenous uses.

Termed *xiuhquilitl* in the Mexica Nahuatl and *ch'oh* among the Maya, indigo was a valuable pigment in pre-colonial Mesoamerica, applied on ceremonial wear, textiles and murals, used to darken hair and as a prophylactic for lice. But perhaps most striking to Europeans was its use on the bodies of sacrificial victims: In Yucatan, a horrified Diego de Landa saw 'those who were to have their heart torn out . . . smeared in blue', and carried, surrounded by crowds, to the raised sacrificial stone, painted blue by the priests before the victim was finally laid upon it.[7] Over a long, and protracted conquest, the new settlers observed similar ritual episodes.

The most revealing, surviving source for the native extraction, use, and application of dyes and pigments is Bernardino de Sahagún's encyclopaedic *Historia general de las cosas de Nueva España*. There are two relevant chapters: from Book 10, Chapter XXI: 'De los que venden colores', with a section on 'tintoreros', and from the following volume, Chapter XI: De los colores, de todos los colores', including the most extensive reference to indigo:

> There is an herb in the hot lands called *xiuhquilitl*, they crush this herb and squeeze out its juice and pour it into vessels; there it dries or curdles. With it they dye it a dark and radiant blue: it is a very appreciated colour.[8]

The accompanying illustration for *tlacehuilli* (Figure 3.1), depicts the process. To the left, a man squeezes the indigo leaves (from what clearly resembles an *indigofera* plant in the background), letting the juice flow into a bowl (a symbolic rendition of a more complex process). At right, a scribe uses the pigment, and in the centre we find a drawing of what appear to be large, circular loaves atop a small basket. The image aligns with Francisco Hernández de Toledo's near contemporary description: indigo, the Court physician wrote, was 'formed into small wheels, hardened by being placed on plates over coals', that could be 'kept for use throughout the year'.[9]

Indigo has been in use since at least 6000 BCE, extracted from several hundred species of *Indigofera* native to Asia, Africa and the Americas.[10] Production methods have varied regionally and over time, but there are broadly three ways of obtaining indigo dye from *indigofera*. The simplest of these is to combine the plant matter with a fermenting agent directly in the dyeing vat, and immersing the cloth, yarn or other, within. Alternatively, the leaves can be composted together and shaped into balls, a method that remained common throughout Asia well into the nineteenth century. The more complex system employed in the colonial processing of indigo yielded the highest-quality dye, and involved the extraction of the pigment through an initial stage of fermentation (soaking the leaves in large vats of water), followed by oxidation (beating the mass), before the resulting paste was scooped out, dried and flattened in racks under the sun. In a final step, the hardened pigment was cut into smaller squares or rectangles – light, durable and ideal for trade.[11]

Whether the description in the Florentine Codex refers to the second or last of these cannot be determined from such a brief text. That the Maya should have used it for their famed blue pigment suggests a higher tinting strength, associated with the third and more complex method of extraction, a conclusion reinforced by the quality of the earliest exports of New World indigo, doubtless produced with (or by) expert, indigenous labourers.

The evidence suggests official interest played a substantial role in the initial surge of enthusiasm for the trade among New World settlers. The Spanish textile industry was uncomfortably dependent on massive imports of French and Azorean woad – the principal and most widely used source of blue in Europe – and the Indies presented an opportunity to reverse this. Over the first half of the sixteenth century, the Crown sponsored repeated attempts to cultivate woad in the Indies; in 1558, just as a last of these ventures failed, news of an alternative reached the Court, and the Council of Castile issued a request for information across the empire's new territories – from New Spain to Buenos Aires – concerning a native dye rumoured to have 'the same effect as woad'.[12] Within a year, samples were sent from New Spain to Seville, and remitted for inspection to the dyers of Segovia, who found that it was indigo.[13]

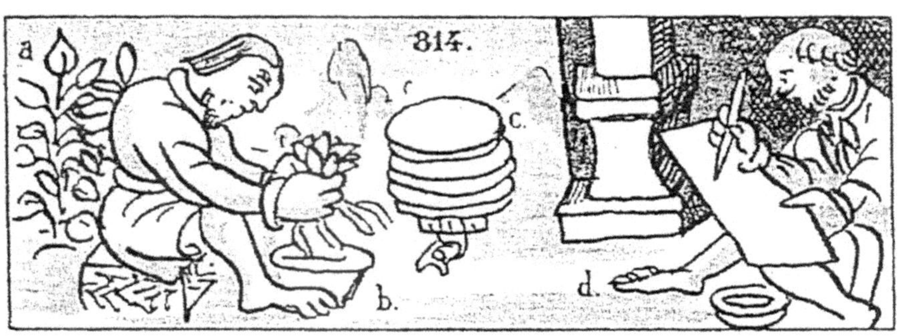

Figure 3.1 'Xiuhqilitl', *Historia General de las Cosas de la Nueva España*, Libro 11, cap. XI, fol. 219R (*World Digital Library*) https://www.wdl.org/en/item/10096/ (accessed 20 January 2023).

As the news reached back across the Atlantic, a small number of settlers began to explore indigo's potential as an export commodity. Principal among these was Pedro de Ledesma, a former conquistador resident in the City of Mexico, who obtained the first and only monopoly over the dye's manufacture and sale between 1563 and 1572.[14] Ledesma made ample use of his privilege, setting up various indigo producing ventures, usually in partnership with local *encomenderos* (who provided the land and labour), instructing each in the making of the dye. Towards the end, however, Ledesma found himself on the defence, attempting to fine an increasing number of indigo producers for infringing his monopoly. Nor was he the only settler to claim the 'invention' of indigo in this period.

Another early enthusiast was Marcos de Ayala, a native of Toledo who had served under Francisco Montejo in the conquest of Yucatan. In the aftermath of a prolonged offensive, Montejo rewarded his soldiers – the region's first Spanish settlers – with *encomiendas*. Yet the absence of significant mineral deposits in the region fuelled perceptions of poverty and stagnation, encouraging a disillusioned group of *encomenderos* to search for economic alternatives.[15]

Ayala received an encomienda in the vicinity of Valladolid, where he set about making silk – breeding silkworms and planting the mulberry trees to feed them – exporting the finished product with some success to Spain and Flanders.[16] Alongside these ventures (with aim at dyeing his silks), Ayala embarked on a focused search for dyestuffs, experimenting with local trees and herbs,[17] claiming the discovery of various colours, 'especially that of indigo and a wood which in this language they call *Ec* (logwood)'. But he had not done this alone: Ayala had 'consulted with the Indians of this land' and hired master dyers from Spain and Flanders, brought at his request to Yucatan.[18]

Around 1561, armed with his recent discoveries, Ayala relocated to Mexico City, where he presented his findings to the Viceroy.[19] Three years later, a letter from D. Diego Quijada, governor of Yucatan, reported that Ayala had 'insisted much on this business' and finally discovered 'how it is done'. The governor sent samples with the letter, observing that several residents had begun cultivating 'these trees' of indigo, and requested licence for the use of indigenous labour in the dye's production. The real news was cultivation: this may well be the earliest reference to the planting of *indigofera* – a Spanish development that seems to have occurred first in Yucatan.[20]

Among the governor's eager residents was Ayala's nephew, Guillermo de Peralta, who had heard of royal interest in the matter from his uncle and wished to communicate his own enthusiasm for the trade.[21] Following Ayala's instructions, received by way of a written report, Peralta had also 'succeeded' in the manufacture of indigo, expertise he would later share with Martín Ruiz de Arce and Juan López de Mena. These two *encomenderos* wrote to the Viceroy on 15 January 1565, in a letter that would include a vivid account of the informal trials and tests that shaped production methods. 'One afternoon', Ruiz de Arce recounted,

> the boys brought some leaves and we left them until the next morning . . . but not as much indigo came out as in the vat the day before. And another day there were too many leaves, and I told Juan López that what was left we should leave to soak

> that night. . . . . And it seems, without a doubt, that by carelessness we hit the point that indigo needs, because soaking it from one day to the next makes it twice as much as cooking it fresh. . . . And this is its 'point', to cook it in such a way that it does not boil, nor gets hotter than in the time of an Apostle's *Creed*.[22]

With his letter, as noted in the marginalia, Ruiz de Arce sent a sample of the indigo produced, only a 'loaf', he clarified, as the rest had been sold. These 'loaves' suggest that they were already employing a more complex extraction system, and at the very least manufacturing a product that could withstand the Atlantic crossing. These brief accounts can give us an idea of how practices and knowledge were created and disseminated among Spanish settlers, validated through empirical testing and circulated (even in writing!) through friends and relatives.

The first Europeans involved in indigo production acknowledged their debt to indigenous techniques, especially when it was convenient for them. In what was ultimately an attempt to boycott Marcos de Ayala's claim to secure a monopoly over indigo production in the Peninsula, the notary Gerónimo de Castro wrote that Ayala could not claim the invention of indigo as his own, because it had been used before his time by the indigenous inhabitants of the land, and later by the conquistadors, for their cloth and silks.[23] What Ayala and others, such as Pedro de Ledesma[24] perceived before their fellow colonists, was indigo's potential as an export product. Within a few years, references to the industry had swelled in correspondence from the region, including detailed reports of evolving production methods.

The earliest description of the manufacturing methods employed in the region is a report by D. Carlos de Arellano, *alcalde ordinario* of the city of Mérida, to the Crown, accompanied by the usual letter of praise for the new industry, with a final request for a decree authorizing *encomenderos* to produce *indigo* within the 'Indian *villages* of their *encomiendas* . . . and with the Indians of their *encomiendas*'.[25] The document is undated, but likely created in 1573, and no later than 1575.[26] Arellano described three methods of extraction, all referring to the most complex of the systems previously described, suggesting that pre-colonial processing techniques may have been no less evolved than those employed in Asia, producing a comparable good in form and quality.

The most relevant divergence in Arellano's methods relates to the temperature of the water used for fermentation, the earliest focus of debate in the manufacture of indigo. Traditional processing methods in pre-Hispanic Mesoamerica seem to have involved heating, but the use of cold water would lower costs. It may not have been a perfect substitute: the best indigo produced in Guatemala was made through hot water extraction, and the same was true of artisanal production in India, where hot water (affordable only on a smaller scale) was associated with a higher-quality product. In 1575, Viceroy Martín Enriquez wrote to the Crown in this regard:

> In what concerns indigo, I sent for Pedro de Ledesma and was informed of the way in which he thought (this trade) could grow and be of such use as he had written. And his resolution was to say that at present it was made in small vessels and with hot water, that he would make it with cold water and in vats that held a greater quantity, and that for this, houses and other mills were to be made.[27]

Two years later, forty-eight indigo mills were operating in Yucatan, producing some 15,000 pounds each year.[28] Yet this growth would also coincide with a period of famine, drought and disease on the Maya Peninsula. Although initially enthusiastic about the new trade, reports of its impact on the health of indigenous labourers moved the Crown to prohibit their employment in the production of the dye in 1563, and again in 1581 and 1601.[29] The need to appease a Crown increasingly preoccupied with the demographic collapse of the native population stimulated the introduction of machinery (particularly for the strenuous beating process) and other, small adaptations designed to alleviate the more labour-intensive stages of production. The evolution of technology for the beating process, a distinctly Spanish-American development, can be highlighted: while smaller estates no doubt continued to use paddles turned manually, larger plantations had wheels inserted into the vat, rotated by mules or water wheels. The influence of these restrictions in the development of new technologies here, before anywhere else, seems evident.[30]

For all these efforts, Yucatan's indigo boom was short-lived: between 1590 and 1600, a mere 700 pounds were shipped from the port of Campeche.[31] Local producers were united in their blame: the Crown's decision to ban indigenous labour had crippled them. Making the dye might have become costly and complicated, but competition had also grown from the South, where an industry was beginning to prosper. By 1620, several hundred *obrajes* had transformed the landscape along the Pacific Coast, from San Salvador to Nicaragua.[32] Here, as in Yucatan, the development of the indigo trade became a moving factor in the acquisition of indigenous land; as Luis de Morales, resident in Guatemala wrote in 1631: 'of those great populations all we see today are the sites were the towns lay, and which are now our principal indigo mills'.[33]

In 1610, the Crown drafted its own instructions for the cultivation and manufacture of *añil*, setting a standard for production in what would come to be known as the 'royal obraje'.[34] Achieving a high-quality dye from *indigofera* requires careful work and expertise, from cultivation to processing – quality will be affected by the timing of the harvest, the plant matter employed, the water used for processing and the labourer's skill at each productive stage, controlling processes of fermentation and oxidation. Indeed, by the first quarter of the seventeenth century, and until the industry waned in the mid-nineteenth century with the development of a synthetic alternative, some of the highest-quality indigo from Guatemala was made by specialized 'poquiteros': small, predominantly mulato producers, as Antonio Fernandez has observed, 'who (whatever their genetic makeup), created their own culture, dissociated from the Indigenous one'.[35] For over two centuries, Mesoamerican indigo maintained a reputation for quality, broadly ranked alongside the Indian varieties, and well above exports from French Saint-Domingue, South Carolina and the British Caribbean.[36] In the seventeenth century, French and Dutch entrepreneurs drew intentionally from the Spanish example in the organization of colonial production.[37]

Taken together, the surviving descriptions of production reveal an uneven trajectory with several discrete stages of innovation between *c.* 1561 and 1578, highlighting the activities of a few, dynamic individuals as they kindled an industry, transforming this new and unfamiliar article into a global commodity. Rather than an excessively linear model of development, therefore, the evidence indicates various practices and techniques coexisted in this period, incorporating elements of indigenous, pre-Hispanic processing methods, and evolving to meet the demands of a nascent export industry. As early as 1577, reports from Yucatan would encourage drying the dye as much as possible, thus reducing its final weight, but also, and more significantly, because this was how it was preferred in Spain: 'they ask that it be blue over purple, and light'.[38] Their geographic isolation notwithstanding, producers in Yucatan were well informed: 'purplish hues' were considered indicators of high quality well into the eighteenth century. Insofar as possible, they responded to merchant demands, modelling their goods accordingly. The example may also highlight the role of merchants in the circulation of knowledge regarding these new goods – as in assessing their potential. At every step, as Prakash Kumar has noted on the topic, 'claims regarding knowledge of improvement had to be validated in the marketplace'.[39]

Agents and networks of trade

In July 1576, Francisco Morovelli, a factor for Simon Ruiz – that 'most famous of all Castilian merchants',[40] wrote to his employer from the port of Seville:

> With these ships from New Spain . . . 500 *arrobas* (12,500 pounds) of indigo have arrived, which they say is much better than what comes from India . . . and many people have sent samples, who write that if it were good business, they could send much more, as the land is good for it.[41]

Settlers were enthusiastic, confidently sending samples for testing in Spain[42] – a first step in the introduction of new Atlantic commodities. Better yet, rumour suggested the new dye surpassed its Indian counterpart: perceptions of quality that would be key to indigo's early success. That September, Morovelli confirmed the recently arrived dye had sold at an exceptionally high 30 reales the pound. Most of it, he explained, had been taken by Portuguese merchants who had come from Lisbon with this objective – 'and they say it is very good, and much better than that of India'.[43]

In this first period, as increasing quantities of indigo reached Europe after 1560, the Portuguese emerged as something of ambassadors for the trade, efforts that would include the dye's promotion abroad and diffusion of proper dyeing techniques. So, for example, in 1577, Pedro Vaz Devora travelled to London, where he instructed representatives of the merchants and dyers of the city in the art of using *aneel* (from the Spanish *añil*, for indigo). The resulting request for permission to use the new dye in England would sum up indigo's advantages over woad: at 40 shillings, 4 pounds of indigo made as much dye as 100 pounds, or 50 shillings worth of woad.[44] But it wasn't just stronger and cheaper: indigo also gave a more attractive, 'oryent' blue. A year later, highlighting the official channels through which knowledge and practices might travel, Dutch officials 'allowed an envoy of the King of Portugal to teach the Leiden textile dyers to dye with *añil de indigo*'.[45]

Interest in the dye was on the rise, yet suspicion and doubt also permeated these early exchanges: 'regarding indigo', wrote Francisco Morovelli in 1578:

> there is such deceit in this commodity, as I have been told, that I have not dared trust anyone, although at 22 and a half *reales* I could have bought more than 300 pounds that a merchant has here . . . it is a matter that must be well understood, and if you should have the intention to buy some amount when the fleet arrives, it would be worthwhile for someone from Segovia or elsewhere to come here, if only to certify it's quality.[46]

Segovia was the 'main centre of the Spanish wool industry' in the late sixteenth century, producing an average 13,000 pieces of cloth per year and specializing in higher-quality dark woollens: violets, blues, blacks and browns, all of which required a base of woad.[47] The city's master dyers were broadly accepted as the best in their field, and it was here that Mesoamerican indigo was first identified for what it was, from samples sent by the viceroy of New Spain.

Simón Ruiz had ample business and an agent in Segovia. For over a decade, Juan de Amarita's activities were mostly concerned with the sale of indigo and textiles assiduously reported back to his employer. The earliest reference to the sale of American indigo in the surviving correspondence appears in April 1572, a decade after reports of a new blue dye had reached the European markets, but Amarita was already selling both new and old world indigo, and keeping separate accounts for each.[48] The dye was dispatched to Segovia from the warehouses of Ruiz in Medina del Campo, packed in crates of an approximate 100 pounds or barrels with a net weight of 150 to 180 pounds.[49] Demand in Castile coincided with the availability of wool, suggesting most sales were destined for the local textile industries – indigo never failed to sell well from June to July, between Saint John and Saint James, after the 'new wools' (*nuevas lanas*) reached the market.[50] By 1578, when Morovelli expressed his misgivings, indigo had appeared in full force on the Castilian markets; 'there is so much indigo it is pitiful', Amarita complained that December – 'there is nothing in Segovia but indigo'.[51]

Records point to an early preference among Castilian buyers for the American varieties; as Amarita insisted throughout 1578: 'I have written many times how, even if the Portuguese indigo is very good, it will not sell well while there is (indigo) from Seville'.[52] Two years earlier, along similar lines, the merchant Luis Alvares Caldeira, writing from Antwerp to Simón Ruiz, was anxious to sell what indigo he had left from Lisbon, despite reports that little had come that year in the *naos da India*, because he had 'seen how much has come from the Indies, and how good, and if it should be sailed here, as soon as it can be had, no other will sell'.[53] A survey of Amarita's letters indicates that in the 1570s and 1580s both the American and Asian varieties sold at similar (if volatile) prices for like qualities. Prices did, however, vary considerably between these. In October 1583, merchants in Segovia were selling the higher-quality indigo from New Spain at 28 to 30 reales the pound, and poorer loads for 18 to 20 R/Lb.[54]

Providing standards for the product's evaluation, quality grades gradually came into use. These classifications evolved from the needs of merchants and producers, from exchanges at the port and through correspondence; in their rawest form, they began taking shape as early as 1574.[55] Describing a shipment of eight barrels of indigo sent to Simón Ruiz, the Portuguese merchant Antonio Gómes wrote: 'the four first are the best, especially numbers 1 to 3; n. 4 is poorer, but the difference is slight. Of the last four, n5 and 6 are very good, and n 7 and 8 are regular'.[56] By 1620, exports from the Audiencia de Guatemala were classified into three grades: *corte*, *sobresaliente*, and the highest quality *flor tizate*.[57]

By the close of the sixteenth century, Castile, at least, had fully embraced the new blue. In 1594, writing to an agent in Lisbon, Simon Ruiz indicated local dyers would buy French woad only if demand remained unsatisfied by indigo.[58] The Atlantic trade grew steadily over the next few decades, becoming increasingly specialized after 1590, with the lion's share of imports controlled by a small group of merchants in Seville. Prominent in this period were the Carmona Tamariz merchants; over the course of three generations and coinciding with this initial boom, the family firm gradually gained control over a substantial share of imports reaching Seville.

A conventional family of Atlantic traders, the Carmonas moved to Seville from Granada, making the transition from silk mercers to Atlantic wholesalers in the aftermath of the War of the Alpujarras. Hernando de Carmona began with a few small investments, and the opportunities were quite real, as Ruth Pike has noted: 'miracles of social mobility were being performed every day'.[59] Within a decade of his arrival, Hernando had consolidated his position in the city, and ensured a promising future for his children: Pedro and Bartolomé would preserve the family firm on the peninsula, while Diego and Andrés took up residence in Puebla, extending the family's social and commercial ties across the Atlantic. Even his only daughter, Isabel, had been successfully wed to a prosperous wholesaler. Two years after the firm received its first shipments of indigo, a last son, Tomé, was sent to Santiago de los Caballeros in Guatemala, signalling the firm's intention to expand its commercial interests in the region. Tomé did exactly this, creating and developing the connections that would, within a decade, allow the Carmonas primacy over the Central American indigo trade.

By 1599, the Carmonas had settled into a variety of shipping patterns in their exchanges through Honduras. The bulk of merchandise was shipped independently, with the risk absorbed in Seville by Hernando, Bartolomé, or both. These consignments were dispatched to Tomé and Miranda in Santiago de Guatemala, who remitted the proceeds of the sale (often in kind) after deducting a commission. Increasingly, the Seville Carmonas began receiving shipments directly from Santiago merchants, which they would handle as encomenderos themselves, remitting the proceeds from the sale on the next departing fleet. Both Tomé and his relatives in Seville, finally, acted as intermediaries in a stricter sense, dispatching or delivering goods to third parties.

Returns for the Seville merchants, as Atlantic historians have stressed, were slow to materialize; cargo shipped to the Indies had to be sold and its proceeds returned, often in kind, before profits could be calculated, a cycle that – because the fleet wintered in the Indies and was usually delayed in the spring – could take up to three years. Above all, the *registros de carga* – shipping registers compiled for each ship, with individual entries detailing the nature of goods sent by a particular merchant, their destination and named recipient – suggest a profound (if justified) preoccupation with risk: shipments sent to each Santiago merchant were systematically distributed among several vessels. So, for example, on the return leg from New Spain in 1606, the firm received 108 *cajones* and 121 *arrobas* (some 14,000 pounds) of indigo spread among sixteen vessels.[60]

By 1603, the Carmonas were shipping goods to at least 18 residents of Santiago, up from 8 in 1601. The tenor of their clients had evolved at a similar pace, to include some of the city's more prominent figures. Records suggest that relations among the principal Sevillian merchants investing in the indigo trade were especially close – it was a small group that conducted business with and for each other, as partners and intermediaries. Evidence for the redistribution of indigo is more limited; under normal circumstances, ordinary sales would hardly have required a notary. Large volumes of indigo were reexported to Italy and the Netherlands – the Carmonas might have had connections in Venice and Florence through their ties to the Fantonis and Federighis,[61] while their Portuguese associations could have given them access to the Dutch markets. More important, however, were the family's lasting ties to the silk industries of Granada and Toledo, which remained stagnant but alive in the early seventeenth century.[62]

Key to their trajectory and success were the social and familial ties maintained over the course of generations, countless connections to the textile sector, friends, relatives and commercial relations that spread across an empire. Their marriage practices remained endogamous to the textile sector – Diego de Carmona married the daughter of a cochineal producer in Puebla, while his brother Tomé wed a cloth merchant's daughter in Santiago. Their example also highlights the strategic moves that could characterize investments in the Iberian Atlantic, keeping trusted agents close to the centres of production from where they might extend the reach of their commercial networks, usually by expanding their social ties – efforts that involved buying public offices and becoming familiars of the Inquisition.

As representative as they were of the possibilities of success, the Carmonas also suffered the constant (and as many unforeseeable) risks of Atlantic commerce – goods damaged or lost at sea, and the legal battles that ensued. In 1614, 524 crates of indigo were lost to a storm after being transferred onto Melchor de Arce's 'Nuestra Señora del Rosario y la Concepción', when the vessel's main mast cracked and collapsed, smashing a hole into the ship's side.[63] Much of the dye had simply leaked out, while the rest arrived wet and unusable. Two trials and several years of legal proceedings ensued.

The fourth generation, to use Enriqueta Vila Vilar's expression, 'lost the commercial spirit'[64] altogether, becoming public officials, entering the clergy and seeking the usual succession of offices and honours. As of 1616, a handful of merchants gradually increased their presence in the *registros de Honduras* and appear to have finally absorbed the Carmona's share of trade around 1620. The Carmonas reached the apogee of their success in Central America between 1606 and 1613, regularly handling an important percentage of the export trade, as even a cursory glance at the *registros de venida* reveals, with page after page of 'Carmona' or 'Carmonas' scribbled on the margins (by the functionary of the Casa de la Contratación, for the purposes of taxation), and the word 'añil' not far removed. In this same period, from 1608 to 1614, total values for indigo exports (including shipments from New Spain and Tierra Firme) exceeded those of cochineal.[65] The close of the family's exchanges with the Indies also coincided with a period of global decline for the indigo trade: a series of locust plagues ravaged Central American plantations, while altercations between the Estado da India and the Mughals precipitated a drop in exports from 1613 to 1618 – signalling an end to this first boom in production and trade.[66]

Into the dyer's vat: Resistance and adaptation

The dyer's craft was the most regulated of all trades in the Spanish textile sector,[67] as might readily be gathered from the length of the section dedicated to this occupation in the *Ordenanzas Generales del Obraje de Paños*, begun under the Catholic Monarchs and issued in 1511, standardizing the production of textiles in the realm.[68] Beautiful, solid colours could make a huge difference in the desirability and sale of textiles, while achieving the same tone consistently on long lengths of cloth required considerable expertise – the use of large equipment, complex dyeing recipes and hazardous, often toxic, mordants. In Spain and across Europe, standard examinations for the dyer's guild involved the dyeing of a long length of cloth to a standard of blue. Dyers were also last in a sequence: a poor job, at this stage, could ruin the work of an entire chain of production. Compounding the risk, a dye's faulty application or the use of lower-quality raw materials might only become noticeable over time – as the cloth was manipulated by merchants or buyers, ruining reputations. Strict regulations thus governed the entire dyeing process, limiting the range of permissible dyes, their application on each fibre and grade of cloth; compliance was another matter.

In Spain as elsewhere, the centrality of blue to the productive process – blues obtained exclusively with woad before the introduction of indigo, cannot be overstated. Spanish master dyers were broadly divided into two categories: those who dealt in an initial base of blue, the so-called *tintoreros de blao*, and the *tintoreros de bullón*, occupied with a second stage when the cloth or wool was overdyed in other colours with the use of mordants[69] – similar divisions could be found across Europe. Dyeing in any shade of green, dark red, brown or purple required a base of blue, achieved through one or more immersions (termed *celestes*) in the large woad vat. Dark greens, for example, were given two *celestes* first, then mordanted with alum and overdyed in yellow weld.[70] Standard samples of these different *celestes* – pieces of cloth in gradations of blue were used by inspectors for the dyer's guilds across Spain in the assessment of quality (see, for example, Figure 3.2, depicting the drapers' guild of Leiden conducting one such evaluation).

Into this world of woad came indigo – a new, superior blue from the Indies, East *and* West. For all indigo's advantages, however, the transition between blues was neither smooth nor immediate. The conflict provoked by the introduction of indigo in Europe – the so-called war with woad, is well known. In the late sixteenth century, as the volume of indigo reaching Europe increased, so did legislation against it. The German Diet at Frankfurt proscribed its use in 1577, 1595 and 1603,[71] as did France in 1599, and again in 1609,[72] restrictions that were only fully lifted in the eighteenth century. In England, the combined use of woad and indigo was authorized in 1581 (after an earlier ban), but declared 'poisonous' that same year, ostensibly as a result

Figure 3.2 The drapers guild, Leiden (1675). *Source*: Jan de Baen, *De Staalmeesters of Leiden* (1675), Museum De Lakenhal, Leiden. Courtesy of the Museum De Lakenhal.

of hazardous attempts to use an Indian dyeing method employing arsenic.[73] This orpiment vat, which posed significant dangers to the dyer, may also have given the dye a difficult reputation in Genoa, where indigo was banned until 1655.[74] Moving this barrage of legislation was a powerful woad sector, but also fear and suspicion arising from technical inexperience. Yet none of these measures were easy to enforce; that they had to be periodically reiterated seems to suggest a failure, rather, to stop the dye's diffusion.

The Spanish textile industries were not entirely unfamiliar with indigo: there is evidence of its use in the royal dye houses of medieval Castile, and within the silk industries of Granada and Valencia, even as they declined throughout the sixteenth century – but it was not employed on wool. Indigo is not even listed in inventories of materials used by Castilian dyers in 1550.[75] Rather, the sector was dominated by woad, where Spain remained an insignificant producer in the sixteenth century, satisfying a large demand with imports from the Azores, or the higher-quality French woad, a profitable trade dominated by Burgalese merchants that fed the large Castilian wool industry.[76] When it arrived, then, indigo was introduced as a product of the realm. Yet the introduction of indigo within the Spanish dyeing sector had larger consequences that generated resistance from the dyer's guild. Six years after the earliest Portuguese imports reached Lisbon, with the publication of the *Ordenanzas* in 1511, indigo was listed as 'false' – a term indicating lack of fastness. The impact of indigo's introduction, however, would not be felt before mid-century, and coincided with both the rise of New World imports and the diffusion of the dye's use on wool.

In 1560, just as indigo began to appear in increasing quantities on the Spanish markets, the dyers of Córdoba condemned this 'evil and false' dye, suggesting the man who introduced it, the Portuguese Pedro Vaez, should be 'banished from the kingdom'.[77] Two years later, the Cortes de Madrid petitioned to ban indigo on grounds that it 'spoiled' the cloth.[78] The reception was warmer in Segovia: a more specific petition by the city's merchants that same year proposed restricting its use on black *velartes* (a popular, high-quality woollen cloth), but the city's dyers appealed this petition, requesting their work to remain undisturbed – that is, to be allowed to continue using indigo – while the Council of Castile investigated the matter further.[79] Highlighting the weight and reach of it all, the dyers of Cordoba would attempt to influence these deliberations in Castile with a report of their own, adamant in the claim that indigo was a fugitive and corrosive dye. Yet Segovia would win this first exchange, and the Council issued a temporary decree allowing the use of indigo for a limited (but unspecified) period of time.

Towards the end of the 1560s, and over the course of the following decade, as the volume of imports increased and purchases became more regular, the nature of complaints changed; it was no longer a question of whether or not to use indigo, but rather of how and by whom it should be applied. In January 1566, the local council issued a first decree regulating its use in the city. 'We have been told that the dye called indigo has recently been brought to these kingdoms, and being new and unknown, there have been different opinions regarding the methods that should be used. And the experiences of a few years have shown the best way to dye with it, and the inconveniences that result from using other methods'.[80] After a diligent investigation, then, the application of indigo had been mastered, and the city now issued a brief set of rules.

Indigo had been introduced as a supplement for woad and gradually added to the vat in higher proportions. But the most visible and practical repercussions of its introduction involved the possibility of using smaller, more affordable vats (*tinacos*) to achieve these blues on wool. Indigo was a much more concentrated dye, 10 pounds dyeing as much as 130 pounds of woad, and could therefore be employed in smaller receptacles.[81] Evidence of a more widespread phenomenon came to light a decade later, in 1578, when the Crown attempted to restrict the use of indigo in *tinacos* and the smaller *tiñajuelas*, *ollas* and *calderillas* (little vats, pots and tubs).

The first to appeal were the drapers of Segovia, no less, suggesting the earlier regulations had been ignored. Instead, as Javier Vela has observed, this document would suggest indigo had revolutionized the sector, 'introducing radical changes in the dyeing process and those qualified to perform it'.[82] As the appeal claimed: with these new materials and techniques 'so much cloth has been made, as to make it cheaper, and *anyone can dye, without having to be a master dyer*' (authors' italics). Many 'poor people', the appeal continued, could work 'who would not be able to if their wools had to be dyed in large vats, because much capital is needed for them, and to dye with the said *tinacos* in indigo and madder no money at all is needed, because each person can dye what they have little by little'.[83]

This last comment on the dyeing of things 'little by little' may be a reference to the indigo vat's endurance: protesting their use of indigo in court, Genoese dyers claimed a combined vat was best, not only because of its 'wonderful effect', but also because mixed in this way the vat could be kept alive for four to six months, whereas the simple woad vat didn't last more than twelve days.[84] The Segovian appeal would also seek to revoke an earlier ban on urine: 'experience has shown that with fresh water (the dyeing) is imperfect, and bleeds, and using urine it comes out perfectly' – all of which is true: urine is an ideal reduction agent for indigo and particularly well suited for wool.[85] Last, 'the city's merchants and makers of cloth', blamed those with interests in the large dye works that used *tinas*, of attempting to restrict the use of *tinacos*. The conflict, then, was between the city's master dyers and a host of others – merchants, drapers and humbler artisans in the dye trade – who stood to gain from this small-scale use of indigo.

The same *tinacos* (small vats) provoked further conflict some years later, when the hatters of Segovia were accused of endangering the population with heated dyeing vats and polluting the main Calle Real with 'foul odours' by pouring the contents of their *tinacos* onto the street.[86] Segovia was the main manufacturer of hats in Spain between the fifteenth and sixteenth centuries, and their wares had a reputation for quality well beyond Spain. More than eighty master hatters (similar in number to the city's dyers) were active in the late sixteenth century, producing a variety of goods in the more fashionable, darker shades of black and blue. As outlined in the *Ordenanzas* of 1567, all woollen 'hats, bonnets, and caps' in shades of 'purple, greens, and blues', required one or more celestes.[87] The increasing availability of indigo, then, made it possible for hatters and artisans with more limited means to conduct their own dyeing. Yet this also implied new dangers (the risk of fires) and forms of pollution. In this sense, the integration of indigo into the productive process would also affect deeply rooted concerns related to urban spaces: medieval Spanish dye works, like other 'foul-smelling, environmentally unfriendly'[88] trades were kept at a distance from towns and cities.

A similar phenomenon took place in Córdoba, where José Ignacio Fortea Pérez has documented a marked increase in non-professionals or newcomers to the dyer's craft encouraged by the introduction of indigo. After 1581, Córdoba's Municipal Council issued numerous licences to 'wool dyers in indigo', usually humbler artisans residing outside the city[89] – again, changes tied to the diffusion of indigo's use on wool.[90] There was a lot more resistance to the diffusion of indigo's use here, however, than in Segovia. As late as 1615, a local ordinance banned the use of indigo on the highest-quality black wools, because it was 'a false dye for them'.[91] Nor were these the only conflicts to arise from the introduction of añil – resistance took various forms, 'growing pains' that New World indigo would experience on a global scale.

Birthing pains of an Atlantic commodity

While this chapter has focused on indigo's introduction to Spain and Europe, the nascent industry would also transform surrounding regions; by the first decades of the seventeenth century, and despite sporadic restrictions on trade between Santiago de Guatemala and Lima, indigo-dyed, ordinary blue woollen cloth had become a staple at the textile *obrajes* of Quito and Cuzco.[92] Across the Americas, and throughout the colonial period, ordinary, indigo-dyed wool and cotton textiles were used overwhelmingly for rough workwear. The only surviving fragment of 'slave' cloth – made in late eighteenth century Penistone, West Yorkshire, and recently discovered in Derbyshire – is a small rectangle of indigo-dyed, deep blue wool.[93] Figure 3.3 illustrates some of the indigo dyeing methods employed in eighteenth-century Peru, from left to right: the dyeing

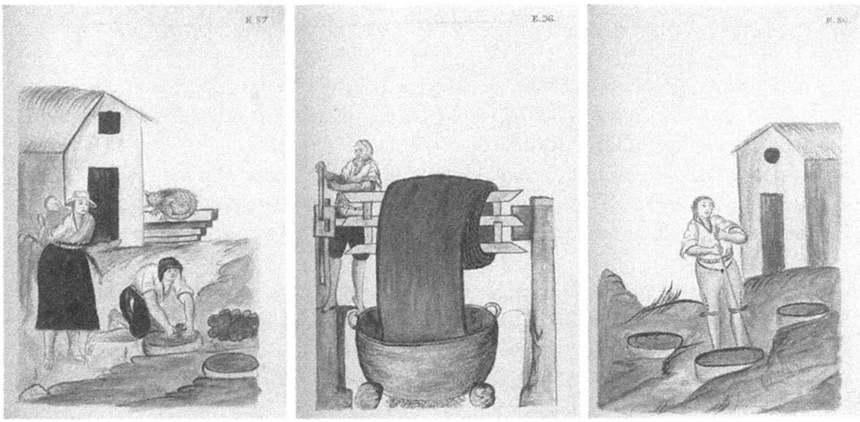

Figure 3.3 Indigo dyeing methods (eighteenth-century Peru). *Source*: (images left to right), *Yndios tiñendo lana*, in Baltasar Jaime Martínez Compañon, *Trujillo del Peru* (1782–5), Vol. II, estampa 87. Biblioteca Virtual Miguel de Cervantes; *Yndios tiñendo ropa*, in Baltasar Jaime Martínez Compañon, *Trujillo del Peru* (1782–5), Vol. II, estampa 96. Biblioteca Virtual Miguel de Cervantes: *Noques de tinte azul*, in Baltasar Jaime Martínez Compañon, *Trujillo del Peru* (1782–5), Vol. II, estampa 86. Biblioteca Virtual Miguel de Cervantes.

of wool skeins in small vats, the dyeing of a length of cloth and an earlier stage in the process, the preparation of the indigo vats.

'It was the range of blues', Luca Molá has observed of the Venetian textile industry, 'that underwent the greatest innovation in the sixteenth century'.[94] New shades, light and dark, proliferated: *fior di lino*, *slattato* (milky blue), *palombino* or *colombino*, a 'deep blue' with 'very little red'. In the words of a local Venetian dyer: 'There is no rule, since these colours have been used only for fifteen or sixteen years'.[95] A closer look at the early trajectory of New World dyes illuminates the broader, social consequences of their diffusion. Without exception, increasing access to these raw materials would incite strong and varied reactions – even the global enthusiasm caused by cochineal was countered by a measure of resistance.[96] For merchants and dyers alike, the introduction of these exotic, superior reds, blues and blacks supposed a learning process that involved experimentation, failures and successes, fuelling the diffusion of new practices and expertise.

Indigo's accommodation into the textile sector in Spain and beyond[97] would affect dyeing sites and technologies; it would require ancillary services and facilities (grinding mills, for example), fostering organizational changes to the industry. Specific episodes of innovation and resistance can convey the complexities of the integration of new raw materials into production processes. Taking a closer look at the early trajectory of New World dyes can illuminate the broader, far-ranging social consequences of their diffusion – from plantations in Yucatan to the streets of Spanish Segovia.

Notes

1 Ana Roquero, 'Materias tintóreas en la época de Felipe II: Introducción en Europa de los Colorantes Indianos', in *Los Ingenios y las máquinas en la época de Felipe II*, coord. Ignacio González (Madrid: Real Jardín Botánico, 1998), 11.
2 Murdo J. Macleod, *Spanish Central America: A Socioeconomic History, 1520–1720* (Berkeley: University of California Press, 1973), 174, 197; Pierre and Huguette Chaunu, *Seville et L'Atlantique* (Paris: S.E.V.P.E.N, 1955), vol. VI-2 ('Arrivées de l'Indigo'), 988–9.
3 For the case of cochineal Luca Mola, *The Silk Industry of Renaissance Venice* (London: Johns Hopkins University Press, 2000), 120–31, for logwood Molá, *The Silk Industry*, 131; and for indigo Carola Ghiara, *L'arte tintoria a Genova dal XV al XVII secolo: tecniche e organizzazione* (Firenze: Giunti-G. Barbèra, 1976), 24.
4 Adrianna Catena, *Enter Xiuhquilitl: Indigo in the Spanish Empire, c. 1550–1700*, DPhil Thesis (University of Oxford, 2015), 84.
5 Antonio Fernández, *Pintando el mundo de azul. El auge añilero y el mercado centroamericano. 1750–1810* (San Salvador: CONCULTURA, 2003), 5.
6 On the knowledge and skills required to make high-quality indigo, see: Jenny Balfour Paul, *Indigo* (London: Archetype, 2006), 117; and Juan de Dios del Cid (attrib.), *El Puntero Apuntado con Apuntes Breves* (San Salvador: ed Isabel Casín de Montes, 1999).
7 Diego de Landa, *Relación de las Cosas de Yucatán* (Madrid: ed. Miguel Rivera, 1985), 51.
8 *Historia General de las Cosas de Nueva España*, Libro 11, ch. XI, fol.219R (*World Digital Library*) https://www.wdl.org/en/item/10096/ (accessed 20 January 2023).
9 Justina Sarabia, *La Grana y El Añil: Técnicas Tintóreas En México y América Central* (Sevilla: EEHA, 1994), 30–1.
10 Aaron Sidder, 'Earliest Evidence of Indigo Dye Found at Ancient Peruvian Burial Site', *Smithsonian Magazine*, 15 September 2016. Retrieved 18 June 2023, from: https://www.smithsonianmag.com.
11 See Balfour Paul, 'Indigo in South and South-East Asia', *Textile History* 30, no. 1 (1999): 103–5.
12 *Consulta del Consejo de Indias*, (4 July 1558), Archivo General de Indias (AGI), Indiferente 738, N.29; AGI, Buenos Aires 1, L.3, fols. 33V-34V; 37R-38R; Santo Domingo 899, L.1, fol. 113V-114V; letter to the Captaincy General of Guatemala in MacLeod, *Spanish Central America*, 426, n.2.
13 Enbióse de la Nueva España una yerva diciendo que hazía el efeto del pastel, y hallose que era añir, y assi se aviso dello al Virrey, 15 de Setiembre [1561], y de aqui parece que comezó su beneficio.' In *Colección de documentos inéditos relativos al descubrimiento, conquista y organización de las antiguas posesiones Españolas de Ultramar* (DIU), vol. xiv, p. 231.
14 On Ledesma and his indigo ventures, see Catena, *Enter Xiuhquilitl*, 49–57, and Sarabia, *La Grana y El Añil*, 26. Documents relative to Ledesma's commercial partnerships in the production and sale of indigo: Archivo General de Notarías (CDMX), Not.1 (Antonio Alonso), Vol.9, fols.179R-182V; Not.1, Vol.9, fol.796V (05.07.1570); Not.2 (Antonio Alonso), Vol.4, fols.83R-84V (28.04.1572).
15 AGI, Mexico 364, 'Carta del Ayuntamiento de Yucatán al Rey' (8 April 1573), fol. 62.
16 AGI, Patronato 64, R.7, fol.2R.
17 'Méritos y Servicios: Marcos de Ayala Trujeque' (1561) AGI, Patronato 64, R.7; and 'Méritos y Servicios: Marcos de Ayala Trujeque' (1591), AGI, Patronato 80, N.3, R.4., s/f.
18 AGI, Patronato 80, N.3, R.4, s/fol.

19 Caroline Cunil and Efrén Torres, 'Experimentando materias tintóreas entre el Nuevo Mundo y Europa: narrativas y muestras en los archivos de la monarquía hispánica (siglo XVI)', *Glocalism: Journal of Culture, politics and innovation* 2 (2021): 16.
20 Macleod, *Spanish Central America*, 178. The governor's correspondence is included in: France Vinton Scholes and Eleanor B. Adams, *Don Diego Quijada, Alcalde Mayor de Yucatán (1561-1565): Documentos Sacados de los Archivos de España*, vol.1, p.92; Diego Quijada to the Crown (20.05.1564).
21 AGI, Indiferente 1092, R.11, D.242; see also Vinton Scholes et al., *Don Diego Quijada, Alcalde Mayor de Yucatán (1561–1565)* (Mexico: Porrúa, 1938), 179.
22 AGI, Indiferente 1092, R. 11, D.242: 'Carta de Juan de Arce' (15 January 1565).
23 'Contradicion' (4 March 1572), AGI, Patronato 64, R.7, fol. 89R.
24 See n.13.
25 'Carta de D. Carlos de Arellano', AGI, Indiferente 1530, N.17.
26 See Sarabia Viejo, 'Intentos de aclimatar productos americanos en Andalucía en el Siglo XVI: el Añil', in *IX Congreso Internacional de Historia de América, Asociación Española de Americanistas* (Sevilla, 1982), 42–3.
27 AGI, Mexico 19, N.166, n.166, n.d.: 'Carta del Virrey Martín Enríquez' (24 October 1575).
28 AGI, Indiferente 1530, N.4, 'Del arbol del hec y del añil' (1577).
29 Robert Smith, 'Indigo Production and Trade in Colonial Guatemala', *The Hispanic American Historical Review* 34, no. 2 (1959): 187.
30 These beating systems were introduced to Guatemala in the late sixteenth century by one Pedro de Valverde, from New Spain. Horacio Cabezas Carcache, *Jiquilite-Añil: Inicio del Barroco en Guatemala* (Guatemala, 2016), 10.
31 Cristina García Bernal, 'El afianzamiento de un precario comercio: los intercambios entre Sevilla y Yucatán (1590–1600)', *Historia Mexicana* I, no. 2 (2000): 207; Nancy Farriss, *Maya Society under Colonial Rule: The Collective Enterprise of Survival* (Princeton: Princeton University Press, 1984), 36.
32 Jacques Heers, 'La búsqueda de colorantes', *Historia Mexicana* XI, no. 1 (1961): 4.
33 Carcache, *Jiquilite-Añil*, 13. The impact of indigo manufacture on indigenous communities has been studied extensively. See: Robert S. Smith, 'Forced Labor in the Guatemalan Indigo Works', *The Hispanic American Historical Review* 36, no. 3 (1956): 319–28; MacLeod, *Spanish Central America*.
34 'Relación de la siembra y calidad del jiquilite', AGI, Guatemala 13, R.4, n.41: 'Carta de Juan Maldonado de Paz, fiscal de la Audiencia de Guatemala' (31 May 1612); AGI, Guatemala 386, L.2, 'A la Real Audiencia de Guatemala' (1 November 1610).
35 Fernandez, *Pintando el mundo de azul*, 348.
36 Ghulam Nadri, *The Political Economy of Indigo in India, 1580–1930: A Global Perspective* (Leiden: Brill, 2016), 141; see also Max Edelson, 'The Characters of Commodities: The Reputations of South Carolina Rice and Indigo', in *The Atlantic Economy During the Seventeenth and Eighteenth Centuries: Organization, Operation, Practice, and Personnel*, ed. Peter Coclanis (Columbia: University of South Carolina Press, 2005), 344–60.
37 Nadri, *The Political Economy*, 141.
38 AGI, Indiferente 1530, N.4, 'Del arbol del hec y del añil' (1577), s/f.
39 Prakash Kumar, *Plantations and Science in Colonial India* (Cambridge: Cambridge University Press, 2012), 8.
40 Elliott, *Imperial Spain 1469–1716* (London: Penguin Books,1975), 197.
41 Archivo Simón Ruiz (AR) Francisco Morovelli to Simón Ruiz (23 July 1576), *c.* 31–252.
42 Catena, *Enter Xiuhquilitl*, 119.

43 AR, Francisco Morovelli to Simón Ruiz (17 September 1576), *c.* 31–256.
44 John Edmonds, *The History of Woad and the Medieval Woad Vat* (Buckinghamshire: John Edmonds, 1998), 38–9.
45 M. F. Van Eikema Hommes, *Discoloration in Renaissance and Baroque Oil Paintings: Instructions for Painters, theoretical Concepts, and Scientific Data*, PhD diss. (University of Amsterdam, 2002), 113.
46 AR, Francisco Morovelli to SR, 17 January 1578, *c.* 42–199.
47 Jean-Paul Le Flem, 'Vraies et Fausses Splendeurs de L'industrie Textile Segovienne (*c.*1460–*c.*1650)', in *Produzione, commercio dei panni di Lana (nei secoli XII-XVIII)*, ed. Marco Spallanzani (Firenze: Datini-Olschki, 1976), 532.
48 AR, Juan de Amarita to Simón Ruiz (10 April 1572), *c.* 16–200.
49 AR, *c.*250–329 'Cuenta' and *c.* 250–367 'Cuenta'.
50 AR, Juan de Amarita to Simón Ruiz (28 February 1578), *c.* 42–94; other examples are cited in the text.
51 AR, Juan de Amarita to Simón Ruiz (17 December 1580), *c.* 59–178.
52 JA to SR, 30 January 1578, *c.* 42–87.
53 Luis Alvares Caldeira to SR, Antwerp, 9 September 1576, in Valentín Vázquez de Prada, *Lettres marchandes d'Anvers* (Paris: S.E.V.P.E.N, 1971), t. II, 217.
54 AR, JA to SR, 5 October 1583, *c.* 84–41.
55 AR, *c.* 250: 'Relaciones de mercaderias y fletes, 1557–1577'.
56 AR, *c.* 250–329: Cuenta de 18 barriles de añir.
57 Smith, 'Indigo Production', 184.
58 José Gentil da Silva, *Lettres de Lisbonne, 1563–1578* (Paris: S.E.V.P.E.N, 1959), t. II, 3.
59 Ruth Pike, *Aristocrats and Traders: Sevillian Society in the Sixteenth Century* (Ithaca: Cornell University Press, 1972), 100.
60 AGI, Contratación 4920: 'Relación de oro, plata y efectos' (1606), s/f.
61 In the 1590s, venetian drapers were using West Indies indigo to dye 15000 bolts of cloth each year. See Molá, *The Silk Industry*, 130.
62 Manuel Garzón, *La Industria de la Seda en España: El Arte de la Seda en Granada* (Valladolid: Archivo de la Real Chancillería, 1972).
63 AGI, Contratación 789, N. 1: 'Autos de Juan de Monasterio Vide, dueño de nao, con los interesados en sus mercaderías, sobre avería gruesa' (1614).
64 Enriqueta Vila Vilar, *Los Corzo y los Mañara: Tipos y Arquetipos del Mercader con Indias* (Sevilla: Universidad de Sevilla, 1991), 205.
65 Macleod, *Spanish Central America*, 174, 197; Statistics in Pierre and Huguette Chaunu, *Seville et l'Atlantique*, and Eufemio Lorenzo Sanz, *Comercio De España Con América En La Época De Felipe II*, 'Relaciones económicas' (Valladolid: Diputación Provincial, 1979), vol. 1.
66 James C. Boyajian, *Portuguese Trade in Asia under the Hapsburgs* (Baltimore: Johns Hopkins University Press, 2008), 130.
67 Paulino Murugarren, *Evolución de la industria textil Castellana en los siglos XIII-XVI* (Salamanca: Universidad de Salamanca, 1974), 180.
68 Preceded by several earlier, more limited versions, the 1511 *Ordenanzas* (recompiled and published during the reign of Philip II: *Nueva Recopilación* repr. facs., Madrid, 1640) set standards for production and quality, focusing on the principal occupations in the productive process (weaving, dyeing, fulling, finishing, shearing), while delegating oversight of the productive process to the guilds.
69 Murugarren, *Evolución de la industria*, 180–1.

70 *Nueva Recopilación de las leyes de estos reinos* (repr. facs. Madrid: ed. Catalina de Barrio Angulo y Diego Diaz de la Carrera, 1640), vol.2, t.XIII, ley.lxxix, https://repositorio.uam.es/handle/10486/3158 (accessed 20 January 2023).
71 W. A. Vetterli, 'The History of Indigo', *Ciba Review*, 55 (1951): 3067; Jamieson B. Hurry, *The Woad Plant and Its Dye, 1857–1930* (London: Oxford University Press, 1930), 268.
72 Hurry, *The Woad Plant*, 267; Balfour Paul, *Indigo*, 56.
73 Hurry, *The Woad Plant*, 259–62; Balfour Paul, *Indigo in the Arab World* (Richmond: Curzon Press, 1997), 85.
74 Ghiara, *L'arte tintoria a Genova dal XV al XVII secolo*, 24.
75 Gabriela Siracusano, *Pigments and Power in the Andes: From the Material to the Symbolic in Andean Cultural Practices, 1500–1800* (London: Archetype, 2011), 57; Murugarren, *Evolución de la industria*, 189; Enrique Otte, *Sevilla y sus Mercaderes a Fines de la Edad Media* (Sevilla: Fundación El monte, 1996), 152.
76 Hilario Casado Alonso, 'El Comercio del Pastel: Datos para una Geografía de la Industria Pañera Española en el Siglo XVI', *Revista de Historia Económica* 8, no. 3 (1990): 523–45.
77 As noted earlier, Portuguese envoys were active in the diffusion of indigo dyeing recipes in this period. José Ignacio Fortea Pérez, *Córdoba en el s. XVI: las bases demográficas y económicas de una expansión urbana* (Córdoba: Monte de Piedad y Caja de Ahorros de Córdoba, 1981), 298.
78 'Petición de las Corts de Madrid' (Alcalá de Henares, 1563), cap. LXXVIII; in Silvia Saladrigas Cheng, Joan Soler i Jiménez (eds), *El Arte Real de Perseguir a los Sombreros: Textos y Documentos para la Historia del Tejido y la Indumentaria en las Monarquías Hispánicas, s. XIV – XVIII* (Barcelona: Museu Textil Barcelona, 2008), 200–1.
79 Javier Vela Santamaría, 'Protoindustria y fábrica difusa: el distrito industrial Segoviano en el Siglo de Oro', paper for *XI Congreso Internacional de la Asociación Española de Historia Económica* (Madrid, 4–5 September 2014), 14–15, n.76. Unpublished, thanks to the author for sharing his manuscript.
80 Vela Santamaría, 'Protoindustria', 15.
81 Illustrating the difference, indigo was sold by the pound, while woad was sold in loads (*cargas*).
82 Vela Santamaría, 'Protoindustria', 15.
83 Vela Santamaría, 'Protoindustria', 16, n.103.
84 Ghiara, *L'arte tintoria a Genova dal XV al XVII secolo*, 24.
85 Vela Santamaría, 'Protoindustria', 18. For a description of the urine vat, see Balfour Paul, *Indigo*, 124–6.
86 Vela Santamaría, 'Protoindustria', 17.
87 *Nueva Recopilación*, vol. II, L.7, t.XIII, Ley 101.
88 Siracusano, *Pigments and Power*, 57.
89 Fortea Pérez, *Córdoba en el s. XVI*, 298.
90 See Molá, *The Silk Industry*, 130.
91 Fortea Pérez, *Córdoba en el s. XVI*, 299.
92 Javier Ortiz de la Tabla Ducasse, 'Las ordenanzas de obrajes de Matias de Peralta para la Audiencia de Quito, 1621', *Anuario de Estudios Americanos* 33 (1976): 875–931; Neus Escandell-Tur, *Producción y comercio de tejidos coloniales: los obrajes y chorrillos del Cusco 1570–1820* (Cusco: Centro de Estudios Andinos 'Bartolomé de las Casas', 1997), 196.

93 'Only surviving fragment of "slave" cloth found in Derbyshire record office', *The Guardian* (2 April 2023).
94 Molá, *The Silk Industry*, 132.
95 Molá, *The Silk Industry*, 132.
96 Molá, *The Silk Industry*, 120–31.
97 For the case of Venice, see Molá, *The Silk Industry*, 132; for Lyon, Gentil da Silva, *Lettres de Lisbonne*, 129–30; for South America, Elena Phipps et al., *The Colonial Andes: Tapestries and Silverwork, 1530-1830* (New York: Metropolitan Museum of Art, 2004), 86–7.

4

A place under the sun

Brazilwood in the Brazilian economy (1500–1875)

José Jobson de Andrade Arruda

The setting for this chapter in Portuguese colonial history in America is the rainforest and its main characters are the Brazilian Indians and the leafy brazilwood (*pau-brasil*). The plot of our story is driven by the trajectory of the colonization process that the European peoples imposed on the lands and populations of South America. The outcome led to the transfer of resources to the dominant centres in Europe as well as to a degree of residual, endogenous accumulation in Brazil, which can be described as a complex process of extended development and underdevelopment. This historical legacy of four centuries of the exploitation of brazilwood is expressed in a vast and tragic panorama of devastated forests and decimated native populations that is in need of much more historical research and debate.

Natural red dyes were part of the original commodity trade that Portuguese merchants brought from Asia from the late fifteenth century, although clearly spices took priority.[1] From as early as the year 1500, after the discovery of Brazil, the first exports consisted of the trunks of a native tree, known as *pau-brasil* which became the source of a very special red dye that soon gained markets in Renaissance Europe. However, in an arbitrary way, most Brazilian historical texts concerning the colonial period do not underline this fact and rather tend to emphasize the beginning of relevant economic activities as the result of installation of the early, sugar industry in 1534, when the first mill was established in São Vicente. But, surprisingly, this neglects the first three decades of colonization, as if there were no commodity that deserved to symbolize the first moment of life as a colony. Hence traditional interpretations have retained brazilwood in the limbo of our history and historiography. This chapter focuses on three aspects of the history of the exploitation of brazilwood from the sixteenth to the nineteenth centuries. It begins with an exploration of the royal monopoly, then delves into the forms of exploitation of brazilwood and native Indians in their interactions with the colonizers. The final part examines the commercial dynamics before concluding with some historiographical remarks.

Exploitation and trade in brazilwood: A royal monopoly

The historical trajectory of brazilwood unequivocally reveals a fundamental aspect of the absolute power of the Portuguese monarchs from the late medieval period down to the early nineteenth century. The Crown established royal jurisdiction over the exploitation of the forests of timber and could make and remake the rules with the sole purpose of increasing fiscal revenue, which the monarch could dispose of at will without any constraints, legal or institutional, just as he had over his domains by right of discovery or conquest. Royal power over the colonies was not subject to the limitations (*limitações concelhias*) that Portuguese continental territory inherited from the Middle Ages. The Portuguese Kingdom and its traditional restrictions were one thing, while the merchant empire was another, since it was constructed through conquest. The king's rights over the valuable commodities were uncontested in fact and in law, and were legitimized by the *Afonsine Ordinances.*[2] Through these laws, overseas trade was considered the property of the Crown as an inalienable possession, but in addition, all discovered or conquered lands, routes, traffic, as well as valuable products effectively became exclusive monopolies (*'estancos'*) that could be granted by contract to third parties without compromising the Crown's sovereignty. This was the case of the salt monopoly as well as of whale fishing and the extraction of brazilwood. In this sense, the Portuguese monarch was both a sovereign and a colonial entrepreneur, in short a merchant king. However, we must recognize that 'management and operation of such monopolies were not entirely left to royal officials' who 'lacked the necessary skills to run them', giving opportunity to foreign merchants to participate in 'various contracts including those under royal monopoly collecting large benefits'. In this sense, as argued by Jorge Pedreira, Portugal was a special kind of merchant state characterized by a monarchical capitalism that linked the interests of the 'monarchy, crown officials and the aristocracy, all reinforced by the role of conquest and plunder in empire building'.[3]

Some of the participation rights for the exploitation of land and natural resources were expanded or reformed as a result of the practical needs imposed by colonization, but without ceding the absolute precedence of the royal rights. This was precisely the case of the rules stipulated in the letters of concession of the Brazilian captaincies that officially demarcated the beginning of colonization in 1531.[4] The *Foral*, the royal letter that established those captaincies, set norms that lasted throughout the colonizing process and specified the rights and duties of the settlers ('the donees'), and, above all, the rights of the king, to whom was reserved a tenth of the products of internal and external commerce, a tenth of fishing, a fifth of all minerals and, above all, the exclusive control over the extraction and commercialization of any and all existing drugs or spices, especially, of brazilwood, allowing the concessionaire (*capitão donatário*) only 1/20 of the value obtained from cutting wood. The extraction and commercialization of the wood without the due royal license was subject to the penalty of loss of all goods by the subject and exile to the island of São Tomé. The only exemptions on cutting were for use on the owner's property. This clause would be eliminated in the Ordinance of 1605, a specific legislative act about brazilwood that would be in force until 1808, when it was reregulated in the forestry plans undertaken by the Court in

Brazil, without changing the exclusive clause of the royal rights over noble woods, particularly brazilwood.[5]

The Portuguese Crown not only strictly legislated in defence of its rights over colonial forests but was also empowered to take over the direct administration of the business, since the use of timber was exclusive to the Crown, and forbidden to private individuals even in areas of private property, except under very special conditions. It was an absolute monopoly, 'the world's most restrictive', because no other country dared to include all the hardwoods existing in private properties as an exclusive right of use by the royal power.[6] The Spanish monarchy recognized this right since it had agreed on the limits established by the Treaty of Tordesillas with the Portuguese, under papal approval. On the other hand, the French ignored the treaty and, from very early on, started to practice trade and piracy on the Brazilian coasts that was carried out by privateers, smugglers and ship-owners, although these were private initiatives rather than governmental ones, since the French Crown attempted to preserve a cordial relation with the Portuguese Kingdom in order to maintain strategic political relations within Europe.[7]

In practical terms, the Portuguese monarch could exploit brazilwood directly, by means of his administration and staff, or transfer it to private individuals, by means of a public auction, on the basis of contracts that specified in detail the terms of the relationship, the amount of wood that could be extracted in the colony per year, the amount to be collected for the royal treasury, the duration of the contract, as well as reserving the monarch the right to cede similar rights to other private parties even during the duration of previous contracts. That is, the Portuguese Crown sliced up his monopoly to make it yield more, retaining the prerogative of denouncing contracts with irregularities at its discretion.

The first contract for exploitation of brazilwood was signed with an association of New Christian (formerly Jewish) merchants from Lisbon, under the leadership of Fernando de Noronha. It had a special meaning because it signalled the beginning of the economic exploration of the territory and more particularly of brazilwood as well as establishing a pattern for future business. Signed in 1501, this first contract lasted for three years, and was renewed in 1503, the year in which the navigator Gonçalo Coelho travelled to Brazil in the service of contractors, from whence he returned with a cargo of 20,000 quintals of brazilwood, which yielded a profit of between 400 per cent and 500 per cent on capital invested, a successful experience that was repeated in 1511. In view of the positive results and hoping to increase the Treasury's revenue by increasing the number of participants in the timber exploitation, the free exploitation of brazilwood was extended to all interested parties upon payment of 1 cruzado (the equivalent of 400 réis) per each quintal of wood to be placed in Lisbon. This figure allows us to assess the value of the monopoly rights transferred directly to the monarch. If we consider that a quintal of brazilwood earned the authorized settler approximately 1,000 réis, the Crown would pocket the corresponding 40 per cent of the value of the timber cut without having to incur any expenses.

At the end of the sixteenth century, the Crown introduced a variation on previous practices. It granted to the Jesuits the monopoly of cutting, gathering and transporting wood to the shipping ports in the region of the Capitanía of Espírito Santo.[8] The purpose was to combat the clandestine exploitation of wood, making use of the Jesuits' privileged access to the native villagers' labour force. However, the Crown retained for itself the monopoly of transport of brazilwood to Portugal, where it would sell the dyewood to the highest bidder, thus creating a bilateral monopoly, although in principle the Jesuits could now better control the voracity of the contractors and protect the Indians from mistreatment and the forest from destruction. This arrangement, however, did not last long due to the resistance of the superiors of the order who determined to end this experiment as they did not consider it appropriate for the religious to get involved in worldly enterprises, distracting them from their main objective which was the conversion of the indigenous populations and religious instruction for all.

Very early in the seventeenth century, with Portugal now under Spanish rule, the policy of contracts returned, whereby the contractors were to pay 40,000 cruzados annually for the right to transport 10,000 quintals of wood, a significant amount compared to the other revenues of the Portuguese mercantile empire. It was this profitability that led King Philip III, of the Castilian Crown, and sovereign of Portugal, to issue the symbolic 1605 *Ordinance for Brazilwood*, using his exclusive right to legislate over his monopoly. It was another stage in the alternating cycle between the regime of direct administration and that of contracts, which attested to the royal precedence in the conduct of the policy of brazilwood exploitation.[9]

The royal decree was a personal disposition of the Crown aiming at the preservation of brazilwood in face of the disorders found in its exploitation, since it was 'a matter of such importance to my Treasury'. It threatened those who violated the rules established in the document with severe penalties, ranging from monetary payments to the death penalty, including loss of property and banishment to Africa, a repetition of the provisions contained in the 1531 letter granting the hereditary captaincies. The distribution of logging areas in all captaincies was to be severely controlled. The authorizations to cut wood would be given to suitable contractors who had not committed any irregularities in previous auctions, and would be confirmed in a document issued by the local royal administrator, and the contractors would be obliged to cut strictly the authorized amount of wood, with the declared objective of controlling forest devastation. In order to preserve the forest, the ordinance urged the contractor to use as much as possible of the felled trees, requiring them to use not only the trunks but also the branches and twigs that produced dye that were commonly thrown away. In the same sense, to ensure the continuity of economic exploitation, the residents were urged to cut down the mature trees, but not to cut down the young trees around them, nor the shoots, to ensure the future renewal of the forest. It was recommended, above all, to avoid cutting down the trees by setting fire to the base of the trees, a practice that led to massive burning and devastation. This set of measures was to be supervised by guards paid by the State, a timely measure, but unfeasible considering the size of the area to be controlled. The signature of the commitment in question was formalized in the presence of the City Chamber officers, the Treasury Superintendent (*Provedor-mor*) and the governor general himself, and then sent to the procurators in the other captaincies for acknowledgement and execution.

An extractive industry in the Brazilian forests

The rainforest that astonished the first navigators who arrived to the Amazon was the result of natural developments that took place over the course of fifty million years, although its historical borders and modern configuration actually date back to between two and four million years. During the very long duration, the forest wrote and rewrote its own history as a palimpsest, retaining on its surface the overlapping layers of its evolutionary experience, representing 'that last hidden empire, ripe for conquest'.[10] The history of the economic exploitation of the Brazilian rainforest and especially of brazilwood, as it was carried out by the Portuguese Crown, cannot be isolated from the concomitant exploitation of the indigenous labour force, the children of the rainforest, transformed in an instant into *blacks of the earth*, who were compelled to destroy the specimens generated during millennia in the heart of the forests, a kind of environmental matricide, whose epilogue could not have been different: the genocide of men and the death of the forest.[11] This image certainly did not enter Amerigo Vespucci's mind on his first trip to the newly discovered land, his *America*, which he described as being devoid of any interest, except for the 'infinite trees of Brazil'.[12] The tropical forest was full of thick, tall, leafy specimens that after being cut down and chopped into logs, revealed a reddish-brown heartwood reminiscent of burning coals, surrounded by sapwood and bark. The brazilwood produced a thick dye ranging from brown to ochre, readily recognized because similar specimens commonly arrived to the docks of Lisbon in ships that came from India. For the first navigators and merchants, who arrived in Brazil, finding these trees rich in dyes, was surely the main reason why the land covered by the forest was worth exploring. With the passage of time, they discovered that many varieties of brazilwood existed along the extensive coastal strip that stretched from Rio de Janeiro to Rio Grande do Norte. The specimens were of distinct qualities, being more numerous and of better quality on the coastal stretch between Itamaracá, Pernambuco and Paraíba, near the equator, but inferior along the southern regions, below Cabo Frio. They were abundantly represented in the maps drawn at the time.[13]

The Portuguese were pioneers in establishing the first large-scale extractive industry in America based on the harvesting of *pau-brasil*, but the vast expanse and density of these valuable trees were not a sufficient condition for beginning exploitation. The massive cutting and extraction of trees from the Brazilian forests could not have taken place at that time without the robust contribution of the labour of the indigenous peoples concentrated on the coast; they saw in the Portuguese possible allies against the warlike hordes of the Tupi tribes that frequently emerged from the backlands and plundered the coast.[14] The remuneration of the native labour force was almost free, considering the standards of the European mercantile economy, since these included trinkets and ornaments to start with, and later, indispensable instruments to continue the work in the forests: these were mainly tools, in this case, knives, machetes, sickles, axes, sledgehammers, wedges and firearms, 'which should be avoided', according to an early testimony.[15] Historian Alexander Marchant argued that the Portuguese in charge of organizing the cutting and shipping did not initially enslave the indigenous peoples. At first, they simply used barter – the usual practice in intertribal relations – to convince local workers to harvest the trees and carry the wood to the shipping location. However, relations were complicated by the intervention of foreign smugglers who were in dispute with the Portuguese for control of indigenous labour, and, as a result, all the settlers were forced to raise remuneration of indigenous labour and also began to ask for shipment of more sophisticated and expensive goods for trade with them.[16]

At the same time, a second form of exploitation of native labour was used to open clearings for the cultivation of subsistence crops, an essential step that preceded the planting of sugarcane and tobacco. Indigenous labour was therefore increasingly valued and, as a result, the agents of the contractors offered them more and more benefits, which led the natives to a kind of satiation with regard to the forms of payment practiced, and soon enough they simply began to refuse to work. But this refusal to adhere to the Europeans' way of life was not simply a material question related to the amount paid for the work. It was, above all, a cultural issue. Their availability as labourers was marked by a limit established in terms of the satisfaction of their vital needs, and, therefore, as soon as this threshold was reached, they saw no reason to continue expending their energy, which, in their conception, could very well be employed in a much more pleasurable way, in hunting, fishing and even in war. Moreover, if the time taken to reach this vital limit was accelerated by the use of more efficient tools, including the use of iron axes which multiplied cutting capacities eightfold, the interest of the indigenous population in work diminished quite abruptly. To change such conceptions would mean to destroy their cultural system, particularly considering that traditionally the work involved in the transformation of nature, in particular agricultural cultivation, was considered a social function of women, and therefore, a gender issue and a feminine attribution, so much so that the harvesting of brazilwood was also performed by native women.

Two absolutely antagonistic worldviews were thus confronted.[17] The Europeans thought of production as a form of accumulation, but this conception was considered absurd by the indigenous people. Saving, accumulating, for what? The preference of the natives of the Brazilian forests was that of immediate consumption to supply physiological needs. When the first sugar cane plantations were launched, they inevitably must have asked why it seemed important to produce so much sugar. Neither the settlers nor the Indians themselves needed it in such large quantities. In these terms, the *engenho* (sugar refinery) was a real absurdity! But they also must have wondered why so much brazilwood should be cut down? What should be done with it? 'They had no interest in profit and no ambition for material wealth.'[18] They disliked physical labour, as it placed them in an inferior position and separated them from the joys of life, something they found absolutely intolerable. The seventeenth-century French missionary Yves D'Évreux, quoted by John Hemming, reproduced in a remarkable passage their worldview so averse to the principles of the capitalist world that the settlers and their enslavers represented:

> I care little about being eaten. Because when you die you don't feel anything anymore. If they eat you or stop eating you, it's the same for a dead man. But I would

> be outraged to die in bed, and not to die like a great man, surrounded by people dancing, drinking, and swearing revenge, before dying, against those who were going to eat me. For whenever I think that I am the son of one of the great men of my land, that my father was feared, that everyone surrounded him to listen to him when he visited the house of men, and seeing me now as a slave, without body painting, without feathers on my head or arms and fists, as the sons of great men are usually paraded in my land, [. . .] when I think of all this, I wish I were dead [. . .] and can only regret being alive.[19]

This was the essence of the ethic of rational non-work and the reverse of capitalist ethics based on seeking profits and accumulation. As long as the tools received in payment for their forced labour were sufficient to produce their subsistence, they had no ambition for any other imported tools that would lead them to systematic work, and 'to give them tools as is the custom' does not seduce them, only, perhaps, 'swords and rifles', because 'once they are fed up with tools, they make themselves meaner than they are, and they get agitated, arrogant and rebellious'.[20]

Under these circumstances indigenous labour was much less productive than that of African slaves, which perhaps explains why the price of Indian slaves was three times lower than that of Africans.[21] But African slaves, initially, were far from being the solution. The most viable alternative was to resort to indigenous slavery, a practice which was only allowed in the case of a *just war* as authorized by the Crown and openly practiced in certain circumstances under licence from the Portuguese monarch, as was the case of Martim Afonso de Souza, the first governor of Brazil, who ended up gathering 3,000 Indians as slaves to work on six sugar mills. The policy with regard to the slavery of the natives was only ratified by the Ordinance for the General Government, in 1549.

This set of circumstances explains the transition from barter to slavery with regard to the Indians. The extraction of brazilwood was a type of work that seemed to fit the profile of the native populations. Cutting, slicing, thinning, transporting and eventually loading on boats were activities closely related to their environment since they were forest dwellers who had accumulated long-lived knowledge about the forests and their management. In fact, an army of men was required to fell trees between 15 and 20 metres high, with trunks that sometime could only be embraced by three men, but whose cutting was exhausting, even with the use of appropriate tools, a difficulty that invariably led to the use of fire, particularly at the base of the trees in the traditional practice of *slash and burn* that was used for the thinning of clearings, followed by the planting of manioc on the ashes of the burnt timber.[22] The outcome could be devastating when the fire spread and destroyed vast stretches of forest, including the youngest trees of brazilwood or their shoots, interrupting the natural reproduction cycle of the forests and leading to their devastation.

Harvesting trees by fire or chopping with an axe was only the first stage in the work of processing the raw material, since, given its high density, the brazilwood did not float, which hindered its transportation by river. Hence the need to peel it, thinning the sapwood and retaining only the heartwood: the red part was segmented into logs of some 1 to 2 metres in length, with a weight of about 30 kilos, allowing for transport on the shoulders of the Indians, who would travel distances of up to five leagues to deposit them in the *feitorias*, precarious warehouses where all the logs intended for shipment to Europe were piled up.[23] One can imagine the impact caused by the introduction of the iron axe in this labour, since it was immediately considered a miraculous instrument as it increased cutting capacity by a factor of eight, allowing the natives to more quickly perform their exhausting tasks, leaving free time for other less strenuous activities.

The form of payment to the Indians for their work proved to be a complicated element in the relations with the Portuguese settlers as well as others interested in the products of their labour. The objectives of fierce disputes with Dutch, English and especially French smugglers of brazilwood went far beyond wood since they also sought possession of the territory, but also and necessarily involved complex relationships with the natives. This circumstance led the French, especially, to cultivate their connections with the natives, and to attract them in order to have them on their side in case they were attacked by the Portuguese, or had to fight back. Fierce confrontations followed, including violent struggles for control of the indigenous labour force, without which the forest resources were worthless. In any case, the settlers' experience in organizing *mutirões* (work crews), for wood extraction, facilitated the adaptation to subsistence production that generally required between four and seven slaves per European settler. Often enough, the Indian slaves could be bought for low prices, at approximately ten *cruzados* per head, being employed not only to harvest the trees, but also to cultivate crops sufficient to provide for their own subsistence as well as that of their owners, as described by Pero de Magalhães Gandavo.[24]

The consequences of this type of labour dynamics are more than evident. Indigenous slavery became widespread when the provision of Indians as negotiated by the Jesuits was not enough to meet the demands not only for the brazilwood extraction industry, but, increasingly in the sugar plantations as the value of sugar production in the middle of the sixteenth century began to match the income generated by the brazilwood monopoly, which ipso facto inflated the demand for native labour for the sugar mills. The result was an increase of enormous pressure on the indigenous population that led to generalized conflict, as the native Indians, who were the true owners of the land, fiercely resisted enslavement. This resistance was a landmark event, a turning point in the colonization process, leading to a transition from the more lenient policies of the initial phase that was accompanied by attempts at pacified coexistence between settlers and natives mediated by Catholic missionaries, to an aggressive policy of conquest by fire and sword. This practice resembled the predatory actions of the Spanish conquerors in America, denounced in many texts of the so-called *Black Legend* published in northern Europe. In contrast, most of the Brazilian historical literature has generally exempted the Portuguese conquerors from these harsh critiques. Furthermore, it should be noted that it was normal practice among the settlers to stimulate conflicts between the indigenous tribes in order to have the opportunity to rescue the defeated Indians, although not to save them from ritual death, but rather to put them to forced work in the extraction of brazilwood or, alternatively, in farming.

The relationships with the indigenous people were never easy and required the Portuguese to strike a balance between exploitation and destruction of the native population. But the problem soon reached saturation points, forcing the adoption of administrative measures that could conjure up the combined threats of resistance by Indians and by the French smugglers. For this reason, the date 1549 is symbolic. This was not only because it marks the installation of the General Government in the Brazilian colony, with headquarters in Bahia, but also because it represented the turning point between the predominance of the extractive economy of brazilwood and the rise of the sugar industry, a time when internal and external tensions had reached a climate of open warfare. The focus of the ordinances that guided governmental functions was the indigenous issue. The fierce resistance of the Indian population to the Portuguese was recognized by the Crown as being the result of slavery itself, but nothing else could be done to solve the problem of lack of manpower. This duality was explicitly presented in the ordinances that guided the policy in relation to the Indians: the meek were to be protected; the resisters who refused to submit were to be eliminated, despite the ideological discourses of the more legally minded administrators who considered them subjects of the Crown, as well as the Jesuits who resisted their enslavement.

The first two governors of colonial Brazil paid some attention to the legal instructions, but Mem de Sá, the third governor, turned them into instruments that legitimized the actions of conquest of the lands dominated by the rebellious Indians and their transformation into slaves. While the royal advisors in Portugal discussed the ethics of the principle of a just war from an ethical and moral point of view, 'Mem de Sá and his immediate successors adopted a practical attitude, organizing powerful military columns to crush the resistance of the Tupi Indians in Rio de Janeiro, Bahia, Pernambuco, Sergipe and Paraíba'.[25] He declared open war on the Caetés to favour the settlers in the acquisition of slaves obtained in a just war, ignoring the numerous conversions made by the Jesuits, who were horrified as they watched 'the Portuguese take men, women and children away from their missions' while at the same time attacking all the Indians they found on the way under the pretext that they were Caetés.[26] In the region of Ilhéus, there was a real slaughter, of which Mem de Sá was proud. He referred to his feat thus: 'I entered the village and destroyed it, and killed all those who wanted to resist, and I went on burning and destroying all the villages that were behind [. . .] so that no Tupiniquim were left alive and they all brought them ashore and put them along the beach in order to take the bodies close to a *legua* (6.5 km)'. But as the Tupiniquim tried to form a confederation of tribes, the Portuguese crushed them and left the 'beaches covered with soulless bodies, and the surface of the sea that washed over them turned the color of blood'; the destruction of 300 villages forced them to accept the terms of the winners, accept vassalage to the king, pay tribute and rebuild the destroyed mills.[27]

In 1567, the governor himself disembarked at Espírito Santo, calmed the peaceful natives but sentenced the resisters to death and restored order. These feats were celebrated by the settlers, praised by members of the church itself and rewarded by the monarch. It was not a discovery, much less an 'encounter', an expression which is the cloak of fantasy. It was a conquest by iron and fire that was in no way inferior in atrocities to those registered in the Black Legend of the Spanish conquistadors and does justice to Jean Parmentier's irony on criticizing the Portuguese in relation to their conquests, when he says of them: 'although this people is the smallest on the whole globe, this globe does not seem large enough to satisfy their greed'.[28]

The northeastern coast of Brazil was quite rapidly controlled by the colonists, but not completely, because there was still an enemy that was equal in violence to the Portuguese, the aggressive tribe of the Aimorés, who had taken refuge in the dense forests that covered the hills near the coast, from where they came to terrorize the settlers and destroy the sugar mills. Fearless, powerful warriors and thirsty for human flesh, their decisive weapon against their opponents was mobility, as they used true guerrilla tactics and elements of surprise and camouflage, fundamental conditions for their victories in the total war declared against the Portuguese. They were so successful that in 1587, in the Capitania of Ilhéus and Porto Seguro – two decades after the raids Mem de Sá – they inflicted a fierce defeat on the settlers, a kind of improbable revenge, which resulted in the death of 300 Portuguese and of more than 3,000 slaves. Moreover, these fearless Indians would continue their bloodthirsty adventures for centuries.

The Portuguese response was military action, designed to effectively take possession of the land and the people settled on it, which soon became a political formula that involved seizing the indigenous populations and taking possession of their lands, extending beyond the imaginary limits assured by the Treaty of Tordesillas – across the dynamic northeastern centre of Brazil, São Paulo, Goiás, Mato Grosso, Maranhão, Pará and Rio Negro – transforming the peripheral areas into reservoirs of Indian labour to be plundered by slaving companies (*descimento, saltos*), whose specific purpose was to hunt down and enslave Indians. This means that not only in its beginnings, but also in its development, the sugar economy was clearly linked to the expansion of Indian slavery. In any case, in all the areas in which the enslavement of Indians was required for the process of colonization, the argument regarding the institution of a *just war* was always essential.[29] The use of this concept varied, depending on the greater or lesser need for compulsory native labour; the opportunity offered by some type of resistance, transformed the Indians – in the view of the Portuguese settlers - into barbarians, savages, brave, gentiles, *tapuias* into targets, that not only could be destroyed but deserved to be so, and even the allied Indians were encouraged to follow suit in their pursuit. This sense of expediency explains the law's vacillations and the differences between the interests of colonization and those of the natives. While the Ordinance of 1549 legitimized the slaughter undertaken by Mem de Sá, the previous law of 1509, dealing with 'the freedom of the gentiles of the land and the war that can be waged against them', declared their unconditional right to freedom, a provision that did not last two years, because a new law reinstated the legality of captivity in case of just war or ransom.[30]

Indigenous resistance combined with the expansion of the sugar economy tended to progressively intensify the use of African labour. This was at a time when the difference between higher cost of African slaves and that of the indigenous labour could be covered by high profits obtained in the sugar enterprises. Contemporary practice revealed that when comparing the two types of slavery, the African was three times more productive, thus justifying the investment. The new slavery was primarily intended for the sugar industry, but it had repercussions on the brazilwood extraction industry. This was because African labour could be displaced from the activities of the sugar mill to the extraction of brazilwood at times of low intensity in the agricultural and industrial tasks, increasing the productivity of slave labour. This also increased the labour intensity in brazilwood extraction, because the African slave was more efficient and also because carts were introduced, pulled by draught animals, to transport the wood, at a time when the timber was to be found as far as twenty leagues away from the coast. Thus, a connection was established between sugar production and forest extraction. In the first case, economic development and internal accumulation benefitted the plantation owners. In the second case, due to the royal monopoly, the transfer to the outside world of almost all the benefits of exploitation left the settlers with only a residual accumulation but also caused enormous forest devastation, whose replacement cost would be incalculable due to the centuries that nature took to form it and to the many other costs that would be necessary to regenerate the vast tree resources.[31]

The balance of losses and gains was extremely unfavourable for the children of the land, although less so for the Portuguese-Brazilians. The indigenous nations were reduced to a minimum expression of what they had been, not only in numerical size, but above all in their self-esteem, because slavery took away their honour and diminished their pride. As a result, the defeated and saddened Indian men and women began to consider the *Carib Indians* in a positive light, and in particular the discourse of their shaman, who offered hope for a better life on their own land, a place of plenty and peace. After the first contacts with the white men, the forest dwellers therefore tended to seek isolation, which was only broken when a coastal expedition or the boats carrying brazilwood arrived. At the same time, those Indians who lived more closely with white men on the plantations, or with the Jesuits in the villages, contracted multiple diseases. This decimated the inhabitants, who, extremely frightened, retreated to their villages in the backlands, contaminating thousands of people, who also died in large numbers. The mortality rate in the villages controlled by the Jesuits ranged between 30 per cent and 75 per cent of the villagers. It was a demographic catastrophe, causing the death of entire nations, companions to the destruction of the hundreds of thousands of felled trees, both thrown to the forest floor in a vast devastation.

Nonetheless, numerous historical texts have contained expressions of euphoria that refer to the apparent economic success of brazilwood exploitation and have extolled this process for having been responsible for the first boom of the colonial economy in the sixteenth century. In this regard, it should be noted that the revenue from exported wood in the 1550s was certainly higher than that from sugar; indeed, up until 1600, when the value of sugar exports reached 2.2 million pounds sterling, the value of brazilwood exports was close behind. In any historical reconstruction it is important not to forget that it was necessary to export millions of trees for this result to be achieved, producing an enormous forest devastation that historian Warren Dean[32] estimated at 8,000 tons per year. A number that would have to be increased by another 20 per cent, due to waste, including discarded wood by contractors who alleged poor quality, as well as the destruction of shoots in the burning and in the harvests, equal to the devastation of 6,000 square kilometres, based on a global estimate of an average of four specimens of brazilwood trees per hectare, which is the typical occurrence in Brazil.[33]

Profitability and accumulation

There is no doubt about the potential profitability of the extraction and commercialization of brazilwood. Still, it is essential to identify empirical data that can help measure the accumulation level generated in the colony, in Portugal, and even in the terminals of European connections, especially in the Netherlands. The list of the shiploads, for example, provide considerable, reliable information. The ship *Bretoa*, which sailed in 1511 and was financed by the holders of an early brazilwood contract, loaded 2,000 quintals of wood in Cananéia and resold them for 2.5 cruzados per quintal in the markets of Flanders, Castile and Livorno, earning 5,000 cruzados.[34] Rondineli, an Italian agent of the brazilwood trade in Europe, calculated that the new contract offered by the monarch to the consortium, proposing the annual extraction of 20,000 quintals of brazilwood against a payment of 1,400 réis for the royal rights, would be very advantageous for the contractors. The 20,000 quintals of brazilwood sold in Europe at 1,000 réis a unit would give 20,000 réis, from which it would be necessary to deduct the expenses of cutting and transport evaluated at 4,000 réis, plus 2,400 réis for random expenses and taxes.

Our calculations of costs would therefore be 6,400 réis, to which must be added the royal monopoly fee, to the amount of 1,600 réis, resulting in total expenses of 8,000 réis for the contractor, which, debited from the gross revenue of 20,000 réis, would produce a net profit of 12,000 réis. This is an astonishing financial result for operations of this nature in the context of the sixteenth century. Of the net profit of 12.000 réis, one fifth went to the king and four fifth to the contractor, rates only comparable to the exceptionally lucrative Indian trade, in which pepper could yield up to 600 per cent on investments by the royal treasury, but with a fundamental difference: the sovereign did not invest a single real in the brazilwood business, since the costs and risks were all the contractor's, but, as we have seen in this quantification exercise, he also received a large share of the final net profit.[35]

Twenty years later, the loading of another ship allows us to reassess the values involved in the brazilwood business. The ship *Pelerine*, fitted out by French merchants in 1531, had the objective of taking control of a Portuguese a trading post for the export of brazilwood that was already installed off the coast of Pernambuco.[36] The case was taken to the Bayonne Court of Arrest, a bilateral court consisting of French and Portuguese judges, whose purpose was to settle conflicts resulting from seizure or piracy involving the two countries. Frictions multiplied as French actions in Brazil intensified and Portuguese retaliations became violent, creating a state of undeclared war between France and Portugal; the French king, Francis I, was not interested in an open war against the Portuguese as it would harm his strategy to maintain good relations with Portugal, which was at war with his enemy, the Spanish monarchy. The cargo of 5,000 quintals of brazilwood seized was estimated at 16,000 réis by the French commissioners, but this was an obvious exaggeration, that was promptly questioned by the Portuguese representatives, as they would be forced to pay a large indemnity. Manipulations that led them to shift the problem to a second front of accusations against the owners of the *Pelerine*, now no longer of brazilwood smuggling but of theft, as they were able to prove that the ship itself had been taken from the Portuguese in recent piracy operations.

This event demonstrated, on the one hand, that brazilwood had become more valuable in international trade and, on the other, revealed the complex nature of commercial exchanges often entangled with piracy, smuggling and diplomatic relations between states, making situations of effective but undeclared maritime warfare frequent.

An official report titled *Livro em que se contem toda Fazenda & Real Patrimonio dos Reynos de Portugal*, presented to King Philip III of Spain, II of Portugal, by his secretary Luiz de Figueiredo Falcão in 1607, specifies the contribution of the empire to the royal property of Portugal, allowing a comparison between the income from India and Brazil and, above all, the income from brazilwood as compared to the income from sugar and other colonial products.[37] In the category of overseas revenue, it is shown that India produced an income of 888,900 cruzados and Brazil 110,000, that is, eight times less than India, but ahead of the Azores and the coast of Mina. In the chapter royal *estancos*, which included the so-called Indian spices, the pepper trade yielded 600,000 cruzados and brazilwood 60,000 cruzados. But in relation to the fiscal income generated by Brazil, brazilwood generated more income for the Crown than sugar at that time. In terms of the global finance of the Portuguese Empire, Brazil produced 12.4 per cent of the royal revenues, 5.6 per cent from sugar and 6.8 per cent from brazilwood. This percentage would tend to change as the colony's economic development continued to favour sugar. In general, the value of the brazilwood trade exceeded the general increase in royal revenues during the first 120 years of its exploitation, going steadily upwards until 1618. During the first decades of the seventeenth century, wood prices fell in the colony, but as the cost of transport remained stable and the resale price in Europe rose, the internal accumulation already established at the minimum level, fell even further, while the king's profits tended to expand.[38]

Brazilwood was undoubtedly the main source of Portuguese fiscal income from Brazil during the sixteenth century. The statistical tables prove that the theories of the *Old Colonial System* are right. The lineage of this historiographical debate harks back to historian Caio Prado Junior who, in a classic text, emphasized the need to seek the meaning of Portuguese colonization in Brazil, which for Caio would be the production of super-profits for the European metropolis.[39] This interpretation was also advanced in empirical and analytical studies by Fernando Antonio Novais, who underlined the function of the colony in the process of primitive accumulation of capital, made possible by the monopoly regime typical of the mercantilist policy, carried out under the aegis of the absolutist state.[40] The prices of raw materials were kept close to or below the cost of subsistence in the colony; transportation costs remained stable and resale prices in Europe tended to rise. In sum, all indicators pointed to minimal income for the colonial actors, maximum profits for the big businessmen who operated the monopoly via contract, and very high profitability for the Crown, which exempted itself completely from the risks of the business, demonstrating that the exploitation of the forest wealth of the colony promoted, above all, the external accumulation of capital. Estimates of the contribution of brazilwood to the public treasury at the beginning of the sixteenth century did not exceed 5 per cent of the total revenue of the Empire, but it was enough to cover a substantial part of the costs of maintaining the territory by the end of the same century.

In any case, brazilwood in European markets was gaining strength. Despite the fact that it was in demand for the dye produced, it also lent itself very well for furniture and housebuilding, so much so that part of wood exported from Portugal returned to Brazil, in part, transformed into fine furniture.[41] When shipped from the ports of Bahia or Pernambuco, the nature of the cargo was meticulously described in a document significantly called a *conhecimento*. The cargo registries were made at the time of unloading in Lisbon and the logs were stored there at the *Casa das Índias*, in the centre of the city, under strict surveillance, waiting to be shipped to the largest consumer market, the Netherlands. The contracting parties were responsible for the cargoes, so they were closely supervised by the royal agents, who controlled the number of logs, thickness, and size, in order to make sure that the conditions stipulated in the contract had been observed, submitting the contracting parties to rigorous demands. The logs were mainly sent to the commercial and industrial warehouses in the Netherlands, whose port cities integrated the North Sea trade routes, including British, Norman and German cities, with ramifications reaching to the Baltic Sea. The most important markets were Antwerp and later Amsterdam.[42] According to Fernand Braudel, Amsterdam was the last 'true empire of trade and credit, without the support of a unified modern State', successfully shipping woollen fabrics, cotton and other articles to the Mediterranean markets, once hegemonic, and expelling the local or rival, imported products.[43]

The links in the chain that connected the Portuguese traders to their confreres in Flanders and the Netherlands were facilitated by the Sephardic diaspora that had begun in the late medieval period, and intensified during the period when the spice trade was practically monopolized by the Portuguese Crown, to the detriment of the older Mediterranean routes. This intricate network of relationships among family clans of new and old Christian merchants, who settled in Portugal and the Netherlands, was an essential condition for the commercialization of brazilwood. Powerful merchant families, such as the Benveniste, are a clear example of the operational mode of the large Sephardic family conglomerates that made fortunes buying spice shipments by cash on delivery or by contracting the royal monopoly for long periods.[44] In the excellent dictionary on Portuguese Sephardim, directed by António Marques de Almeida, in the entry on Diogo Mendes Benveniste, an exact picture is provided of the historical density of these transnational characters, whose activities were decisive in the process of accumulation of mercantile capital, that was also applied to the financing of the unstable public treasury of Portugal.[45] Sustained by solid fortunes, amassed in the spice trade, the Benveniste family was projected to the condition of Antwerp's first fortune, receiving the nickname of *the greatest of merchants*. The powerful mercantile networks, of which Benveniste was a part, played an active role in the first moments in the contracting of the brazilwood monopoly and operated according to practices established in the spice trade, as a result of which the mercantile capital circulated at both ends of the Atlantic axis, from the south to the north, from the Brazilian coasts to the Flemish cities, a fact that gives full reason to the testimony of Friar Vicente do Salvador in the seventeenth century, when he stated that the main destination of Brazil and its economy was oceanic and European, following exactly the same direction of capital accumulation.[46]

When the brazilwood arrived in Amsterdam, the logs of up to 2 metres were reduced to chips and then scraped until they formed a thick powder. The work was conducted by prisoners convicted for minor crimes in the prison of Amsterdam. The arduous effort earned the convicts some income and, above all, an opportunity of re-socialization through work, because these special prisons had the purpose of rehabilitating young offenders, while creating a less hostile environment in prison, where they became scrapers, *rasphuis*.[47] This is the reason why the prison governors of Amsterdam claimed the monopoly of the process of transforming brazilwood into the essential raw material for the production of red dye, although this monopoly was also desired by the prisons of other Flemish cities, such as Rotterdam and Leiden, which took disputes to the courts. It was also the reason why the municipalities would buy the raw material from the merchants associated with the Portuguese, so that in the prisons it could be manufactured according to contractual rules approved by the burgomasters. The teams formed by pairs of prisoners were needed because they had to move the equipment to scrape the wood; the labourers positioned at each end of the log were obliged to produce a certain amount of dye powder per day of work, which varied between 8.69 and 13.04 kg per day, depending on the physical capabilities of the prisoner-workers, who received for their work a wage plus a productivity premium.

Frédéric Mauro saw in this process an incipient division of labour, since some inmates specialized in scraping the wood, while others worked on maintenance of the blades that made up the sharp tools or in distributing food to their companions, ending the process with the bagging of the product to be sent to the dyers. The bags were duly identified with the arms of the municipality or with the initials of the prison that produced them.[48] Indeed, it was a miracle that the mercantile economy was able to accomplish, that is to say, to transform a house of social correction into a strategic link in the textile industry of which its creators were proud, to the point of stamping their arms on the packages destined to the dyers. In any case, around 1655 this process underwent a revolutionary change in its standards. This was the invention of a windmill that was erected in the city of Amsterdam. It was a revolutionary technical innovation in the production process, aimed at accelerating the time it took for the prisoners-workers to reduce the wood to sawdust, thus eliminating the most arduous and time-consuming part of the entire production process. The innovation significantly increased the productivity of the work, which went from 13.04 kilos/day to 65.21 kilos/day, a performance that suggests the adoption of hydraulic energy to drive the gears, a technical transformation that undoubtedly reflected the expansion of the consumption of Brazilian dye in Europe, being an invention produced by a social need.

But the production line was longer, for the bags containing brazilwood dust were distributed to the main industrial centres in Europe, where the second phase of the industrial process took place, obtaining the desired reddish colour dye. Lamarck states that the dye obtained from brazilwood was a false red colour, false in the sense of fleeting, for not retaining the colour for a long time. The fading of the colour led Camilo Torres Barragán to title his master's thesis on the theme *Um Rojo Fugitivo*.[49] In fact, the Brazilian dye 'evaporates easily and cannot be employed without alum and tartar'. Usually the dyers dissolved 0.92 grams of the powder in two buckets of water, to which an alkaline mixture was added after being reasonably homogenized, a process that required weeks of fermentation. The shade of colour obtained depended on the ingredients added, and, above all, on the quality of the wood, the type, the place where it was extracted, the care taken in storing it and the degree of humidity in the holds of the ships that transported it. The colour expected from the best brazilwood, usually from Pernambuco, was the carmine obtained by adding acid ingredients, which could reach purple or brown tones, as well as a liquid lacquer for miniatures or a reddish chalk for painting in general, 'although brazilwood never gave a dye comparable to *granum*', of mineral origin, used by the best dyers to dye luxury fabrics red.[50] The difficulty of the dyers in obtaining a dye with constant intensity and hue practically prevented the use of *brazilina* in the most sophisticated fabrics, being, for this reason, applied in the dyeing of common cloth, of lower quality and price, and therefore, accessible to a more elastic consumer market and, certainly, with greater growth potential in Europe.[51]

The commercialization of brazilwood became more complex in the first half of the seventeenth century due to the war between Spain and Holland, when the Dutch invaded the Brazilian Northeast, which had become an important sugarcane-producer on coast, exactly in the regions where the best quality of brazilwood was found. The timber was now sent directly to the processing centres in Holland, while in the remaining areas of forest exploitation, from the coast of Rio de Janeiro to Bahia, the traditional routes through Lisbon were preserved and from there the red logs were shipped to Holland through traditional mercantile networks or redirected to the Mediterranean markets, including Florence and Livorno. The profitability of this commodity trade allowed for the expansion of this business, even with declining prices. The price of goods shipped to the port of Lisbon was four times higher than in the ports of shipment in Brazil. Even adding transportation costs, which corresponded to 30 per cent of the value of the commodity, boarding and disembarkation costs, and eventual losses from shipwrecks or corsair actions, the brazilwood business continued to be a good investment for the contractors, for the intermediaries in the colony, and especially for the king, who made no capital contribution to the operation, conditions that became even more advantageous after the expulsion of the Dutch from Brazil in 1654, as the Portuguese regained control over the most valued areas of the production of brazilwood.

In any case, by this time the golden century of brazilwood was reaching its close. This was an age when the most powerful rival European states – the Dutch, English and French – advanced into the Caribbean region, aiming at breaking Portugal's monopoly on the sugar business and to break the Spanish monopoly in the region as well, creating trade centres and trade routes outside the control of the Habsburg Empire. At the same time, there had begun the exploitation of brazilwood and logwood in other regions extending from southern Mexico and the Yucatan peninsula as far as to the viceroyalty of Nueva Granada (modern-day Colombia and Venezuela) where Spanish settlers exploited the valuable timber, which was exported to Europe. At the same time, English corsairs established a bridgehead in so-called British Honduras, concentrating on the exploitation and trade of logwood wood with high market value. There they resisted the onslaught of local Mayan populations and Spanish military contingents, and, protected by international treaties, eventually gave birth to the autonomous state of Belize, now a member of the *Commonwealth of Nations*, which, in the creative image of Wayne Palomares Armstrong, formed with Brazil a pair of nations born of the redwood.[52]

Even before the withdrawal of the Dutch, the Portuguese government moved in 1649 to reorganize the commerce in Brazil, and created the *Companhia do Comércio do Brasil*, which initially held a monopoly of the brazilwood business in regard to cutting and transportation. This represented an effort to restore the economic advantages of the metropolis in the terms of the colonial system, a domain that had been crumbling in the face of multiple local interests, especially those of the large merchants from Recife. It was an innovative experiment intended to solve the frequent frictions between the settlers and the agents of the monopoly contractors. Through this new modality, the Crown reserved for itself 6 cruzados per quintal of wood sent to Portugal. On the other hand, the monopoly required the company's administrators to deal with the royal administration in the colony, including officials and officers of the kingdom who, out of excessive zeal or in order to sell difficulties and reap advantages, created obstacles to the consummation of the cutting and transport operations to the port of shipment. In these terms, the company was left negotiating between the king, with whom it signed the contract, and its own administration, which was the company's supplier of timber. These tensions were aggravated by the strong resistance of the poor local residents who demanded the right to extract brazilwood for their own purposes, as they said this was their last resort against poverty.[53] Historian Frédéric Mauro considered brazilwood exploitation impressive for its variety, in time and space, as well as the multiple systems of cutting and transport adopted. Many actors participated in this activity: isolated explorers, not infrequently doubled as smugglers; seasonal explorers operated in the dead seasons of sugar activity; monopoly explorers, from the Companhia de Jesus (the Jesuits); agents of the Companhia de Comércio do Brasil and even some Dutch contractors.[54]

In the eighteenth century, as a result of gold discoveries in Minas Gerais and large shipments of gold to Portugal,[55] interest in brazilwood decreased. Indeed, the profitability produced by the net export of metals was incomparably superior to the time-consuming and expensive procedures to transform brazilwood into finished dyes, carry out sales and obtain payments. But with the decline of gold mining after the mid-eighteenth century, brazilwood contracts were once again taken into account by the Portuguese prime minister, the Marquis of Pombal. It was in this context that a new modality of exploitation of this timber was established, being highly valued by the expansion of the European textile industries, mainly the English ones.

A practice was established which included direct extraction and transportation of the dyewood to the Casa da Índia in Lisbon. By means of a contract that established limits and prices for the shipments, the buyers of the dye took it from there to resell it in the main European market, thus establishing a powerful, international network of traders involved in the brazilwood business. The first contract signed by the Crown, with Pombal's approval, was negotiated with a group of merchants based in England, *Purry, Mellish and Devisme*, who monopolized the purchase and resale of brazilwood for twenty-seven years, taking responsibility for contracts that spanned the period from 1757 to 1784. The originality of these contracts is to be noted in the first place to the fact that they were destined specifically to ship this commodity to the Lisbon market. In the second place, the contracts made it obligatory to restrict acquisitions of the timber from the North of Brazil, specifically in the colonial territory that extended from Pernambuco to Ceará, the area where the best quality dye wood was found. The third intention of the contract was to help transform Pernambuco into a dynamic centre of export operations, to which all captaincies in the demarcated area would be economically and politically subordinated.

This new arrangement designed by Pombal for the Brazilian Northeast led historian Thiago Alves Dias to use the expression *indirect monopoly* in his study of the subject.[56] Alves argues that the new form of brazilwood extraction was an improvement over the traditional form of the royal monopoly practiced until that moment. The king was now authorized to extract, transport and commercialize the dyewood under license, a new and significant innovation from the point of view of the management of mercantilist economic policy in the colony. But in contrast to the traditional monopoly there were other advantages to the new policies. In the first place, the area of dominion and exploitation was strictly demarcated; second, the extraction itself was exclusively to be carried out through the port of Recife; third, the Recife-Lisbon transport route and market stopover was administered by contractors, led by David Purry.[57] That is to say, with the passage of time, the royal monopoly, instead of being softened, was hardened and modernized for the benefit of the king and his international partners and to the detriment of colonial interests.

The value of the brazilwood monopoly increased in the first decade of the nineteenth century as new fiscal needs came to the fore, since declining gold exports were no longer in a position to meet them. The most critical moment was the forced transfer of the Portuguese royal family to Brazil in 1808.[58] This was a symbolic date for the Brazilian colony and also for its first product of importance in the international market, since this is the date when the French chemist, Michel-Eugéne Chevreul, synthesized the artificial colouring of brazilwood, establishing a limit to its secular trajectory. But it still had significant value. With the aim of providing capital for the newly created Banco do Brasil (1808), rights were granted in its statutes with regard to exploitation of brazilwood wherever it was found on the Brazilian coast. In fact, the brazilwood resources – after the proclamation of independence in 1822 – were destined to the payment of the debt contracted by the Portuguese with the British banks. In effect, these funds were intended to be a kind of compensation for the recognition of political emancipation by Portugal, and, complementarily, to cover the necessary expenses for the installation of diplomatic representations of the new Brazilian state abroad.[59]

Resorting to resources from state rights over brazilwood was nothing new in the financial history of this commodity for, on the contrary, it had often been used as a panacea for many problems of the royal treasury. At a critical moment, in 1804, Napoleon demanded the Regent D. João to break neutrality with England, threatening to invade the kingdom, unless 16 million francs were paid to prevent the invasion, a huge amount for the time. D. João instructed his banker, Baron Porto Covo da Bandeira, to gather the funds with several Dutch bankers and pay the required subsidy in sixteen annual instalments of 1 million francs, mortgaging the income from the brazilwood contract as the guarantee, a financial operation which was still pending in 1843, when the Baron's heirs requested the Portuguese government to settle the existing liabilities.[60]

Table 4.1 Profit and Profitability of the Pau-brasil Monopoly (1796–1807) (in Quintals* and Mil réis)

Year	Quantity (in Quintals)	Portugal		Pernambuco				Profit (%)
		Re-export Price	Revenue	Extraction Cost	Transportation Costs	Expense	Operating Profit	
1796	2.939	4.400	12.931.600	1.000	400	4.114.600	8.817.000	68.18
1797	1.801	4.400	7.924.400	1.000	400	2.521.400	5.403.000	68.18
1798	7.474	4.400	32.885.600	1.000	400	10.463.600	22.422.000	68.18
1799	1.030	4.400	4.532.000	1.000	400	1.442.000	3.090.000	68.18
1800	12.149	4.400	53.455.600	1.000	400	17.008.600	36.447.000	68.18
1801	27.740	15.000	416.100.000	1.000	400	38.836.000	377.264.000	90.67
1802	12.426	14.000	173.964.000	1.000	400	17.396.400	156.567.600	90.00
1803	22.742	15.000	341.130.000	1.000	400	31.838.800	309.291.200	90.67
1804	27.401	15.000	411.015.000	1.000	400	38.361.400	372.653.600	90.67
1805	19.544	15.200	297.068.800	1.600	640	43.778.560	253.290.240	85.26
1806	5.459	18.000	98.262.000	2.000	800	15.285.200	82.976.800	84.44
1807	9.199	18.000	165.582.000	2.000	800	25.757.200	139.824.800	84.44
1808	-	20.000	-	-	-	-	-	-
1809	-	16.000	-	-	-	-	-	-
1810	-	14.000	-	-	-	-	-	-

Source: Balança Geral do Commercio do Reyno de Portugal com seus Domínios e Nações Estrangeiras: Anos 1796 a 1811, Biblioteca Nacional do Rio de Janeiro, Instituto Nacional de Estatística de Lisboa. Seção de Manuscritos.
*The quintal was a weight of approximately 100 pounds.

Concluding remarks

In the final years of its domination of the Brazilian colony, the Portuguese government was still struggling to reorganize the brazilwood extractive business.[61] One of the concerns was to the need to reform the 1605 Ordinance, considered by many to be excessively rigid in defence of the royal monopoly and responsible for the settlers' lack of interest in its exploitation, since it yielded them little gains. Proposals for reform were outlined in the Resolution of 30 September 1818, whose terms were maintained until 1822 when the Portuguese royal monopoly came to an end as well as the Brazilian monopoly.[62] But despite these changes, the fate of brazilwood was ultimately to be determined by the chemical revolution underway in Europe, most notably in Germany. The long period when brazilwood was *royal timber*, which had begun in 1501, finally ended in 1875, when the export tax that weighed on it was extinguished by law, and brazilwood became equivalent to other national timber.

One can conclude that the central problem with the Portuguese policy of exploitation of brazilwood was the extremely monopolistic form imposed by the Crown, which ended up engaging the settlers themselves in the service of forest devastation, not to mention the role of the royal contractors who were already naturally committed to this task. While the monarch claimed one-fifth of the gold production, he demanded complete and absolute ownership of the best trees reserved for shipbuilding and dyeing. Indeed, all the conservationist precepts incorporated in the various ordinances had the sole purpose of reserving for the Crown the natural treasure stockpiled for future use. The best trees were not the property of the Portuguese state, but of the sovereign, which could only be cut down by the settlers to be sold to the agents of the contractors, being themselves prevented from cutting them down for their own use, subject to severe punishments that could even devolve in a risk of execution and death. The local settlers were therefore prevented from transforming a natural capital into social capital. They could not harvest their trees, sell them, and capitalize themselves to invest in rural activities as the British forestry policy in North America allowed. The forest effectively became useless to the grantees and settlers and when possible led them to destroy the trees to open up land for other purposes. This absurd forest policy is summarized in an expression of historian Shawn Miller: 'neither harvested nor exploited, they annihilated'.[63] Only a tiny portion of the available forest wealth was transformed into domestic wealth. Local capital accumulation did not surpass more than 1 or 1.5 per cent of the total value of colonial exports, and therefore did not contribute significantly to endogenous development.

In fact, the dyewood ended up having virtually no value for private individuals, neither for use nor for exchange, and ended up being even a hindrance, because on the properties where there was an abundance of brazilwood, agriculture was hampered because the land could not be cleared. The main complaint of the first governor, Duarte Coelho, in a letter sent to the king, was addressed to the agents of the brazilwood monopoly contractors, for 'their manner' of conducting the relationship with the Indians, who were seduced to work by offering them even dangerous instruments as payments, such as swords and muskets, a behaviour that increased their aggressiveness against the settlements, besides removing them from the agricultural tasks, which were the pillars of the colonial economy.[64]

Complaints against the monopoly echoed throughout the centuries, but in the twilight of Portuguese rule in Brazil, local inhabitants openly positioned themselves against the monopoly without fear of reprisals: 'The brazilwood monopoly, under the present circumstances, is without contradiction an obstacle to farming throughout the vast extension in which the forests are found' and 'in this state of things, the farmers or landowners consider it a misfortune that brazilwood exists on their lands'.[65] By the end of the eighteenth century, in the wake of revolutionary ideals, anonymous pamphlets preached the destruction of the forests. In this context, we must remember that instead of easing the restrictive laws on forests in order to meet local and immediate needs, they became increasingly rigid. According to the royal regulations that governed the Brazilian captaincies from the 1530s, the owner had been allowed to use the wood in general and brazilwood, in particular, for his personal needs, a concession removed by the 1605 Ordinance, and only reintroduced in the resolution of 1818. In fact, the monopoly regime became stricter as time went by, indicating its indispensable contributions to royal fiscal resources, exactly the opposite of what was happening in Spanish America with the brazilwood trade, which was increasingly captured by illegal trafficking networks, leading to the transfer of control over the forests to the local settlers in the wake of the political upheavals that ended Spanish rule in America.

This means that the settlers in Brazil, the would-be conquerors of the land, had an inferior situation to the Indians regarding the exploitation of the forest, as regulations turned the trees into forbidden goods for them. With the monopolization of the forest, the rational management of the forest became extremely difficult, according to Shawn.[66] Although historian Warren Dean appeared to agree with this interpretation, stating that the goal of the Portuguese Crown 'was to accumulate capital not in the colony but rather in the metropolis', he concluded that the destruction of the forest could be attributed, as many do, to colonialism and capitalism. Yet this reflects an ideological bias, which led him to argue on behalf of the Crown, that the 'colonial authorities not infrequently showed themselves more enlightened and progressive than their subjects in relation to nature', probably in reference to preservationist clauses inscribed in the 1605 Ordinance.[67] But the fact is that virtually all arguments in defence of the forest that were registered in official regulations during the entire colonial period had no real intention of promoting the defence of the environment *in and for itself*. One cannot, therefore, recognize any enlightened and, much less, progressive intention in actual practices employed.[68] In fact, the conservationist discourse of the time was an explicit defence of the royal patrimony. It was not *preservationist*; rather, it was *reservationist* and designed to guarantee the king's present income and that of his future heirs. This fact did not go unnoticed by his contemporaries, so much so that Brandonio, in the *Dialogue on the Grandeurs of Brazil*, wrote: 'Brazilwood is his commodity [. . .]. All Brazil yields to His Majesty's treasury without any expense'.[69] What could we expect from a sovereign who, when referring to the public patrimony, does not think of the greater interest of the State, nor of his subjects, nor of the settlers, because he addresses them in these strictly personal terms: 'Brazilwood [. . .] will belong to me, and everything will always be mine and my successors'.[70]

In the face of the historical evidence, it becomes impossible to concur with the historiographical tradition that affirms that there were no significant benefits for the European states from colonization. It is even less valid to accept the interpretation that Portugal did not obtain any advantages from its maritime empire, nor from Brazil, with the argument that benefits were quickly transferred to other European states, which would imply a passive inert behaviour by the Portuguese Crown that contrasts with its fierce defence of its patrimony. Faced with these arguments, it is necessary to recognize the benefits that both the Portuguese monarchy and the other European powers received and the costs they imposed on the vast territorial spaces they kept under custody for such a long time. More specifically, what were the *benefits* to the Brazilian colony resulting from the exploitation of brazilwood? The answer is: barely none. And how so for Portugal? There were clearly many benefits. And as for Portugal's partner nations? Quite a few. In contrast, we may ask what were the *costs* for the colony in the exploitation of these precious trees? All of them! Brazilwood has no place even in the historical memory of Brazilians because it left nothing, no ruins, only devastation, unlike sugar and gold, which contributed to the building of villages, towns and cities.

In summary the costs and benefits in processes such as the colonial exploitation of brazilwood over the centuries have to be confronted with the fact that a great portion of the extensive forests of many regions of Brazil are long gone, a fact that suggests the need to consider the darker side of this chapter, which is titled *A Place in the Sun*. The purpose of this study has been to illustrate the place of brazilwood in the colonial economy and society and on the trajectory of this valuable dye, thinking of the size of the trees that stood out in the forest and shone in the sun like gold on the Brazilian historical scene. But, it is also essential – after a concentrated effort of research – to ask what is left of the splendour of the forests of brazilwood? What happened to this magnificent historical character and to its ephemeral glory as the first of all Brazilian exports? The virtual disappearance of the brazilwood trees and their death on the forest floor reflect a harsh reality that is relevant for contemporary concerns on the environmental history of Brazil.

Notes

1 This term designates various species related to the production of red-brown dyes, according to Nicholas Eastsugh, Tracey Chaplin, Valentine Walsh, and Ruth Sidhall, *The Pigment Compendium. A Dictionary of Historical Pigments* (Amsterdam: Elsevier, 2004), 60–1. Several of the dyes mentioned came from western India and the Malabar coast according to A. O. Baião, 'O Comércio do Pau Brasil', in *História da colonização portuguesa do Brasil*, ed. Carlos Malheiro Dias, Vol. 2 (Porto: Litografia Nacional, 1921–1924), 323.

2 *Ordenações Afonsinas*, liv. 2°, tit. XX, n°226. Chancelaria de D. João III, liv. 21, fls. 73, in Arquivo Nacional da Torre do Tombo.

3 Jorge Pedreira, 'To have and to have Not: The Economic Consequences of Empire: Portugal (1415–1822)', in *The Costs and Benefits of European Imperialism: from the conquest of Ceuta, 1415, to the Treaty of Lusaka, 1974*, ed. Patrick Karl O'Brien and Leandro Prados de la Escosura (Madrid: Marcial Pons, 1998), 104.
4 António Vasconcelos de Saldanha, *As capitanias do Brasil. Antecedentes, desenvolvimento e extinção de um fenómeno atlântico* (Lisbon: CNCDP, 2001).
5 Bernardino José de Souza O, *Pau-brasil na história nacional* (Rio de Janeiro: Companhia Editora Nacional, 1939), 150. And Shawn William Miller, *Fruitless Trees. Portuguese Conservation and Brazil's Colonial Timber* (Stanford: Stanford University Press, 2000), 49.
6 Miller, *Fruitless Trees*, 5–47.
7 Charles André Julien, et al., *Les Français en Amérique pendant la première moitié du XVIéme siècles* (Paris: PUF, 1946), 6.
8 Frédéric Mauro, *Portugal, o Brasil e o Atlântico, 1570–1670* (Lisbon: Editorial Estampa, 1989), 173.
9 See the document titled: *Regimento do Pau Brasil*, Rio de Janeiro, AN, col. 60, f. 1065. The analysis of the document reveals the intention of producing order in the exploitation of brazilwood, but in fact it tended to generate conflicts by focusing on the royal monopoly which was widely criticized by the owners of colonial properties. See Maria Isabel de Siqueira, 'Considerações sobre a ordem em colônias: As legislações na exploração do pau-brasil', *Clio-Revista de Pesquisa Histórica* 29, no. 1 (2011): 1–17.
10 Warren A. Dean, *A ferro e fogo: A história e a devastação da Mata Atlântica brasileira* (São Paulo: Cia das Letras, 2004), 35–7.
11 *Blacks of the earth* was a name proposed in the book by John Manuel Monteiro, *Negros da terra-índios e bandeirantes nas origens de São Paulo* (São Paulo: Cia das Letras, 1994).
12 Americo Vespucci, *Cartas de Viaje* (Madrid: Luciano Farmisano, 1986), 13. Since the twelfth century the wood of trees that produced red pigment was known in Europe and designated as *brasile*, a derivation of the Latin that originated the name of Brazil, which was adopted among popular sectors as the designation of the Portuguese colony as early as the 1530s; moreover, the wood merchants were called *brasileiros*.
13 Yuri T. Rocha, Andrea Presotto, and Felisberto Cavalheiro, 'The Representation of Caesalpinia Echinata (Brazilwood) in Sixteenth-and-Seventeenth-Century Maps', in *Anaís da Academia Brasileira de Ciências* 79, 4 (Brazil: Anaís da Academia Brasileira de Ciências, 2007), 751–65.
14 John Hemming, *Ouro vermelho. A conquista dos índios brasileiros* (São Paulo: Edusp, 1995), 67.
15 Jean de Léry, *Viagem à terra do Brasil* (São Paulo: Martins Editora, 1951), 210.
16 Alexandre Marchant, *Do escambo à escravidão. As relações econômicas de portugueses e índios na colonização do Brasil, 1500–1580* (São Paulo: Editora Nacional, 1943), 95–9. A unique variety of goods was brought for barter by Palmier de Gonneville, one of the first French to reach the southern coast of Brazil, including 300 fabrics; 4,000 axes, shovels, reapers, cleavers and pitchforks; 2,000 spears; 50 dozen mirrors; 6 quintals of beads; 8 quintals of Rouen iron instruments and 240 dozen machetes.
17 Julie Kapyrka and Mark Dockstato, 'Indigenous Knowledges and Western Knowledges in Environmental Education: Acknowledging the Tensions for the Benefits of a "Two-Worlds" Approach', *Canadian Journal of Environmental Education* 17 (2012): 97–112.

18 Hemming, *Ouro vermelho*, 85.
19 Hemming, *Ouro vermelho*, 85.
20 Letter from Duarte Coelho to the Portuguese king, Olinda, 20 December 1546, HCP, 3:13.
21 Stuart Schwartz, *Segredos internos. Engenhos e escravos na sociedade colonial* (São Paulo: Cia das Letras, 1988), 72–3.
22 Lamarck described the typical brazilwood, the *brasilete*, as having the following characteristics: 'The trunk wood is red, but is covered by a very thick sapwood. It is very heavy, very dry and brightly burns, with almost smokeless flame due to its dryness. It works well on the lathe and receives polishing well'. See Jean-Baptiste Lamarck, *Dictionaire Encyclopédique de Botanique* (Paris, 1789), in Mauro, *Portugal, O brasil e o Atlântico*, 167. The term 'Ibirapitanga' used by the Indians for brazilwood became, for the botanists, *Caesalpinia echinata* in honour of the botanist and physician of Pope Clement VIII, Andrea Cesalpino. See Ângelo C. Pinto, *O pau-brasil e um pouco de história brasileira* (Rio de Janeiro: Instituto de Química, UFRJ, 1999).
23 To load 5,008 logs weighing 125 tons onto the ship *Bretoa*, it took seventeen days for the Indians, probably assisted by some sailors, which gives an average of almost 300 logs a day. The French took the wood directly to Europe, while the Portuguese concentrated it in authorized ports for export. Copies of some contemporary engravings of the natives working in the forest can be found in Robert C. Smith, 'The Wood-Beach at Recife: A Contribution to the Economic History of Brazil', *The Americas* 6, no. 2 (1949): 209–33.
24 Pero de Magalhães Gandavo, *Tratado da terra e da gente do Brasil*, Vol. II (São Paulo: Editora Nacional, 1978), 152–3, describes this experience: 'Some fish and hunt, others make food and farm, and thus little by little men become rich [. . .] because the same Indians of the land seek food for themselves and for the masters, and in this way men do not make expenses with their slaves in food nor with their people'.
25 John M. Monteiro, 'The Crises and Transformations of Invaded Societies: Coastal Brazil in the Sixteenth Century', in *The Cambridge History of Native Peoples of the Americas v. III, South America*, ed. Frank Salomon and Stuart Schwartz (Cambridge: Cambridge University Press, 1999), 1005.
26 Hemming, *Ouro vermelho*, 146.
27 Hemming, *Ouro vermelho*, 149.
28 Hemming, *Ouro vermelho*, 80.
29 Carlos Alberto de Moura Ribeiro Zeron, *Ligne de foi: La Compagnie de Jésus et l'esclavage dans le process de formation de la sociéte coloniale en Amérique portugais (XVe-XVIIe siècles)* (Paris: Honoré Champion, 2009).
30 Guida Marques, 'Do índio gentio ao gentio bárbaro: usos e deslizes da guerra justa na Bahia seiscentista', *Revista de História* 171 (2014): 26.
31 The forest was a gateway for Portuguese mercantilism and provoked the profound alteration of the forest ecosystems, as is argued by Marcelo Henrique Dias, 'A floresta mercantil: exploração ladeira na capitania de Ilhéus no século XVIII', *Revista Brasileira de História* 30, no. 59 (2010): 2.
32 Warren Dean, *With Broadax and Firebrand The Destruction of the Brazilian Atlantic Forest* (Berkeley: University of California Press, 1997), 64–5.
33 Quoted by F. F. A. Aguiar and R. A. de Pinho, *Pau-brasil. Caesalpinia echinata Lam* (São Paulo: Instituto de Botânica, Folheto 18, 1986).
34 Charles André Julien, *Les voyages de découverte et les premiers établissements* (Paris: PUF, 1948), 25–9.

35 Carta de Pietro Rondinele, Sevilla 3 de octubre de 1502, in *Comissione Colombiana. Raccolta de Documenti e Studi*, Vol. I (Roma: Ministero della Pubblica Istruzione, Parte III, 1893), 121.
36 Michel Mollat, *Le commerce maritime Normand à la fin du Moyen Age* (Paris: PUF, 1952), 250.
37 *Livro em que se contem toda a fazenda, & real patrimônio dos Reynos de Portugal, Índia, ilhas adjacentes de sua Coroa & outras muitas particularidades.* Feito e ordenado por Luiz de Figueiredo Falcão, 1607. See José de Souza, *O, pau-brasil na história nacional*, 81.
38 Mauro, *Portugal, o Brasil e o Atlântico*, 186–9.
39 Caio Prado Junior, *A formação do Brasil contemporâneo (Colônia)* (São Paulo: Editora Brasiliense, 1961).
40 Fernando A. Novais, *Portugal e Brasil na crise do Antigo Sistema Colonial (1777–1808)* (São Paulo: Hucitec, 1979).
41 Eddy Stols, 'A "nação" flamenga em Lisboa (séculos XV-XVII)', in *Flandres e Portugal. Na confluência de duas culturas*, ed. J. Everaert and E. Stols (Lisbon: Edições Inapa, 1991), 121. In the course of the fifteenth century the term *nation* was used to designate the trade associations of a certain nationality. See Frédéric Mauro, 'Merchant communities, (1350–1750)', in *The Rise of Merchant Empires. Long-distance Trade in the Early Modern World 1350–1750*, ed. J. D. Tracy (New York: Cambridge University Press, 1990), 262.
42 Oscar Gelderblom, 'Organization of Long-Distance Trade in England and the Dutch Republic, 1550–1650', in *The Political Economy of the Dutch Republic*, ed. Oscar Gelderblom (Surrey: Ashgate Publishing, 2009), 226–7.
43 Fernand Braudel, *Civilização Material, Economia e Capitalismo. Séculos XV-XVIII. O Tempo do Mundo*, Vol. III (São Paulo: Martins Fontes, 1996), 157–9.
44 Vitorino Magalhães Godinho, *Os descobrimentos e a economia mundial* (Lisbon: Editorial Minerva, 1981).
45 A. A. Marques Almeida (dir.), *Mercadores e gente de trato* (Lisbon: Campo da Comunicação, 2009), 412–18.
46 Friar Vicente do Salvador wrote the first *History of Brazil* in 1627.
47 Kenneth Harold Dobson Haley, *The Dutch in the 17th Century* (London: Harcourt Brace, 1972), 159. Also see J. Thorsten Sellin, *Pioneering in Penology: The Amsterdam Houses of Correction in the Sixteenth and Seventeenth Centuries* (Pennsylvania: Pennsylvania University Press, 1944).
48 Sellin, *Pioneering in Penology*, 199–200.
49 Camilo Alfonso Torres Barragán, *Un rojo fugitivo: extracción, circulación y consumo del palo de Brasil de la Nueva Granada (1770–1850)* (Colombia: Universidad de Los Andes, Master's Thesis, 2013), 47.
50 According to Frédéric Mauro, the red dye from *brazilwood* (C16 H14 O5) when dissolved in water and in contact with air, or with an alkaline solution, becomes brick red, and gives rise to *brasilene* (C16 H12 O5 H20), which is precisely the colouring matter used by dyers. See Mauro, *Portugal, o Brasil e o Atlântico*, 199. Also *Granum* are rock crystals that can be treated to produce red dyes. See Harold Benjamin Johnson and Maria Beatriz Nizza da Silva, *O Imperio Luso Brasileiro 1500–1620* (Lisbon: Editorial Estampa, 1992).
51 Marchant, *Do escambo à escravidão*, 42. And Arthur Ernest Everest and Arthur George Perkin, *The Natural Organic Coloring Matters* (London: Kissinger Legacy Reprints, 1918), 345–62. Also W. S. Stanfield, 'Half a Century in the Dyeing and

Finishing of Flannel Garments, Wools and Linens', *Journal of the Society of Dyers and Colorists* 50, no. 4 (1934): 139–69.

52 Wayne P. Armstrong, *Logwood and Brazilwood: Trees that Spawned 2 Nations* (1992). https://www.waynesword.net/ecoph4.htm.

53 Uiá Freire Dias dos Santos, *Negociação e conflito na administração do pau-brasil: a capitania de Porto Seguro (1605–1650)* (Salvador: Universidade Federal da Bahia, Master's Thesis in History, 2015), 120–206.

54 Dias dos Santos, *Negociação e conflito*, 175.

55 Pinto, *O pau-brasil e um pouco de história*, 116–17.

56 Thiago Alves Dias, *Monopólio Indireto: Colonização Mercantil no Norte do Estado do Brasil (c. 1710–c. 1780)* (São Paulo: Universidade de São Paulo, PhD Thesis in History, 2017), 10.

57 T. A. Dias, *Monopólio Indireto*, 254.

58 José Jobson de Andrade Arruda, *Uma colônia entre dois impérios: a abertura dos portos brasileiros 1800–1808* (Brazil: Cátedra Jaime Cortesao/EDUSC, 2008), 45–185.

59 The loan to which the commitment referred was contracted by Portugal with England in 1811, when it gave as a guarantee of payment the income from the *estanco* of hardwoods and part of the revenue from diamonds, ivory and orchil. See José de Souza, *O, pau-brasil na história nacional*, 81.

60 'Decreto de 2 de maio de 1804 para assistir ao Governo Frances com 16 milhões de francos', quoted by António Alves Caetano, *A Economia Portuguesa no tempo de Napoleão: constantes e linhas de força*, (Lisbon: Tribuna da História, 2008), 115–17.

61 Resistance to the policy of exploitation of the Brazilian forest was sporadic and sparse during the entire colonial period, but at the turn of the eighteenth century there emerged enlightened reports on the damage caused to the forest. See Jose Augusto Pádua, 'Annihilating Natural Productions: Nature's Economy, Colonial Crisis and the Origins of Brazilian Political Environmentalism (1786–1810)', *Environment and History* 6, no. 3 (2000): 255–87.

62 Resistance to the royal monopoly intensified after the Court settled in Brazil. Plans to reformulate forest policy were frequent but culminated with the end of the Crown monopoly of brazilwood purchases, assuring the right of the landowners to open the forest and sell the wood extracted from their properties, except in the *Sesmarías* near cutting sites. See F. W. O'Morton, 'Royal Timber in the Late Colonial Bahia', *Hispanic American Review* 88, no. 1 (1978): 59–61.

63 Miller, *Fruitless Trees*, 10.

64 Marchant, *Do escambo à escravidão*, 95.

65 *Coleção cronológica das leis, decretos, resoluções de consultas, provisões (. . .) do Império do Brasil desde o ano de 1808 até 1831, coligidas pelo conselheiro José Paulo de Figueiredo Nabuco, Legislação brasileira ou Coleção Cronológica* (Rio de Janeiro: Arquivo Nacional, 1836–1844).

66 Miller, *Fruitless Trees*, 11.

67 Dean, *With Broadax and Firebrand*, 133.

68 On this point, see Stuart McCook's review of Miller, *Fruitless Trees*, in H-LatAm (November 2000). And Jose Augusto Drummond's review of Dean, *With Broadax and Firebrand: The Destruction of the Brazilian Atlantic Forest*, in *H-Environment* (June 2000).

69 de Magalhães Gandavo, *Tratado da terra e da gente do Brasil*, 146–7.

70 Letter of Donation of the Captaincy of Bahia to Francisco Pereira Coutinho, in Ignacio Accioli de Cerqueira e Silva, *Memórias históricas e políticas da Província da Bahia*, Vol. 2 (Salvador: Bahia, 1925). There was no lack of contemporaries who advocated a more restrictive monopoly that would avoid including the brazilwood forests in the land concessions for new colonists (*sesmarias*), conceded by the Crown. See José Carlos Mayrink da Silva Ferrão, *Memória sobre o pau-brasil, 1805*, Vol. 7 (ANRJ, Codex 807): 213–19. And José Carlos Mairink da Silva Ferraz, 'Memória sobre o pau-brasil', *Revista do IHGB* 416 (2002): 385–7.

5

Mexican cochineal and the material history of art (sixteenth to nineteenth centuries)

Georges Roque

The history of Mexican cochineal is of enormous interest not only for the understanding of the role dyes play in the manufacture of luxury textiles since the sixteenth century, but also for a comprehension of their contribution to art, in particular painting, since pigments are the material basis with which painters provide colour to their canvases. A number of museum exhibitions have been dedicated to particular colours, such as blue or red, and specialized art journals have also published dossiers on some hues, but it is crucial not to forget that to be interested in a hue as such usually means paying no attention to the specific way in which it has been obtained, that is to say, its materiality.[1]

Red, for instance, is an abstract concept subsuming everything that is red; as a matter of fact, there are thousands of different reds. Hence our emphasis on the distinction between *hue* and *pigment*.[2] In sum, what interests us in these pages is to contribute to a material history of art, taking into account a given colourant/dye, its properties, functions and meanings, an approach that traditional art history has largely concealed since the Renaissance.[3] This neglect is largely to be explained by the prejudice regarding the pigments used by an artist often considered a raw material and, as such, virtually devoid of meaning. Our goal here is precisely the reverse in order to emphasize the meanings that colourants and pigments can have.[4]

If studying colour is so complex, it is because it encompasses so many areas. Art historians and critics often see colour as an aesthetic component of some works of art. However, the aesthetic value of colour cannot be dissociated from other values from the moment one takes into account its materiality, which also includes its economic value and sociopolitical significance. In fact, these aspects are inseparable, and cochineal is an extraordinary example of how they are all entangled.[5]

Let us start with the economic aspects of the history of Mexican cochineal, an extraordinary dye, the most expensive in the world for centuries. If we ask ourselves the reason for the exceptional success of Mexican cochineal in Europe, the answer is the strong demand of the earliest European textile industry (often described as by historians as *protoindustrialization*), which was one of the central axes of the economic dynamics that made possible the gradual emergence of European capitalism from the end of the Middle Ages. This explains what might seem surprising, namely that cochineal was the second most important of the products exported from *New Spain* (Colonial Mexico) to Spain after silver, from the mid-sixteenth century to the end of the eighteenth century.[6] The letters of wealthy European merchants provide a generous source of information about the millionaire cochineal business. Already in 1585, two of them (from Florence and Burgos) made an agreement to try to buy the entire stock of cochineal available in Seville, Italy, France and Flanders, with the purpose of monopolizing supply and obtaining important capital gains after prices rose.[7] And the same type of mercantile operation was repeated on at least one other occasion two centuries later, in 1787, by London merchant bankers.[8]

We turn now to the sociopolitical value of cochineal. If the enormous European success of cochineal is due to its use by dyers, the artisans themselves depended on the strong demand from the wealthy for elegant textiles dyed with cochineal. Several factors are involved here. Firstly, the very high price of this dye transformed fabrics which incorporated cochineal into luxury commodities that were unattainable for ordinary consumers. Furthermore, it should be emphasized that the extraordinary dyeing properties of Mexican cochineal gave very saturated and stable results, in the sense that cochineal dyed clothes retained their saturated tone without fading as quickly as most dyes. Thirdly, the colour red had symbolic importance and is usually related to 'wealth' or 'elegance', because of a precedent: the colour purple formerly had the same sociopolitical value of a colour reserved for the elite because of its price. Consequently, cochineal could easily substitute the natural purple dyes that had been so important from as far back as the age of the classic Roman Empire.

Finally, we come to the aesthetic values related to certain dyes. Since the Middle Ages, textiles that gave a sense of chromatic intensity, such as scarlet or crimson, were appreciated as elegant, refined and luxurious. In other words, what was appreciated was the vividness of the highly saturated tone due to the concentration of colourant. For this reason, cochineal came to signify prestige and wealth and became a fashion phenomenon.[9] This also explains why the study of cochineal cannot be dissociated from its materiality, as well as why we dedicate one strand of our analysis to the relationship between textiles and painting, and another to cochineal as the colour of power.

Textile and plastic arts

A consideration of the importance of the textile industry in the development of the cochineal business in Europe permits a first hypothesis: one of the main reasons for painters' interest in Mexican cochineal was its use by dyers, which produced both stimulation and emulation, since painters wanted to obtain shades as rich, saturated and brilliant as those produced by the dyers. Such a hypothesis has significant consequences in terms of art history: it contributes to raise the question of whether an 'autonomous' history of painting is a valid approach, in which it is argued that the use of given colours actually developed independently, without owing anything to the dye craftsmen. This was the traditional approach of the old Academies of Art, which argued that painting is a liberal and non-mechanical art. But, in fact, it was not by chance that painters used cochineal where it was readily available. I provide three examples:

- **Seville** was the main city where cochineal shipments arrived from the Mexican port of Veracruz during the sixteenth and seventeenth centuries. In the sixteenth century, the two main centres of the silk industry in Spain were Toledo and Granada, while in Seville, the textile industry developed mainly from the last third of the seventeenth century.[10] The Sevillian painter and treatise writer Francisco Pacheco wrote in his treatise *El arte de la pintura* (1649) that he had 'made with this color [carmine] some well imitated velvets' but praised another painter, Alonso Vázquez, for his unparalleled way of using cochineal to imitate velvet fabrics.[11] It is also well worth noting that two Sevillian painters who used cochineal came to Mexico to continue this tradition, namely Alonso Vásquez, already mentioned, and Andrés de Concha. But many other Sevillian painters also used cochineal. To this list must be added Pacheco's son-in-law (and pupil) Diego Velázquez, as well as Zurbarán and Murillo. In fact, several Sevillian painters bought their cochineal directly upon the arrival of the fleets.[12] In Spain, this trend continued throughout the Golden Age. Furthermore, we also know that subsequent Spanish painters of the eighteenth century (Goya, among others) bought cochineal.[13]

- **Venice**[14] was the seat of a very important textile industry from the late medieval era, and also a centre for the sale of pigments to artists.[15] It is worth noting the speed with which the use of cochineal spread among Venetian painters. The first shipment of cochineal arrived in Spain at the end of 1523, but in the following decade other shipments arrived in Seville, probably with small quantities for the purpose of experiments.[16] From 1554, there is evidence of increased shipments of much cochineal but also with a high price.[17] We now know that Titian began to use it at the beginning of the 1550s,[18] and in Spain, it has been found from the following decade in the paintings of Luis de Morales.[19] This means that in less than thirty years after its introduction in Europe, the famous pigment was already being used in painting. There is no doubt that the textile industry played a major role in accelerating this process due to the sumptuousness of the textiles dyed by the very important Venetian textile industry, which had been using cochineal since the 1540s.[20] In fact, from the second half of the sixteenth century, we find cochineal in the palette of most of the great Venetian painters such as Titian, Veronese, and Tintoretto. We should also add Bassano as well as Lorenzo Lotto, who considered himself Venetian. El Greco, who studied in Venice, served as a link between Venice and Spain, and later Tiepolo and Canaletto also did so.

- **Antwerp** and **Amsterdam** were other centres of an important textile industry, where, for the same reason, cochineal shipments arrived by sea from Seville and later from Cádiz.[21] It is no coincidence that these are ports. This also helps to explain the importance of the use of cochineal among Flemish and Dutch painters (Martin de Vos, Rubens, Van Dijk, Rembrandt and ter Borch).

A second hypothesis that arises from considering the importance of the textile industry for contemporary painters is that the first pictorial use of cochineal was precisely the representation of fabrics, cloths and curtains. This is very clear in the works of Venetian painters and in Tintoretto, in particular, since these artists were concerned with assuring that their representations of velvet fabrics could have the glow of textiles dyed with cochineal. Such is the case of two works by Tintoretto. The first one is *The Deposition of Christ*,[22] very interesting because of its date, since it was painted at the end of the first decade of the use of cochineal in painting, that is, the 1550s. The other is *Judith and Holoferne*. In both cases, the laboratory analysis of the paintings detected cochineal in the red areas of the canvases represented. In the case of Tintoretto, the link between textile and painting becomes even more evident, recalling the well-known fact that he was the son of a dyer; hence his nickname.

The idea that the first use of cochineal in painting was the reproduction of fabrics can be verified if we take into account a few recommendations found in the art treatises. I have already pointed out that Francisco Pacheco praised Alonso Vazquez for his way of reproducing velvet fabrics in his paintings. It is worth noting that Pacheco also considers that 'the main difficulty of the coloring is in the imitation of the flesh', and then adds that 'It is convenient to know how to imitate the colors of the cloths, silks and golds of all qualities with such skill that the tenderness or hardness can be seen more or less according to the condition and variety of each thing'.[23]

In both cases, Pacheco recommends the use of cochineal.[24] Moreover, in the case of cochineal dyed fabrics, he adds:

> However, if the intention is to make a crimson, or maroon, or velvet cloth, the vermilion and carmine must be tempered together, making an equal cheerful color, from which the light ones must be extracted by mixing it with the albayalde [white lead] more, or less, as intended; and if the pure carmine is not enough to darken it, it will be helped with a little black.[25]

It would appear to be logical that if the cloth to be reproduced in oil was dyed with cochineal, the same cochineal, alone or mixed with another red pigment, could be used to obtain a similar tone on a canvas. In the eighteenth century, Palomino had exactly the same opinion: 'Now if we consider the clothes of carmine [. . .] these fabrics when they are well dried, are also bathed in fine carmine, and make a very beautiful crimson color'.[26] A very interesting case in this regard is that of the Spanish painter Francisco Zurbarán, who was probably aware of Pacheco's recommendations through the artistic circles of Seville. In his painting *Penitent Magdalene*, he reproduces in detail one of the two-colour brocades that were widespread at the time and generally made of linen with decorative wefts in silk dyed with cochineal. Now, the museum laboratory analysis has allowed us to determine that it is precisely in the reproduction of the brocade that Zurbarán used cochineal for the reds. We can assume that the cloth he reproduced was dyed with cochineal. Such is also the case of a seventeenth-century Spanish brocade analysed at the Metropolitan Museum in New York, similar to the one reproduced by Zurbarán. In this sense, it is clear that the *Penitent Magdalene* is not an exception, since the presence of cochineal in many other paintings of the age is also inspired by a textile.[27]

It should be added that the use of cochineal in the pictorial representation of textiles helped obtain particular effects such as glazes and transparencies that are very useful in this type of painting. There are in fact numerous uses of cochineal to reproduce fabrics as can be seen in another example, also from the seventeenth century, but this time from France, in the case of Eustache Le Sueur. In his painting *Polyphilus before Queen Éleuthérilide*, cochineal was found on the dress of the female character on the right side of the painting.[28]

Now if we consider painting in Mexico during the viceroyalty, we find, not surprisingly, a similar concern. An interesting case, since it serves as a link between the Old and the New World, is that of Andrés de Concha, a Sevillian artist who arrived in Mexico in 1568. In *El martirio de san Lorenzo* (1590) cochineal lacquer was found in the representation of the shirt of one of the wood-carriers.[29] Among other examples of paintings which used cochineal is Cristóbal de Villalpando's *Los desposorios de la Virgen* more than half a century later. Here again, it is in the representation of such fine and elegant clothing, as befits the nuptials of the Virgin, that the presence of cochineal was confirmed by laboratory analysis. But to move beyond painting, it is also worthwhile noting that in the sculpture of colonial Mexico, a subject of particular interest is to be found in the importance of the 'estofados', polychrome wood sculptures imitating textiles.[30] We already knew that the *estofados* made in Spain contain cochineal;[31] yet an investigation of the *estofados* in the Franz Mayer Museum in Mexico City found cochineal in three of them. The dye is particularly visible in the Virgin and Child where the Virgin's dress has a strong crimson hue painted with cochineal. Therefore, it is no surprise if Pacheco also advised the use of cochineal in the *estofados*,[32] a fact which reinforces our emphasis on the importance of the textile paradigm at that time.

Price, prestige, luxury and power

The excellent results obtained by the great dyers in Italy, Flanders, Spain, France and England gave cochineal an enormous prestige, as it produced scarlets and crimsons of exceptional quality. The use of Mexican cochineal very quickly became a fashionable phenomenon, since it was used to dye the most precious fabrics in the royal courts and in the palaces of the nobles (clothes, curtains, chairs and armchairs, rugs, tapestries), as well as in clothing for religious, civil and military purposes. However, to speak of a fashion phenomenon is not enough, insofar as fashions are fleeting, while this phenomenon lasted three centuries, and covered the main European countries.

The speed with which cochineal spread after being introduced into Europe is striking. An early example is a crimson velvet chasuble (liturgical vestment) from a church in south-west France, dyed with Mexican cochineal, and already presented as old and worn when it had to be restored in 1589. As far as the decorations in the courts and mansions of the nobles are concerned, an impressive example is the Croome Court Tapestry Room ordered to be created by the Earl of Coventry for his mansion in Worcestershire in England, which was later reinstalled in the Metropolitan Museum of Art in New York City. The best artists and designers of the time were involved (among others, François Boucher), and the tapestries were made in Gobelins. Furthermore, it should be noted that even at the beginning of the nineteenth century, the uniform coat of a general of the English army was dyed with cochineal, a privilege reserved for the high command of the army, while the uniforms of the soldiers were dyed with a much cheaper dye, madder, which provided duller results.

Cochineal thus appears as a colour of power, and not simply of fashion. However, the expression 'color of power' requires some comments. It can indeed give rise to inadequate interpretations, such as the widespread idea that red is the 'color of power'.[33] In reality, psychologists have argued that red has a multitude of symbolic meanings, so that the association of the colour red with power is one among many possible relations.[34] If purple, and later cochineal, came to symbolize power, it is not because of its red dye, but because of its colouring properties and its high price. It is worth noting that the word 'crimson' came to signify the maximum saturation and purity of a given hue and not the hue itself.[35] This also explains why in the seventeenth century, for example, people spoke of crimson blue, crimson purple or crimson brown.[36]

The prestige of these shades and tones has, of course, to do with their price. It is interesting to note in this regard that to help the dyers who used cochineal, Louis XIV, king of France, decided in 1712 to waive the tax that had to be paid by the dyers of Languedoc (South-West) to import this dye, which was by far the most expensive.[37] However, a century later, the price of cochineal was still very high. A significant example can be found in the accounts of the dyeing workshop of the famous state textile establishment of the *Manufacture des Gobelins*. In 1810, the detailed list of purchases of this factory mentions all the ingredients and prices of dyes. In this regard it is interesting to note that Campeche logwood was quoted at 2 francs per kilo, and cochineal at between 130 and 136 francs.[38] This confirms the exorbitant prices that cochineal still reached until the beginning of the nineteenth century, a commercial reality which implied that owning clothing or furniture dyed with cochineal was a very clear symbol of wealth.

The combination of economic, sociopolitical and aesthetic factors explains the enormous success of cochineal. In France, the Sun King, Louis XIV, set an example at Versailles, asking that the armchairs, as well as the curtains of royal and princely beds, be dyed with cochineal.[39] For this work, the *Manufactures Royales des Gobelins*, founded mainly to provide the furniture and decoration of the royal residences, were of great help. It is also interesting to know that, at the beginning, this manufacture belonged to a family of dyers, the Gobelins, who earned a great deal of money and fame thanks to their dyes obtained from cochineal which, significantly, was called for a long time 'scarlet of the Gobelins'.[40] Similarly, we know from inventories of 1658 and 1659 that the decoration of the Real Alcazar of Madrid, the royal palace of the Spanish monarchy, was dyed with cochineal.[41] This phenomenon also may help to explain why so many painters from the main European countries used cochineal in their palette. In the case of Velázquez, it has been suggested that, if he used cochineal in *Las Meninas*, it is because he knew that the royal palace was decorated with cochineal stained decorations, so that, when depicting the royal couple in the palace, he considered that the best way to highlight it was through this prestigious pigment.

However, this very frequent use of a very expensive pigment poses a problem and calls into question what the great art historian Michael Baxandall had established about the transition between the Middle Ages and the Renaissance, namely that during the Quattrocento, 'while precious pigments become less prominent, a demand for pictorial skill becomes more so'.[42] Later, he insists on this idea: 'As the conspicuous consumption of gold and ultramarine became less important in the contracts [for works of art], its place was filled by references to an equally conspicuous consumption of something else - skill'.[43] Baxandall thus highlighted a 'dichotomy between quality of material and quality of skill'.[44] However, an examination of the pictorial use of cochineal shows that there is no dichotomy but rather a complementarity between skill and the use of an expensive pigment, the two being mutually reinforcing, since the aim was to imitate velvet and silk as well as possible, and at the same time to obtain this effect by means of cochineal, a very expensive pigment.

All this leads to a stimulating reflection that allows to realize that, in reality, the use of cochineal kept alive a rather medieval concern, that is to say, the urge to highlight illustrious personages by the use of the most prestigious pigments, a practice which continued until the nineteenth century. This concern was already reflected in contracts at the beginning of the sixteenth century, specifying that the Virgin's dress had to be painted with the most expensive ultramarine blue.[45] We see here again to what degree an approach based on the materiality of the works calls into question a certain idea of art from the Renaissance onwards.

There are indeed numerous examples of the use of cochineal to represent the Virgin and Christ in their clothing from early on. This is the case of the works of the painter Luis Morales, nicknamed El Divino, in *La Virgen con el Niño* (1570, Museo del Prado) or *Ecce Homo* (1560–70, Museo del Prado), where we find early examples of the use of cochineal in Spanish painting.[46] Later, the same concern is evident, for example, in El Greco's *Head of Christ* (1600). A similar situation prevails in art of colonial Mexico, a fact which need not surprise us. What has been said above about cochineal as the colour of power in Europe is also valid in New Spain, where the representative of the king of Spain, the viceroy, was in principle his living image, as the social and political institutions of colonial society were similar to those which prevailed in Europe.[47]

Of course, the tradition of using cochineal to represent the most sacred characters of the church was not limited to the Virgin and Christ. We also find it in the pictorial representation of the saints. This is the case of *Saint John writing the Apocalypse* by Marteen de Vos and in a youthful work by Poussin, *Saint Denis*.[48] Similarly, we find examples in Mexico, for example in the *Saint Michael the Archangel* by Luis Juarez, whose beautiful cape was painted with cochineal. In summary and to conclude with our commentaries on religious painting, it is worth mentioning the prelates. Indeed, the hierarchy of the church did not escape the wish to highlight their power and prestige through the use of cochineal. An excellent example is the stupendous work painted by the Spanish painter Diego de Velázquez, *Portrait of Archbishop Fernando de Valdés y Llano* (*c.* 1645, National Gallery, London), which was probably realized posthumously, since the archbishop died in 1639. He is represented all in black dress and is seated in front of a sumptuous curtain painted with cochineal.

The portrait of the powerful is a particular pictorial genre, to which Titian contributed in a spectacular way, due to his proximity to power. Few artists have depicted so many of the most important figures in the social hierarchy. In fact, he was appointed, at a very young age, as official painter of the Venetian government, and later, by Charles V as official painter of the Spanish Court. If his portraits are so appreciated, it is due to Titian's talent for highlighting not only the features of the characters, but also their social and political function, to such an extent that several of his portraits are considered true 'icons of power', particularly those painted for Charles V and his son Philip II.[49] The *Portrait of a Man in Armor* is an excellent example of this achievement. It has been suggested that the model was Ferrante Gonzaga, a young *condottiere* of very high rank, who made a brilliant military career, appointed captain general of the cavalry in the army of the Habsburg emperor, who decorated him with the Order of the Golden Fleece.[50] In the painting he wears a particularly sumptuous armour, while his right hand rests on his baton, symbol of his power. In this portrait, as in others like it, Titian developed a specific iconography, based on the heroic image of a warrior. The power enjoyed by the military nobleman is thus celebrated. While this work (*c.* 1530) probably does not contain Mexican cochineal, which had just arrived in Spain, we know that he did use cochineal in his portrait of *The Vendramin Family venerating a relic of the Holy Cross* (The National Gallery, London). This work is a very interesting case of transition in dyes, as it was begun in the early 1540s, using kermes for the main character's ceremonial costume, but also cochineal.[51] It would be no surprise that some other of his fascinating portraits of the powerful from after 1550 also may have been painted with cochineal. Indeed, how better to depict people of great power than by highlighting the richness of the fabrics they owned, and the textiles of their clothing.

Titian's portraits of power had an enormous influence on this genre.[52] For example, Van Dyck, who specialized in portraits, made several copies of Titian's and bought the *Portrait of the Vendramin Family*. Therefore, one should not be surprised if, like Titian in this painting, he used cochineal in some of his own works. In the research carried out for the exhibition *Mexican Red*, we found cochineal in two of his paintings, *Portrait of King Charles Louis* and *Portrait of John Hamilton, Lord Belhaven and Stenton and His wife Margaret*, in both cases in the red fabrics of the clothes and curtains.[53] Furthermore, it should be noted that Peter Paul Rubens, Van Dyck's master, also copied many of Titian's portraits that he saw in Madrid, in particular the head of Charles V from the very famous portrait of the emperor on horseback, now in the Museo del Prado, and he also collected works by Titian. Rubens remembered this portrait of Charles V, dressed in his armour, when he painted his *Portrait of Cardinal Ferdinand of Habsburg*, also dressed in armour. In addition to a pose that confers distinction and nobility, Rubens added a red veil that gives more dynamism to a static pose. It is in the lower part of this veil that cochineal has been found.

A century later, Titian's influence was still present, as can be confirmed, in particular, in the work of the great English artist Joshua Reynolds. The British painter made a copy of Titian's *Portrait of the Vendramin Family*, and highly praised his art, not only for its colours but also for his portraits. He stated that by his way of giving his models 'nobleness and simplicity of character [. . .], he undoubtedly stands in the first rank in this branch of his art'.[54] The assimilation of Titian is evident in his *Portrait of Sir James Hodges*, a high official of the city of London. This work is an excellent example of how, even in the second half of the eighteenth century, the best way to highlight an important man was, of course, through the pose, but also through all the attributes of wealth, and in particular the crimson red present in the sleeves of his elegant lace shirt, in the armchair in which he is seated, as well as in the curtain behind him. It is precisely on the curtain that cochineal has been found; however, since the shade of crimson is the same, it is very likely that he also used cochineal to paint the fabric of the armchair and the shirt of his model.[55] Moreover, it is also well known that he used cochineal in several works, especially in the representation of fabrics.[56] The presence of this same crimson gives, of course, a chromatic unity to the painting, but it is also a testimony of the richness of the character.

Final comments

It is beyond the scope of this chapter to examine the changes that occurred during the nineteenth century in the pictorial use of cochineal.[57] In some cases, the issue of cochineal as the colour of power remains actual, for example in Sir Thomas Lawrence's *Portrait of Sir Nicholas Vansittart* (1825) or Renoir's portrait of *Madame Léon Clapisson* (1883), but these cases are isolated, as cochineal got a more negative connotation, being associated with colonialism abuse and excess. Yet the major change was the decline of the cochineal trade in the nineteenth century due to its replacement as a textile dye by the introduction of synthetic dyes made from aniline and alizarin.[58] Surprisingly, however, painters continued to favour cochineal. In the case of French painters of this new age, we are faced with a paradox. On the one hand, the palette of modern painters in France was characterized by intense use of the new synthetic pigments,[59] and on the other hand, almost all these artists continued to use cochineal. The list is truly impressive, particularly as far as the Impressionists are concerned, as it includes Manet, Monet, Renoir, Degas, Cézanne, Pissarro, Guillaumin and Caillebotte. And then Seurat, Gauguin, Van Gogh and even Matisse. To understand this paradox, since cochineal could be replaced by other red pigments, the only way is to consider that it had preserved its ancient prestige, and that painters wanted to try a pigment that so many important painters had used for three centuries.

As a way of conclusion, it is worth noting that a new and dramatic shift occurred in the twentieth century, when scientists realized that the synthetic red colourants widely used in cosmetics and food were dangerous and carcinogenic. As carmine was a safe natural alternative, causing few, if any, allergies, it quickly substituted the synthetic red colourants. As a result, the production of natural cochineal increased and reaches now the same level it had in the second half of the nineteenth century. Peru produces nowadays roughly 85 per cent of the world's cochineal, that is, more or less 850 tons annually. Most of this production is used as colourant in the food-processing, cosmetic and pharmaceutical industries, and partly in contemporary craft and fashion.[60]

Notes

1 For example, *Del rojo al rosa mexicano*, special issue of *Artes de México* (no. 111, November 2013).

2 For additional information, see my book *La cochenille, de la teinture à la peinture. Une histoire matérielle de la couleur* (Paris: Gallimard, 2021), 63–98. Let me add that the present essay is an original text and not a translation or extract of the latter.

3 On this issue, I refer to my texto 'La cochinilla, entre pintura y tintura', in *Rojo mexicano. Coloquio internacional sobre la grana cochinilla en el arte*, ed. G. Roque (Mexico: Museo del Palacio de Bellas Artes, 2017), 80–7. An expanded version is also available in French, in Roque, *La cochenille*, 99–112.
4 For a semiotic approach of the meaning of pigments, see Roque, *La cochenille*, 83–98.
5 See the stimulating book edited by A. Feeser, M. Daly Goggin, and B. Fowkes Tobin, *The Materiality of Color. The Production, Circulation and Application of Dyes and Pigments, 1400–1800* (Burlington: Ashgate, 2012).
6 P. Chaunu, *Séville et l'Atlantique (1504–1650)*, 2nd part, t. VIII, Vol. 1 (Paris: S.E.V.P.E.N., 1959), 771.
7 F. Ruíz Martín, *Lettres marchandes échangées entre Florence et Medina del Campo* (Paris: École pratique des hautes études, VIe section 1965), cxxviii.
8 F. Braudel, *Civilisation matérielle, économie et capitalisme. XV-XVIII siècles*, Vol. II, *Les jeux de l'échange* (Paris: Armand Colin, 1979), 371–2.
9 For the importance of textile fashion, see Fernand Braudel, *Les Structures du quotidien* (Paris: Armand Colin, 1979), 275–6 and 282.
10 See J. Pérez Morera, 'El tejido brocado en México virreinal: Telas ricas, españolas, italianas y francesas', in *Vegueta. Annuario de la Facultad de Geografía e Historia* (Universidad de la Las Palmas de Gran Canaria, No. 19, 2019), 694–7.
11 Francisco Pacheco, *El arte de la pintura* [1649] (Madrid: Cátedra, 1990, facsímile edition), 485.
12 Rocío Bruquetas, 'El Carmín de Indias', in Roque, *Rojo mexicano*, 143.
13 Rocío Bruquetas, 'Sources for the Study of Pigments Used by Goya and other 18th-century Spanish painters', in *The Artist Process. Technique and Interpretation*, eds. S. Eyb-Green et al. (London: Archetype, 2012), 138–46.
14 On the importance of Venice as a commercial centre, cf. F. Braudel, *Civilisation matérielle, économie et capitalisme. XV-XVIII siècles*, Vol. III, *Le Temps du monde*, 95–113.
15 L. C. Matthew and B. H. Berrie, '"Memoria de colori che bisognino torre a vinetia": Venice as a Center for Purchase of Painter's Colors', in *European Trade in Painter's Materials to 1700*, ed. J. Kirby (London: National Gallery of Art, 2010), 245–51.
16 R. A. Donkin, 'Spanish Red: An Ethnogrographical Study of Cochineal and the Opuntia Cactus', *Transactions of the American Philosophical Society*, new series 67, no. 5 (1977): 23.
17 Donkin, 'Spanish Red', 37.
18 J. Kirby, '"One of the Most Beautiful Reds". Cochineal in European Painting', in *A Red like No Other. How Cochineal Colored the World. An Epic Story of Art, Culture, Science and Trade*, eds. Carmella Padilla and Barbara Anderson (New York and Santa Fe: Skira Rizzoli/Museum of International Folk Art, 2015), 177.
19 Dolores Gayo, 'El uso de la grana cochinilla en la pintura española de los siglos XVI y XVII', in Roque, *Rojo mexicano*, 138.
20 J. Dunkerton and M. Spring, 'Titian after 1540: Technique and Style in his Later Works', *National Gallery Technical Bulletin* 36 (2016): 21.
21 On the importance of these cities at the time, see Braudel, *Le Temps du monde*, 118–29, and 145–56. And on the importance of the pigment market in Antwerp, see F. Vermeylen, 'The Colour of Money: Dealing in Pigments in Sixteenth-Century Antwerp' and R. Fabri, '"Eenen remenant van verf ende pinselen": Some Aspects of the Materials used by Seventeenth-Century Cabinet Painters in Antwerp', both in Kirby, *European Trade*, respectively 356–65, and 366–73.

22 See the detailed analysis of the work in Hugh MacAndrew et al., 'Tintoretto's "Deposition of Christ" in the National Gallery of Scotland', *Burlington Magazine* 127 no. 989 (August 1985): 501–17. The authors highlight Tintoretto's closeness to the Venetian textile world.
23 Pacheco, *El arte de la pintura*, 400.
24 For the incarnations, Pacheco, *El arte de la pintura*, 499. Palomino will say the same in his own treatise: see A. Palomino de Castro y Velasco, *El museo pictórico y la escala óptica* [1715–1724], Vol. 2 (Madrid: Imprenta de Sancha, 1797), 58–9.
25 Pacheco, *El arte de la pintura*, 484–5.
26 De Castro y Velasco, *El museo pictórico*, 69.
27 This is the case of Zurbarán's *Misa del Padre Cabañuelas* (1638), inspired by a brocade from Castile, Toledo, in the last third of the sixteenth century. See figures 6 and 7 of the article by Jesús Pérez Morera, 'El tejido brocado en el México virreinal: sedas orientales y criollas', *Cuadernos de arte de la Universidad de Granada* 49 (2018): 181. I thank the author for all the information he provided me about these brocades. It is also worth noting that this painting has been analysed in laboratories and extensive use of cochineal has been found, see R. Bruquetas and M. Gómez, 'Carmine of the Indies. Cochineal in Spanish Painting and Sculpture (1551–1670)', in Padilla and Anderson, *A Red like No Other*, 197.
28 Cochineal was also found in reproduction of fabrics in *Alexander and his Doctor* (National Gallery, London). See J. Kirby et al., *Natural Colorants for Dyeing and Lake Pigments. Practical Recipes and their Historical Sources* (London: Archetype Publications in association with CHARISMA, 2014), 83.
29 This work was analysed by Elsa Arroyo et al., 'Paños labrados de carmín en la pintura novohispana', in Roque, *Rojo mexicano*, 175. This interesting paper drew my attention to the pictorial use of cochineal to represent red cloths and fabrics and is therefore one of the main sources of the hypothesis suggested here.
30 See P. F. Amador Marrero and P. Díaz Cayeros (eds.), *El tejido polícromo. La escultura novohispana y su vestimenta* (Mexico: UNAM, 2013).
31 See Bruquetas and Gómez, 'Carmine of the Indies. Cochineal in Spanish Painting and Sculpture (1551–1670)', 190–9.
32 Pacheco, *El arte de la pintura*, 462.
33 See, for example, M. A. Brown, 'The Color of Power. Red in the Portraiture of the Spanish and British Empires', in Padilla and Anderson, *A Red like No Other*, 166–71.
34 E. Heller, *Psicología del color: Cómo actúan los colores sobre los sentimientos y la razón* (Barcelona: Gustavo Gilli, 2009), 53.
35 See also C. Buss, 'Silk Gold Crimson', in *Secrets and Technology at the Visconti and Sforza Courts*, ed. C. Buss (Milan: Silvana Editoriale, 2009), 48–9.
36 Dominique Cardon, *Le monde des teintes naturelles* (Paris: Belin, 2003), 489–90; A. Mollard-Desfour, *Le Dictionnaire des mots et expressions de couleur au XXe siècle. Le rouge* (Paris: CNRS Éditions, 2000), 236–7.
37 See the anonymous manuscript published by Dominique Cardon, *Mémoires de teinture. Voyage dans le temps chez un maître des couleurs* (Paris: CNRS Éditions, 2013), 9 and her comments on the very expensive price of cochineal, 217–22.
38 Besides my research in the Archives of the Mobilier national, see also a master's thesis, G. Le Reste, *L'Utilisation de la cochenille d'Amérique dans l'atelier des teintures de la Manufacture des Gobelins* (Paris: École du Louvre, 2016–2017), vol. of annex, 48–9.
39 F. Delambre and B. Guineau, *Les matériaux de la couleur* (Paris: Gallimard, coll. Découvertes, 1999), 74.

40 Ch.-Er. Guignet, *Les Couleurs* (Paris: Hachette, 1889), 144.
41 B. Ellsworth Hamann, 'The Mirrors of Las Meninas: Cochineal, Silver and Clay', *The Art Bulletin* 92, no. 1/2 (March–June 2010): 8–19.
42 M. Baxandall, *Painting and Experience in Fifteenth-century Italy. A Primer in the Social History of Pictorial Style*, 2nd ed. (Oxford and New York: Oxford University Press, 1988), 14.
43 Baxandall, *Painting and Experience in Fifteenth-century Italy*, 15.
44 Baxandall, *Painting and Experience in Fifteenth-century Italy*, 16.
45 J. Gage, *Colour and Culture: Practice and Meaning from Antiquity to Abstraction* (London: Thames and Hudson, 1993), 129–30.
46 D. Gayo, 'El uso de la grana cochinilla en la pintura española de los siglos XVI y XVII', 138.
47 See A. Cañeque, *The King's Living Image. The Culture and Politics of Viceregal Power in Colonial Mexico* (New York: Routledge, 2004), as well as M. Schreffler, *The Art of Allegiance. Visual Culture and Imperial Power in Baroque New Spain* (University Park: The Pennsylvania State University Press, 2007).
48 Samples could not be taken from the Saint's cloak, but from a very similar shade used by the artist for the clothing of the seated figure in the lower left of the painting; this is where the analysis allowed the detection of cochineal. It is interesting to note that in another work from the same early period, *The Death of the Virgin* (1621), cochineal was also found; see S. Saverwijns and J. Sanyova, 'De Schildertechniek van de Dood van de Maagt van Nicolas Poussin', *Monumenten en Landschappen* 25, no. 5 (2006): 46–56.
49 See M. Costantini, 'Les visages du pouvoir', in the catalogue of the exhibition *Titien. Le pouvoir en face*, (Paris: Musée du Luxembourg, 2006), 69–77.
50 See the entry devoted to this work in the catalogue of the exhibition *Titien. Le pouvoir en face*, 128.
51 Dunkerton and Spring, 'Titian after 1540', 21–2.
52 On this point, see J. M. Fletcher, '"La vraie ressemblance". Les portraits de Titien', in the catalogue of the exhibition *Titien. Le pouvoir en face*, 49–50.
53 Roque, *Rojo mexicano*, 46.
54 Sir J. Reynolds, *Discourses on Art* [1797] (New York: Collier Books, 1966), Discourse 4, 63.
55 I am very grateful to Joyce Townsend who provided me with this valuable information (email 13 May 2017).
56 A. Gent et al., 'Practice Makes Imperfect: Reynolds's Painting Technique', *National Gallery Technical Bulletin* 35 (2015): 28.
57 The reader interested in this topic might refer to my book Roque, *La Cochenille*, 199–241.
58 P. Ball, *Bright Earth. Art and the Invention of Color* (Chicago: The University of Chicago Press, 2001), 147ff.
59 See my article 'Les pigments synthétiques de la palette impressionniste: de la modernité au modernisme', forthcoming in the catalogue of the exhibition *Léon Monet, frère de l'artiste et collectionneur* (Paris: Musée du Luxembourg, 2023).
60 N. Chávez, 'Carmine and Earthly Delights. Cochineal in Cosmetics, Contemporary Craft, Fashion, and the Strawberry Frappuccino', in Padilla and Anderson, *Red like No Other*, 266–75.

6

Logwood, masterless men and British interests in Yucatan and Central America

Karl Offen

Logwood and the masterless men who acquired it directed British interests in Yucatan and Central America during the seventeenth and eighteenth centuries. Across the seventeenth-century Caribbean basin, masterless men – including ex-indentures, sailors, turtlers, cattle hunters (*boucaniers*), pirates, debtors and fugitives – routinely violated imperial authorities to pursue livelihoods of plunder, illicit trade, and the procurement of natural resources.[1] By the 1670s, many of these men had turned their attention to acquiring logwood, specifically the heartwood of *Haematoxylum campechianum*, a dyewood tree found around the Yucatan Peninsula (Figure 6.1). In so doing, these masterless men commanded the attention of British officials on both sides of the Atlantic and shaped the origins and developments of several Anglo-Spanish conflicts in the early modern period. It was this broad Atlantic impact of logwood that inspired one historian to state that 'with the exception of sugar no single commodity has played a greater part in Caribbean history than logwood', and another to add that 'logwood cutting was arguably the most profitable illicit activity in the history of the Atlantic world'.[2]

Known in Spanish as *palo de Campeche* or *palo de tinte*, logwood is a multi-stem, thorny, and deciduous tree ranging in heights from 5 to 12 m. It grows in marshy and seasonally flooded limestone soils found in patches around the Yucatan Peninsula. Its native range includes portions of the modern Mexican states of Tabasco, Campeche, Yucatan and Quintana Roo, as well as northern Guatemala, and northern Belize. In the early eighteenth century, ship captains brought logwood seeds to Jamaica and other Caribbean islands, but only wild-sourced trees from the Yucatan and what is today Belize supplied the early modern markets of Europe.[3] Today the tree has a wide distribution in the tropics and subtropics around the world, and many industries employ its chemical compound haematoxylin to dye natural fabrics, for inks, and especially in histology.[4]

From about 1660 to 1750, logwood was the most commercially valuable and the politically important wild natural resource extracted by the subjects of Great Britain anywhere in the American tropics. The wealth that logwood generated contributed to the early development of Jamaica, and the dyewood's extraction and transportation

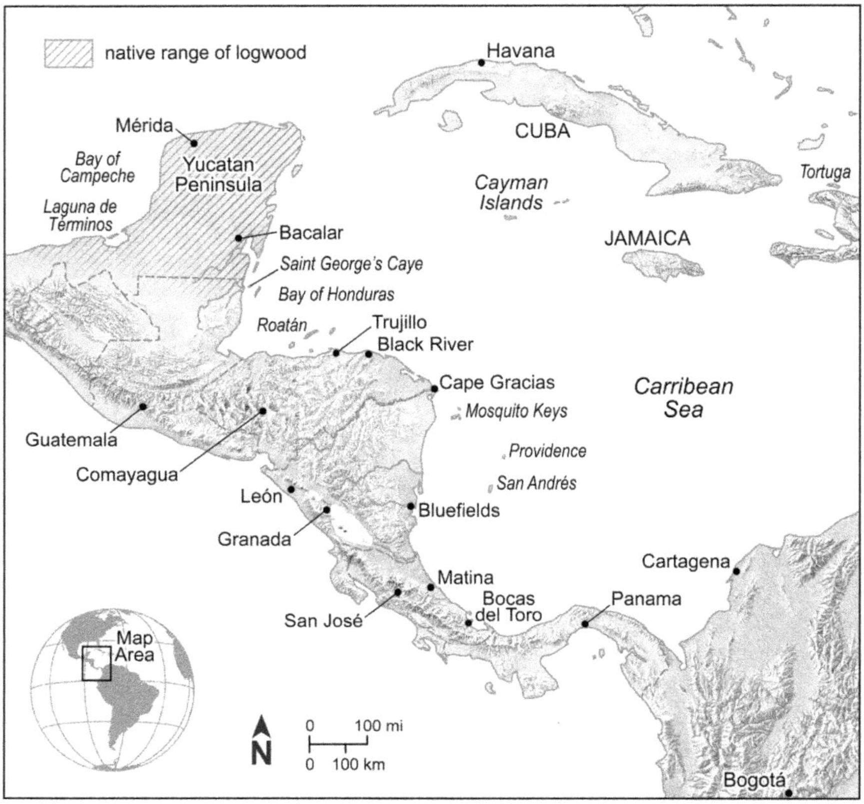

Figure 6.1 Map of the Captaincy General of Guatemala and the western Caribbean showing the native range of logwood around the Yucatan Peninsula.

employed thousands of loggers, sailors, and dockworkers around the north Atlantic, various trade workers in Europe, and provided significant customs duties to Great Britain. Although logwood was not available along the Mosquito Shore, a Miskitu-Indian dominated British enclave to the east of Yucatan – 'for there is no logwood to the south of the river Belize'[5] – settler activities and prospects there remained closely tied to those of the logwood cutters in the Bay of Honduras, or the eastern side of the Yucatan Peninsula.

The basic historical outline of logwood is well known.[6] The Mayan people of Mesoamerica used the heartwood of the tree for millennia to colour cotton fabrics, fibres, ink and paint. They also used the leaves as a medicinal. Spaniards took notice and during the early sixteenth century, they shipped small quantities of the wood to Spain. English, French and Dutch buccaneers first discovered the value of the dyewood via salvage operations or by finding it serendipitously in captured Spanish ships. Its high price and, critically, the suppression of privateering and piracy in 1670 following the Treaty of Madrid inspired these masterless men to initiate their own cutting ventures at locations around the north and western coasts of the Yucatan Peninsula. Although

Spain protested these interlopers, the British government temporarily condoned logwood cutting in 1672. For the next several decades, Great Britain and continental Europe received several thousand tons of logwood per year via Jamaica, Britain's colonies in North America, and the Dutch island of Curaçao, but all of it originated in the Yucatan. In the late seventeenth century, Spanish assaults on the logwood camps in the Bay of Campeche inspired the logwood cutters to concentrate their activities along the Belize River in the Bay of Honduras on the opposite side of the Yucatan Peninsula. Although demand for logwood declined slightly in the 1750s, its commercial viability never waned, and the erstwhile logwood enclave in the Bay of Honduras demanded the attention of the British government into the nineteenth century. Anglo-Spanish peace treaties in 1763, 1783, and 1786 gave baymen, or the logwood cutters of the bay, use-rights to forest resources within defined geographic limits, but a more formal and externally regulated colonial settlement did not take root until after 1798, the symbolic beginning of the Crown colony of British Honduras, or Belize.

Dyewoods in the transimperial Caribbean basin

Logwood was only the most important of several tropical American dyewoods that entered the Atlantic economy in the early modern period. In addition to logwood – unique to the Yucatan Peninsula – the greater Caribbean and Central America furnished European markets with brazilwood, brasiletto, Nicaragua wood, and fustic. All these dyewoods originated from three plant genera: *Haematoxylum*, *Caesalpinia*, and *Chlorophora*. The genus *Chlorophora* is best known for its species *tinctoria*, called fustic in English and *mora* in Nicaraguan and Costa Rican Spanish. The genus *Haematoxylum* is represented in Mesoamerica by two species, *H. campechianum*, or logwood, and *H. brasiletto*, often referred to a 'Nicaragua wood' in English and *brasilleto* or *palo de brasil* in Central America. The genus *Caesalpinia* contains several New World species and have a wide-ranging geographic distribution. None of the various *brasil* woods in Central America or the Caribbean should be confused with the dyewood found in the Atlantic forests of Brazil (*Paubrasilia echinata*). Both *Caesalpinia* and *H. brasiletto* contain the chemical compound brazilin, while logwood contains the compound haematoxylin, which is chemically close to, but distinct from, brazilin. Logwood produces an orange-red dye in boiling water, becoming yellowish when cooled, and turning violet and black when oxidized over time. With different mineral mordants to fix the dye, logwood produced a range of colors and, in addition to dyes for textiles, served in the manufacture of inks.[7]

On his second voyage Columbus brought back dyewoods from the island of Hispaniola, but Spanish interest in their potential was limited.[8] In 1501, the Portuguese returned from Brazil with *Paubrasilia echinata*, initiating a traffic that rose to 1,200 tons a year quickly thereafter, and reaching up to 12,000 tons annually by the 1550s.[9] After discovering a Mayan dyewood called *ek*, the Spaniards maintained a monopoly on logwood into the mid-seventeenth century. This was partly due to limited demand in northern Europe, as dyers could not fix haematoxylin. Indeed, an English Act of Parliament in 1581 specifically prohibited the importation of logwood to England because it bled and faded.[10] This law was not overturned until 1662 after dyers used an oxidizing mordant technique in their process. As early as 1650, however, English tradesmen claimed that drapers, dyers, feltmakers and salters – whether using cotton, linen, wool or leather – could not carry on their work 'without the Use of Logwood'.[11] The change in the wood's legal status, and its perceived value in Britain, opened the door for masterless men to begin searching out the wood by hook or crook.

English interest in vegetable dyes native to the Caribbean basin began with their earliest forays into the region.[12] Their specific interest in dyewoods from the western Caribbean began with the puritan colony on Providence Island (1629–41) and its auxiliary encampment on the adjacent mainland of Central America.[13] On Providence, investors hoped to establish a yeoman farming community, but a profitable crop proved elusive. In response, shareholders instructed settlers to seek out potential commodities on the mainland.[14] From their fortuitous interactions with the Miskitu Indians, settlers sent home samples of many natural resources, including fustic.[15] When investors in Providence expanded their initial charter to ensure it included the small island of Tortuga off the northwest coast of Hispaniola, settlers there immediately procured dyewoods with the labour of enslaved Africans.[16] In 1638, Providence's originator colony at Bermuda got permission to ship 100 tons of logwood that settlers salvaged from a Spanish wreck. After the wood was valued between £8,000 and £10,000, twice the annual worth of Bermuda's tobacco crop, Bermudians scoured the Bahamas for fustic and brasiletto. By 1650, they sent ten tons of brasiletto worth £124 to endow Harvard College. By the 1660s, they shipped so much dyewood to Britain that they flooded the market and the price dropped to £10 per ton in Bermuda and £20 per ton in England.[17]

The charter of the Providence Island Company in 1629 gave proprietors the right to trade, extract resources, and settle in lands not occupied by any 'Christian prince' in a wide swath of the western Caribbean.[18] English writers justified such charters and their infringements into domains claimed by Spain through appeals to the Roman Law concept of *res nullius* and the doctrine of effective occupation. The implications of this legal discourse was embodied in the Americas by woodsmen, pirates, itinerant settlers called 'merooners', and other masterless men who shaped imperial interpretations of the legal concepts in turn.[19] Such developments inspired Atlantic and legal historians such as Laura Benton to highlight the spatial unevenness of imperial territorial control and the geography of activities by masterless men, Indigenous peoples, maroons and others responsible for the patchiness of empires. These new patches of autonomy, or what Benton calls 'micro-regions' of the Americas, created liminal spaces outside effective Spanish control and helped redefine what territorial dominium meant in practice and law.[20]

Laying the foundation: Logwood cutting in Yucatan

Following the English capture of Jamaica in 1655, the relationship between masterless men and logwood grew significantly. Before the signing of the Treaty of Madrid in 1670, Jamaican officials commissioned ship captains as privateers to defend the island from Spanish efforts to retake it. Such letters of marque generated unprovoked assaults on Spanish ports and shipping, and produced new knowledge about the Spanish mainland.[21] Following the removal of the ban on logwood importations to England in 1662 and a belief that the wood fetched up to a £100 a ton, buccaneers from Tortuga and privateers from Jamaica led assaults on Spanish shipping and logwood depots in the Bay of Campeche. They also explored the coast of Yucatan for trees they could cut themselves.[22] Although initial woodcutting occurred around Cape Catoche on the northeastern side of the peninsula, the most important site of logwood extraction was several miles up creeks that emptied into the Laguna de Términos in the Bay of Campeche. Although the British trade in logwood was officially discouraged in 1679, and then, fully restricted by royal edict in 1682, the commerce continued with profound impacts for the region.[23]

With the termination of officially sanctioned privateering by the end of 1671, about a dozen former privateering vessels based in Jamaica began participating in the logwood trade, selling their cargoes for £25 to £30 per ton.[24] The total value of all natural dye products coming into London in this period grew substantially, from £3,000 in 1663 to £7,100 by 1669, and reaching £85,000 per annum at the turn of the eighteenth century, and much of this increase came from logwood.[25] By 1675 there were around 270 Englishmen in the Bay of Campeche, and between 1671 and 1684 some 200 vessels supported the logwood trade, generating exports to England of some 2,000 tons per year via Jamaica and North America, exclusive of the significant portion of logwood that cutters sent clandestinely to continental Europe.[26] North American ships carried near half of all logwood to Boston, New York and Charleston for re-exportation to Britain, usually on a return voyage that brought provisions to the Caribbean. As an indicator of this intra-American traffic, in just three months between September and December 1675, seventeen ships from New England visited Port Royal, Jamaica, on their way to Campeche to load logwood before returning home. One geographer claimed that logwood was 'the second most valuable [North] American export' shipped to Britain in the early eighteenth century.[27] Before 1680, logwood comprised over half Jamaica's exports and was an important source of capital for the development of plantation agriculture.[28]

William Dampier spent almost three years between August 1675 and April 1678 as a logwood cutter in the Laguna de Términos. Dampier stated the tree was found in stands outside the tidal zone in calcareous soils whose thin uppermost surface was rich in organic matter deposited by annual floods. Logwood cutters looked for sizeable stands located 300–600 metres from creeks, as dragging the dense heartwood to the water was laborious. The men cut the trees into lengths convenient for hauling and used gun powder to blow up trunks too big to cut. They chipped away the wood's outer sapwood, and stacked up the remaining heartwood during the dry season (approximately September to May). During the rainy season (approximately June to August) the ground flooded, and the men loaded the wood into shallow-bottomed boats and brought their quarry down to a port on the Island of Triest or Trist (Isla del Carmen) where ship captains exchanged cash and provisions to acquire it. During the cutting periods the men lived on platforms raised above the flood zone, and slept under tent-like pavilions covered with a coarsely woven cloth called osnaburg that served as a mosquito net.[29] Ship captains paid Dampier £5 per ton for the wood he hauled out, which was almost twice the monthly wage as a second mate or Newfoundland fisherman, and four times that of a common labourer in England.[30]

In their spare time, the men hunted feral cattle on the surrounding savannas. According to Dampier, they also 'made Sallies out in small Parties among the nearest Indian Towns; where they plundered and brought away the Indian Women to serve them at their Huts, and sent their Husbands to be sold at Jamaica'.[31] Indeed, buccaneers had been organizing slave raids against the Maya from Tortuga since the 1640s and continued the practice while in residence. The kidnapping and forced concubinage of Mayan women was so pronounced that when the Spaniards temporarily rooted out the logwood cutters in 1680 they found seventy-four mestizo children.[32] These facts belie claims made by Jamaican governor, Thomas Modyford (1664–71), who maintained that sailors traded peacefully with the Maya who provided logwood and tortoise shell in exchange for knives and beads.[33]

The logwood trade inspired a broader British vision in the metropole about the commercial possibilities in the region. Following the Treaty of Madrid, maps such as Robert Morden and William Berry's 'A New Map of the English Plantations in America' fired Britain's ambitions in tropical America (Figure 6.2). Printed in 1673, this map was among the earliest to emphasize the location of Jamaica vis-à-vis the Spanish mainland.[34] One of only four textual glosses on the map highlights 'Port Royal and English Colony' in the Bay of Campeche, and included a symbol that may represent the flag of St George. Morden and Berry drew attention to the 'English Colony' to imply

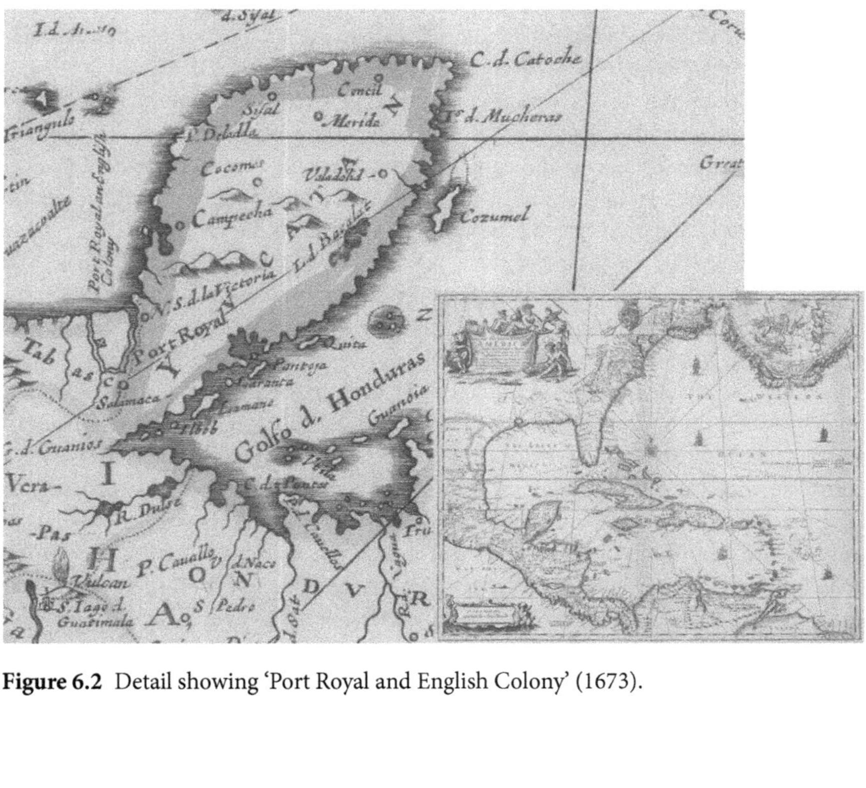

Figure 6.2 Detail showing 'Port Royal and English Colony' (1673).

that the logwood camps pre-existed the 1670 treaty and therefore had every right to demand governmental support. In this way, the printed map sought to both document and project British durability in the Bay of Campeche, and to communicate intentions to the Spaniards.[35]

The question for British authorities, as opposed to private mapmakers, was to establish whether their citizens had a legal right to logwood in the Bay of Campeche or not. English officials in London debated this question with the Spanish court and with the Jamaican governor between 1671 and 1682. Although article VII of the Treaty of Madrid guaranteed the subjects of Great Britain the 'right of Sovereignty, Dominion, Possession, and Propriety' to lands in the Americas that they 'do at present hold and possess', it obligated them to end privateering and rein in piracy.[36] When Sir Thomas Lynch replaced Thomas Modyford as governor in 1671, he took this latter charge seriously, but to occupy the masterless men prone to autonomous mobility he needed to ensure they had a right to cut logwood. In several letters, Lynch beseeched officials in London to 'give your commands about the Logwood'. The governor's conundrum was deep, and he quickly argued that the Treaty of Madrid protected the logwood settlements because cutters had resided in the Laguna de Términos before the peace was promulgated, and because cutting occurred only in 'desolate places'. He also informed his readers that the logwood trade was so profitable that without sanctioning it, he could not occupy the privateers and pirates he was charged with subduing. He further pointed out that logwood cutting would be advantageous to England's cloth trades, make England the storehouse for all Europe, and be worth £100,000 per annum in customs dues.[37] Lynch claimed that cutters saw neither Spaniards nor Indians, and that they worked in 'morose and swampish grounds' where it was impossible to plant, inferring the land had no other uses.[38] The argument was effective, and the Council for Foreign Plantations supported the governor's position on logwood cutting so long as it occurred in 'desolate and uninhabited places'.[39]

The issue of actual rights of access to the logwood forests, however, remained unsettled. In 1674, Charles II sought advice from his Council for Foreign Plantations, and its secretary, John Locke, and from William Godolphin, the English minister responsible for negotiating the 1670 treaty. Godolphin reported that the Spaniards viewed unoccupied commons as part of their dominion, as England would, regardless of their use. But he left the door open for ways to connive at logwood cutting: 'if the English in Cutting Wood of Campeche, would restrain themselves to that alone . . . and without making inroads or other depredations on the Country, It may be adviseable for His Majesty tho not to authorise yet to connive at their so doing till it be discerned to what degree [the Spaniards] will resent it'.[40] For his part, Locke argued that through their labour and entrepreneurial spirit the English logwood cutters improved nature and, in effect, turned wasteful commons that the Spaniards had ignored into productive land as Protestant theology commanded.[41] While the king ignored these ideas in favour of appeasing Spain, these written opinions proved critical to the recommendations of an influential report by the Board of Trade in 1717 that strengthened both popular and official support for British access to logwood in the Bay of Honduras.

Locke's imperial vision hinged on turning unproductive commons and wastes into productive lands via the dynamic labour of Protestant Englishmen. This, he argued, justified the establishment of British government and dominion. The writings of the Jamaican governor had supported this very idea, noting in 1675 that without government regulating access to the commons, the logwood trade could not receive investment, could not expand, and that the English could not exclude illicit shipping to continental Europe. This was the same alleged problem of the commons that Garrett Hardin famously identified as a 'tragedy' inherent to all commons three centuries later.[42] Locke's coffee house drinking buddy, Dampier, echoed his view by noting that the 'Logwood-Trade in general . . . resembles that of Newfoundland [fisheries], since what arises from both, is the product of bare Labour, and that the Persons imployed herein are supported by the produce of their Native Country'. The pirate-savant also pointed out that the Spaniards received more damage from the masterless men when they are not occupied in the logwood trade than when they were.[43]

Dampier wrote this opinion in hindsight informed by developments that he had participated in. From the moment the English king instructed the governor of Jamaica to curtail the logwood trade in 1679, piracy along the isthmus picked up substantially. Without logwood to occupy them, masterless men such as Joseph Coxon, Bartholomew Sharpe, and Dampier returned to their illicit activities of plundering Spanish shipping and soon almost 200 men had crossed the isthmus at Darien and opened a new phase of piracy.[44] This uptick in piracy coincided with an effective 1680 Spanish assault on the logwood camps in the Laguna de Términos.[45] This military victory did not end the logwood trade in the Bay of Campeche, but it did initiate a gradual shift in British cutting from one side of the Yucatan to the other.

Logwood in the Bay of Honduras

Despite successful Spanish raids on logwood cutters in the Laguna de Términos, the logwood trade remained strong, but so too did Spain's commitment to interdict logwood shipping. In 1699 alone, the Spaniards allegedly captured twenty-four British logwood ships in the Bay of Campeche, most originating from North America.[46] Still, at the turn of the eighteenth century 1,200 tons of logwood valued at £20,000 annually reached Britain via Jamaica alone.[47] During the War of Spanish Succession (1702–14), all forms of illicit activities increased along Central America's littoral, and by 1715 one historian suggested that contraband trade 'dominated all forms of commerce in Central America'.[48] Illicit trade also hurt the British interests, as logwood from Yucatan made its way clandestinely to continental Europe via Curaçao and North America. One contemporary estimated that England lost £1,500 per annum in customs due to this illegal trade.[49] Meanwhile, the few masterless merooners living around Black River on the Mosquito Shore enticed the Miskitu with arms and goods to raid Indigenous peoples in the Bay of Honduras. These raids combined with earlier buccaneer abuses to displace Mayas in the Bay of Honduras, opening access to logwood stands along the Belize River.[50] By the 1720s, tributaries of the Belize River provided the majority of logwood entering world markets, prompting the Spaniards to build Fuerte San Felipe in nearby Bacalar (see Figure 6.1).[51]

Shortly after combatants signed the various peace treaties of Utrecht (1713–15), Spain notified Britain that they would forcibly remove the logwood cutters in the Laguna de Términos.[52] The varied interpretations of these treaties ensured that Anglo-Spanish relations remained 'envenomed' over the logwood trade between 1713 and 1730, and one contemporary reported British logwood shipping and cutters suffered £1,239,000 in damages from Spanish depredations.[53] Still, with logwood valued at £13–20 per ton between 1715 and 1730, and providing £4 per ton duties to the king, the valuable dyewood continued to reach Britain whether government officially sanctioned it or not.[54] Jamaican officials continued to beg their counterparts at home to help ensure access to Yucatan logwood because they feared what would become of the many masterless men roaming around Kingston and Port Royal.[55]

In 1717 the Board of Trade addressed the post-Utrecht confusion about logwood rights. Their lengthy report summarized the opinions of Lynch and Locke discussed earlier. The document was copied and circulated widely and became among the most referenced reports by British diplomats and politicians in the various Anglo-Spanish peace treaties affecting this region in the eighteenth century. The report established, at least in the British mind, that the presence of logwood cutters in the Laguna de Términos pre-dated the 1667 and 1670 treaties, and that the areas where the men worked were 'in a manner wholly desolate and uninhabited'. The board provided figures showing the economic importance of the trade, with imports to Britain averaging 3,741 tons per year between 1713 and 1716, or not less than £60,000 per annum. The report urged the king not to 'return' the territory to Spain, and that Britain's right to logwood was 'absolutely granted and confirmed by the Treaty of Commerce made at Utrecht'.[56] Such policy documents relied on a legal discourse whose foundation was greatly informed by the multiple on-the-ground activities of masterless men for several decades.

By the time the Jamaica-based merchant Nathaniel Uring spent five months among the baymen in 1720, and the naval surgeon John Atkins visited the Bay of Honduras two years later, the Belize River supplied the majority of logwood entering world markets. From the two narratives of these men, we learn some 500 men, including 'slaves', resided part of the year at Barcadares some 42 miles up the river (Figure 6.3). These embarkation points are shown on two maps drawn by Uring, one small-scale map showing the Bay of Honduras (with east up), and another showing the lower course of the Belize River in a large-scale map that locates the Barcadares at the very top. Although Uring does not mention the presence of enslaved labour, Atkins's reference is unequivocal.[57] Scholars tend to date the beginning enslaved labour in the acquisition of logwood to Atkins's observations dating to 1724, noting that by 1745 slaves outnumbered freemen in the bay, and by 1779 slaves in the Bay of Honduras outnumbered freemen 3,000 to 500, or 6 to 1 – some of the slaves were of Indigenous ancestry.[58] Besides the enslaved Mayan women mentioned earlier, we know that at least a few enslaved African men had worked cutting logwood earlier in the Bay of Campeche.[59]

According to Uring the land around Belize River was flat and full of morasses, and 'with several large Lagunes, which in the Rain-times are almost all over-flowed'. During much of the rainy season, the men resided at the Barcadares, but in the dry season they searched out suitable stands of trees some 30 miles above the Barcadares, which

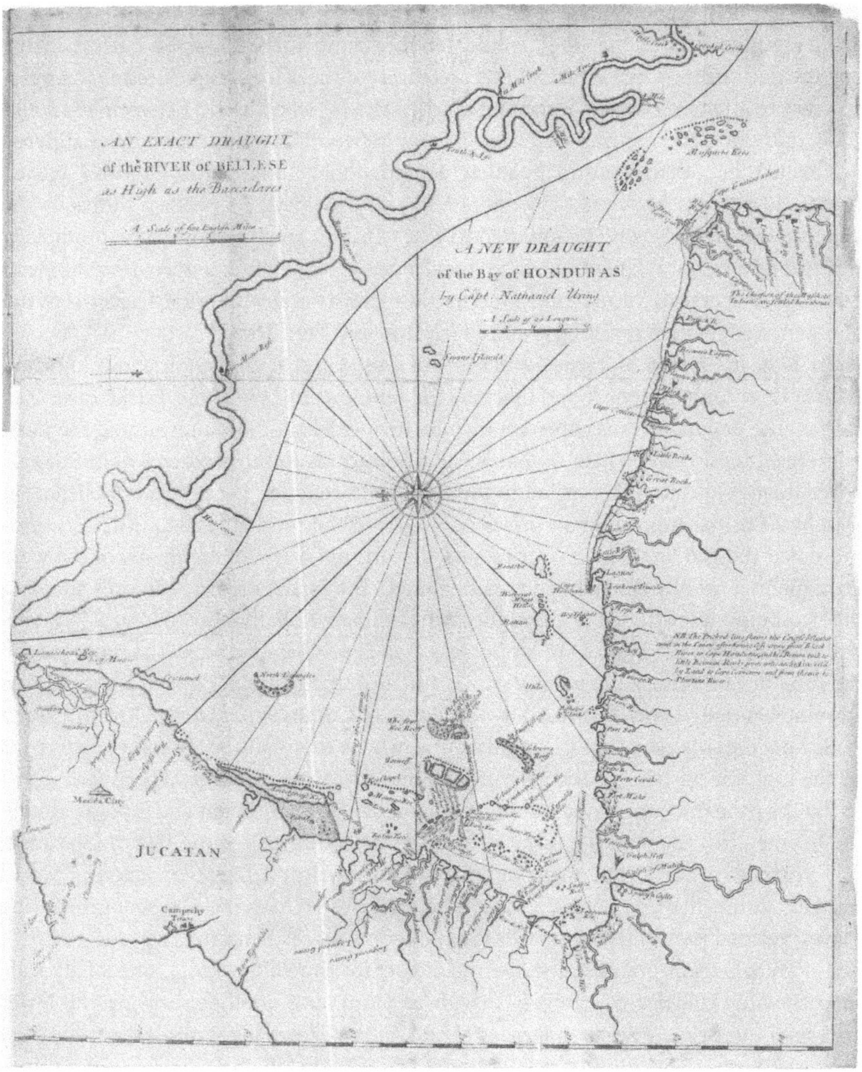

Figure 6.3 Nathaniel Uring, 'An Exact Draught of the River of Bellese as High as the Barkadores. A New Draught of the Bay of Honduras, by Capt Nathaniel Uring' (1726).

Atkins claimed ran 'in a Line or Vein (like Minerals of the Earth) of some miles'. Like Dampier, Uring stated the men built temporary huts by the stands, cut the trees down, chipped off the bark and sap, lay the wood in heaps, made paths to the creeks, and with the floods loaded the wood into canoes and brought it down to the Barcadares where the wood fetched £5 per ton Jamaica money. For his part, Atkins claimed merchants hired the logwood men at a rate of £10 a ton of logwood per month. He added that the masterless men chose 'a King from among their Body, and his Consort is stiled [*sic*] Queen, agreeing to some Laws by common Consent, as a Guide to them'. With their earnings, the men bought liquor, provisions, small arms, powder, shot, axes, cutlasses and osnaburg cloth.[60]

More than half of all ships loading logwood in the eighteenth century came from North America. During just three months in early 1700, six ships carried 260 tons of logwood to Charlestown alone, and Boston received 2,572 tons in 1714.[61] A particularly strong relationship developed between Boston and the logwood cutters in the Bay of Honduras. The latter famously donated in 1727 over 100 tons to the benefactors of Christ Church so they could build a spire that could be seen from the sea. In their acknowledgement, the Anglican church's committee voted 'That a Pew be expeditiously built next to the Pulpit and lin'd handsomely for the use of the Gentlemen of the Bay of Honduras who have been or shall be Benefactors to this Church'.[62]

The switch from the Bay of Campeche to the Belize River as the site for masterless men to acquire logwood was gradual, but the latter location became the primary source for logwood entering world markets after 1720 and for the next century. The 1717 Board of Trade document helped British cutters retain access to logwood forests – a document based on ideological interpretations of common property access, effective occupation, and nature's improvement written forty years earlier in response to what masterless men were actually doing on the ground. The document also helped British officials condone the trade. The relative insecurity of logwood cutting, however, helped ensure that it remained mostly in the hands of masterless men who had little to lose from Spanish assaults into the early eighteenth century, but this was about to change.

Maintaining access to logwood

By the middle of the eighteenth century slave-owning white men increasingly controlled access to logwood labour, preferred cutting locations, and shipping, and their needs began to drive British policy towards the region. Meanwhile, Spanish seizure of logwood ships in the 1730s was an underlying cause for the Anglo-Spanish conflict known as the War of Jenkins Ear (1739–42). During the war, protecting access to logwood in the Bay of Honduras justified the establishment of a military presence along the Mosquito Shore and for the occupation of the island of Roatán off the coast of Honduras. Subsequent wars and a diplomatic ignorance of the geographic distinction between the Bay of Honduras and the Mosquito Shore produced confusing Anglo-Spanish peace treaties concerning the region in 1748, 1763 and 1783. Specific questions concerning the last of these treaties were only resolved with another treaty in 1786 in which Great Britain gave up claims to the Mosquito Shore to retain woodcutting rights at Belize. By this time, however, most masterless men found themselves disadvantaged by an emerging woodcutting oligarchy whose enslaved labourers had begun to displace them.

The successful capture of logwood and other merchant ships in the Caribbean by Spanish coast guard helped precipitate the War of Jenkins (1739–42).[63] Following Admiral Vernon's successful attack at Portobelo, Panama, at the beginning of the conflict, Jamaican governor Edward Trelawny sought to advance the war effort by expanding the role of the British government on the Mosquito Shore. Although much of the shore had been inhabited by merooners since the early eighteenth century, it was not until the Bermudian William Pitt settled at Black River, on the northwestern edge of the shore, that anything approximating a permanent settlement took root. When Spaniards attacked the logwood cutters at Belize in 1730, the men fled to safety at Black River. When the Spaniards withdrew, the men returned to their cutting up the Belize River, but Pitt remained at Black River. There, he built a small mahogany, sarsaparilla, and illicit trade empire with assistance from the Miskitu Indians and his own enslaved labours until his death in 1771.[64] The safety and security of Black River as a point of retreat insured a strong connection between the logwood cutters at Belize and the Britons residing on the Mosquito Shore. British protection at Black River relied on the maintenance of friendly relations with the Miskitu, but also a dangerous and shallow river bar that prevented all but small sloops and pirogues from crossing it. As one contemporary put it, this made Black River 'the readiest and safest retreat for our Logwood Cutters who were driven from the Bay of Honduras [by the Spaniards] in 1730 and 1754'.[65] Another contemporary opined that 'Was it not for Black River no English Subjects could ever think to venture to cut Logwood in the Bay of Honduras'.[66]

To initiate his plan of expanding the official British presence at Black River, Governor Trelawny sent Captain Robert Hodgson to the Mosquito Shore to mobilize the 1,000 Miskitu men and the 100 merooners.[67] Hodgson soon provided Trelawny with four reasons why greater British support for the shore and for taking Roatán was a good idea: (1) the logwood trade in the nearby Bay of Honduras (mostly in control of Dutch) could be secured for Britain; 2) the logwood traders who are frequently harassed by the Spaniards from the sea could now be protected by warships stationed at Roatán; (3) illicit trade with the Spaniards at Trujillo (also in Dutch hands) could now be secured for Britain, and new branches of trade could be opened along the rivers near Black River; and (4) Britain could prevent the many islands in the Bay and along the Shore from becoming a retreat for pirates 'and disorderly people averse to all Government'.[68] As this last point intimated, more government and a greater military presence would mean more slaves and a reduced influence of masterless men, or 'people averse to all government'.

Admiral Vernon and vocal baymen supported Trelawny's efforts.[69] When Spanish ships cleared the logwood men out of the Bay of Honduras in 1747–8, Roatán provided them refuge. The slave-owning baymen leveraged the historical moment to petition the Jamaican governor that an official be sent to the bay to keep order.[70] Although, their plea failed in the short term it marked an important shift in power from the masterless men whose livelihoods depended on access to logwood (and the liberty to sell it to the highest bidder, including the Dutch), and slave-owners who sought to control and profit from the trade. Most slave-owning baymen lived at Saint George's Caye while their male slaves worked the interior on their own. At these interior locations, as one slave-owner put it without a hint of irony, 'slaves need to have maximum freedom to acquire game and provisions, or people would starve'.[71] Such relative autonomy in remote settings encouraged many enslaved people to flee across the Spanish frontier in hopes of becoming free, a situation that increased after a Spanish royal decree in 1739 granted British slaves their freedom if they accepted baptism.[72] The open frontier with Spanish Guatemala and Yucatan, in the words of historian Jennifer Anderson, 'necessitated a relatively flexible form of bondage' to retain the loyalty of woodcutting slaves.[73] Rather than tolerate this necessity, however, slave-owners sought to direct the power of the colonial state to protect their investment in human chattel.

Following the Treaty of Aix-la-Chapelle in 1748, Britain returned Roatán to Spain and removed its military stores and some of the soldiers to the Mosquito Shore. The logwood men returned to the Bay of Honduras and by 1751, they were shipping 8,000 tons at £20 per ton.[74] But political problems remained, as the treaty did nothing to address Spanish concerns about the bay. As one historian put it, 'between 1750 and 1755, the most serious New World problem that tended to muddy the Anglo-Spanish diplomatic waters was that presented by the activities of British loggers in the area of the Honduras Bay'.[75] Following a Spanish assault on the logwood cutters in 1754, the Jamaican governor Charles Knowles (1752–6) sent the engineer Richard Jones to access the best location to build a defensive structure to protect the logwood trade. Knowles argued that the proposed fortification provided sufficient security for the baymen, and he strongly rejected any additional support for the Mosquito Shore. He considered Trelawny's expanded vision for the shore 'chimerical', and initiated a Jamaican policy to downgrade the Mosquito Shore in favour of the Bay of Honduras, a policy that culminated with the complete evacuation of the shore in 1786.[76]

At the end of the next Anglo-Spanish conflict in 1763, Article 17 of the Treaty of Paris illustrated the geographic confusion of British diplomats.[77] The article stated that Great Britain would demolish 'all the fortifications which his subjects shall have erected in the bay of Honduras, and other places of the territory of Spain in that part of the world'. In exchange, Spain would allow British subjects 'or their workmen' [e.g., slaves] to labour unmolested 'in their occupation of cutting, loading, and carrying away logwood'.[78] Not surprisingly, the Spaniards assumed this language included the Mosquito Shore, but since there was no logwood anywhere along the Mosquito Shore, British residents there felt the treaty did not concern them. They claimed that their rights to occupancy derived from the free and independent Miskitu nation.[79] Still, as with Jamaican governor Knowles before him, Governor Lyttleton did not think highly of the shore, and he demanded their fortifications also be dismantled.[80]

Before the ink had dried on the Treaty of Paris, the baymen tested the prescribed territorial limits and began cutting logwood along the Río Hondo, outside the prescribed northern limits, generating protests by Spain. Some historians have noted that the treaty was essentially obsolete because mahogany cutting in the bay had eclipsed logwood.[81] This claim, however, overstated the situation because during six months in April–October 1764, 77 vessels carried away 7,449 tons of logwood and 401,235 ft. of mahogany, the lot valued in Britain at £24,863 sterling.[82]

Meanwhile, settlers on the Mosquito Shore continued to justify their presence through their links to logwood cutters in the bay. In 1768, the Mosquito Shore population consisted of 203 whites, 189 freemen of colour, 3,000 Miskitu, and 914 'negroe' slaves. Natural resources acquired by settlers on the Mosquito Shore included mahogany, other hardwoods and pine, sarsaparilla, shell of the hawksbill turtle, deerskins and small quantities fustic, Nicaragua wood, cotton, seal oil, silk grass (flax), gums, balsams, wax and honey. Via contraband trade with the Spanish settlements of the interior, British settlers received silver bullion, indigo, cochineal, cacao, coffee, mules, horses and black cattle, the majority of which they sent to Jamaica.[83] Still, their most important justifying argument to retain governmental support centred on logwood in the Bay of Honduras, 500 kilometres to the northwest.

Mosquito Shore settlers made their case in two ways. First, they continued to emphasize the importance of Black River as refuge for the logwood cutters at Belize.[84] Second, they noted that the logwood trade at Belize had no future: the trees were cut out and slaves escaped to the Spanish. In contrast, the Mosquito Shore had land to transplant and raise an infinite amount of logwood, and that the Miskitu would help control slave flight.[85] Clever shoremen worked around the contradictory implications of these two positions by stating that logwood was

> an Article absolutely necessary to our Dyers, and even if the Spaniards should [accept] our Right to cut it in the Bay of Honduras . . . there will soon be no more to cut. But on the Mosquito Shore any required Quantity may be raised, the low Lands of that Country being of the very same quality as the neighbouring Bay of Honduras where the Wood comes to its growth from the seed in the less than fifteen years.[86]

From 1779 to 1782 the region was again at the centre of Anglo-Spanish conflict in western Caribbean. On the eve of the successful Spanish sweep of Saint George's Caye on 15 September 1779, one writer reported slave-owners residing at the Caye had instructed their slaves to cut logwood and raise their subsistence crops along the 'Rio Hondo, New River, Rowley's Right, Northern River, Balize River, Chaboon [Sibun] River, and Manatee Lagoon, for 100 miles and upwards – the banks of the Balize River in particular are settled above 200 Miles'. The English, he said, numbered 500, of which 200 could bear arms, and their slaves numbered 3,000, of whom '500 may be depended upon'.[87] During their successful assault, the Spanish captured 140 baymen and 250 of their slaves – suggesting many fled to Black River – ending woodcutting in the bay until after the Treaty of Versailles in 1783.[88]

The peace treaty of 1783 required the evacuation of British settlers from all parts of the 'Spanish continent', except for a narrow territory bounded by the Belize River north to the Río Hondo. The disputed phrase 'Spanish continent' immediately became a new source of discord. Once again, British settlers along the Mosquito Shore argued vigorously that their home was not part of the Spanish continent because it had never been conquered or settled by Spain, contained no logwood, and that the Miskitu were a free and independent nation. In 1784, Robert Liston was appointed Minister Plenipotentiary to Madrid to resolve the dispute. His solution was to give up the shore for an expanded area in the Bay of Honduras to accommodate the additional settlers who would need to relocate there. In writing home to the foreign secretary, Liston stated that when he first broached this idea to his Spanish counterpart, 'his Excellency [Condé de Floridablanca] appeared to have difficulty to contain the marks of his dissatisfaction within the bounds of moderation'. Floridablanca stated, 'That the known intention of the Court of Spain was that we should remove from the whole of the continent of South America [*sic*], except the Districts marked out for the cutting of logwood; that he would much rather have continued the war than have made peace on any other conditions'.[89] The two sides eventually came to terms.

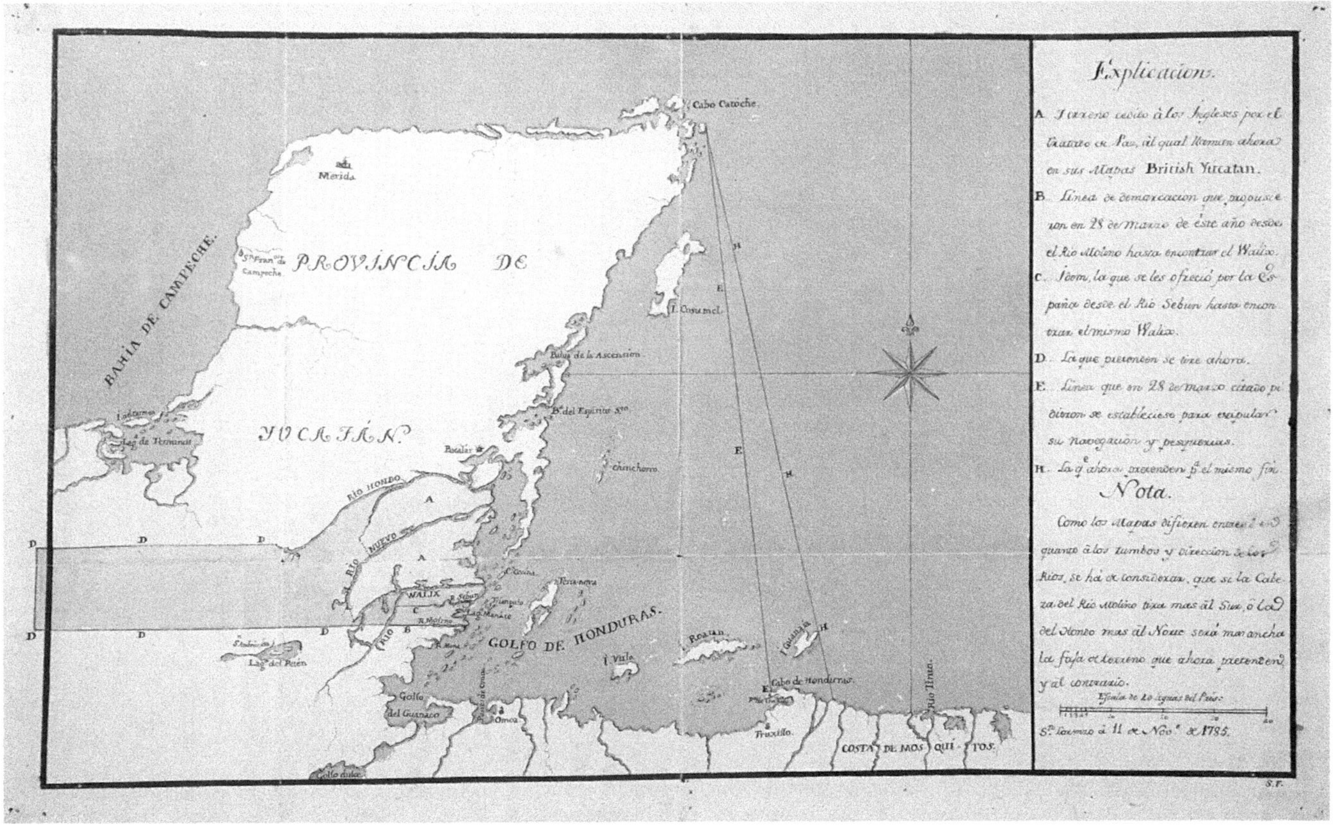

Figure 6.4 'Mapa de la península de Yucatan, y porción de costa comprendida entre la laguna de Términos, en la bahía de Campeche, y golfo de Honduras hasta la costa de Mosquitos' (1785).

A Spanish map drawn on 11 November 1785 illustrates the original 1783 treaty boundaries, those proposed by Great Britain, and those countered by Spain (Figure 6.4). The text on the map specifically refers to areas granted to cut *palo de tinte*, or logwood, and does not refer to mahogany or generalized woodcutting, illustrating that diplomats continued to assume the main dispute concerned logwood as it had a century earlier. After the Convention of London was signed in 1786 to enact Liston's proposal, the royal geographer William Faden published a map with the relevant text from the 1783 and 1786 treaties (Figure 6.5). The map shows the expanded British forest access: a southern increase to the Sibun River; the eastern increase to the 'Source of said River', and then north in a straight line until it intersected with the 'River Wallis', or Belize River. Two things germane to this study happened quickly thereafter. First, slave-holding white men of means grabbed up the best woodcutting areas, and sought to elevate their political power vis-à-vis the mostly poor and mixed-race freemen who came from the Mosquito Shore. Second, all cutters, but especially those excluded from political power pushed the territorial bounds almost immediately. Despite the increase in mahogany cutting after the 1786 convention, logwood continued to be an important commercial item.[90] As late as the 1880s, Belize was exporting 15,000 tons of logwood annually, and it shipped 3,000 tons per year between 1911 and 1914.[91]

Conclusion

Logwood and the masterless men who acquired it directed British interests in the Yucatan and eastern Central America in the early modern period. An expanding European cloth and textile industry created an insatiable demand for natural dyes. Because of its versatility and application in producing many colourants, the chemical compound haematoxylin – unique to the heartwood of the logwood tree native to the Yucatan Peninsula – was in high demand for more than a century spanning the middle of the seventeenth and eighteenth centuries and beyond. Even after 1679 – when Britain first moved to curtail logwood cutting to conform with Spanish ultimatums – masterless men supplied several thousand tons of logwood per year to European markets. Sporadically before, but especially after the Board of Trade wrote an influential report in 1717 that defended British access rights to logwood forests around the Yucatan Peninsula, the British government 'connived' at the illicit logwood cutting operations. By the 1740s, the British government fully backed their subject's right to cut logwood in the Bay of Honduras. Logwood remained a viable commercial activity for masterless men of the Caribbean, even as their ability to retain logwood access diminished vis-à-vis wealthy and politically connected slave-owning white men after the 1750s. The cutting and shipping of logwood shaped Anglo-Spanish relations throughout the late seventeenth and eighteenth centuries, and British decisions to invest blood and treasure in Yucatan and Central America revolved around their desire to maintain access to logwood in the Bay of Honduras. Until it was eventually eclipsed by mahogany in the nineteenth century, logwood was the most politically and economically important wild natural resource extracted from the American tropics.

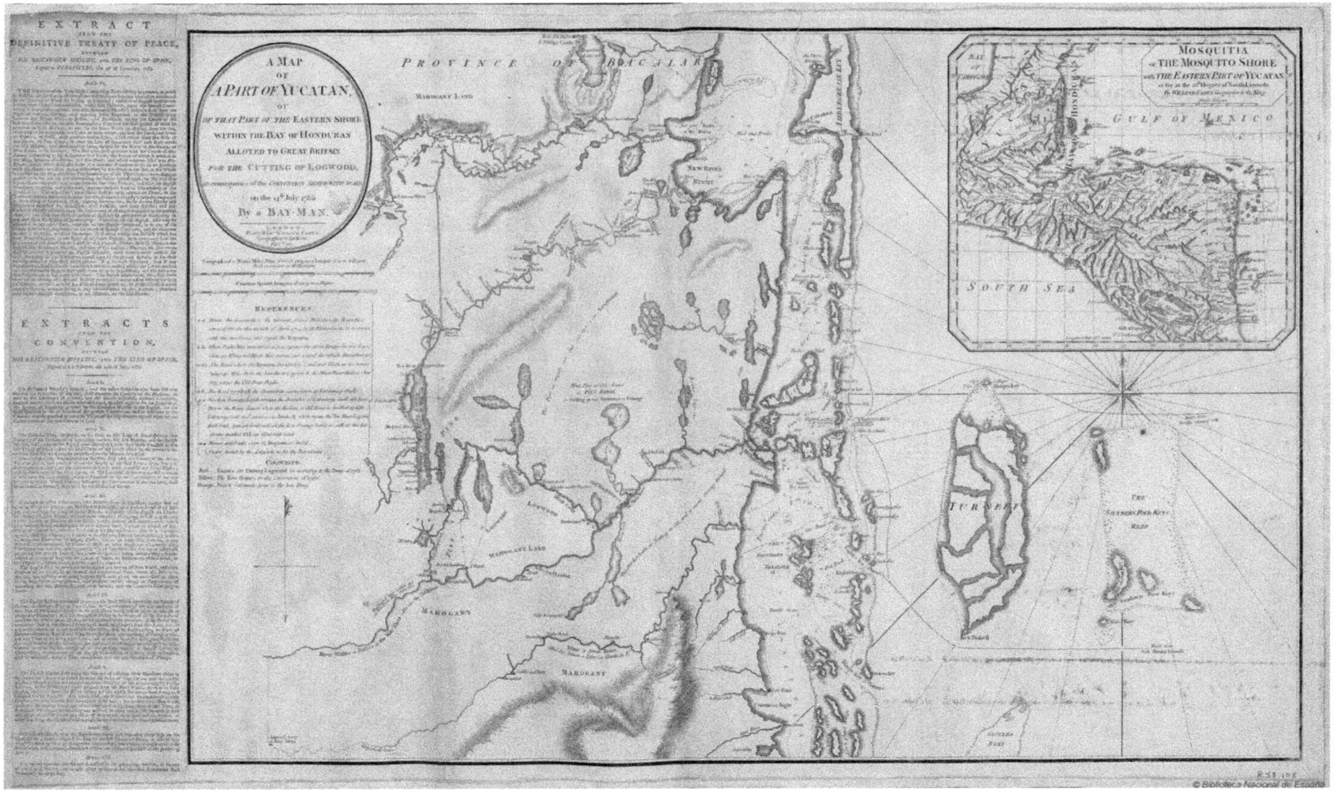

Figure 6.5 William Faden, *A Map of a Part of Yucatan* (1787).

Notes

1 Julius S. Scott first used the term 'masterless men' in his dissertation, published as *The Common Wind: Afro-American Currents in the Age of the Haitian Revolution* (New York: Verso, 2018). See also Isaac Curtis, 'Masterless People: Maroons, Pirates, and Commoners', in *The Caribbean: A History of the Region and Its Peoples*, ed. Stephan Palmié and Francisco A. Scarano (Chicago: University of Chicago Press, 2011), 114–62; Jesse Cromwell, 'More than Slaves and Sugar: Recent Historiography of the Trans-imperial Caribbean and Its Sinew Populations', *History Compass* 12, no. 10 (2014): 778.

2 Arthur P. Newton, *European Nations in the West Indies, 1493–1688* (London: A&C Black, 1933), 144; Kevin P. McDonald, '"Sailors from the Woods": Logwood Cutting and the Spectrum of Piracy', in *The Golden Age of Piracy: The Rise, Fall, and Enduring Popularity of Pirates*, ed. David Head (Athens: University of Georgia Press, 2018), 50.

3 The word Belize originated from a corruption of a Mayan word that Europeans initially applied only to the Belize River, and then, to a territory only in the nineteenth century. I have avoided using the neologism 'Belize' to refer to the logwood cutting locations of the eighteenth century. Mathew Restall, 'Creating "Belize": The Mapping and Naming History of a Liminal Locale', *Terrae Incognitae* 51, no. 1 (2019): 5–35.

4 On current uses of hematoxylin, see Carlos Ortiz-Hidalgo and Sergio Piña-Oviedo, 'Hematoxylin: Mesoamerica's Gift to Histopathology: Palo de Campeche (Logwood Tree), Pirates' Most Desired Treasure, and Irreplaceable Tissue Stain', *International Journal of Surgical Pathology* 27, no. 1 (2019): 4–14.

5 R. Temple, 'On British Honduras, Its History, Trade, and Natural Resources', *Journal of the Royal Society of the Arts* 5, no. 217 (1857): 118; Karl H. Offen, 'British Logwood Extraction from the Mosquitia: The Origin of a Myth', *Hispanic American Historical Review* 80, no. 1 (2000): 113–35.

6 Seminal studies include Arthur M. Wilson, 'The Logwood Trade in the Seventeenth and Eighteenth Centuries', in *Essays in the History of Modern Europe*, ed. Donald C. McKay (New York: Harper & Brothers, 1936), 1–15; Alan K. Craig, 'Logwood as a Factor in the Settlement of British Honduras', *Caribbean Studies* 9, no. 1 (1969): 53–62; Gilbert M. Joseph, 'John Coxon and the Role of Buccaneering in the Settlement of the Yucatán Colonial Frontier', *Terrae Incognitae* 12 (1980): 65–84; David Morgan McJunkin, *Logwood: An Inquiry into the Historical Biogeography of Haematoxylum campechianum L. and Related Dyewoods of the Neotropics*, PhD diss. (University of California, Los Angeles, 1991); Offen, 'British Logwood Extraction', 113–35; Jesse Cromwell, 'Life on the Margins: (Ex) Buccaneers and Spanish Subjects on the Campeche Logwood Periphery, 1660–1716', *Itinerario* 33, no. 3 (2009): 43–71.

7 Susan Fairlie, 'Dyestuffs of the Eighteenth Century', *The Economic Historical Review* 17 (1965): 490–2; Raphael Zon and William Sparhawk, *Forest Resources of the World*, 2 vols. (New York: MacGraw Hill, 1923), 2: 978; McJunkin, *Logwood*, 6–14; Offen, 'British Logwood Extraction', 117–21.

8 Carl O. Sauer, *The Early Spanish Main* (Berkeley: University of California Press, 1966), 92–3.

9 Warren Dean, *With Broadax and Firebrand. The Destruction of the Brazilian Atlantic Forest* (Berkeley: University of California Press, 1995), 45–6.

10 'The Development of the Textile Industries', *Journal of the Royal Society of the Arts* 64, 3293 (1915): 132–4.
11 House of Commons Journal, 20 June 1650, *Journal of the House of Commons*, Vol. 6, 1648–1651 (London, 1802), 426–7; Ortiz-Hidalgo and Piña-Oviedo, 'Hematoxylin', 5.
12 Karl Offen, 'Puritan Bioprospecting in the West Indies and Central America', *Itinerario* 35, no. 1 (2011): 18–20.
13 On the puritan colony, see Karen Ordahl Kupperman, *Providence Island, 1630–1641: The Other Puritan Colony* (Cambridge: Cambridge University Press, 1993), 18. This chapter follow's David Armitage's suggestion that 'British' provides a better category than 'English' for much of the period before the Act of Union in 1707 because the ideology informing the British Empire developed in the sixteenth and seventeenth centuries. David Armitage, *The Ideological Origins of the British Empire* (New York: Cambridge University Press, 2000).
14 Kupperman, *Providence Island*, 33, 83–7, 94, 109; Offen, 'Puritan Bioprospecting', 25.
15 Instructions to Capt. Sussex Camock, 1 July 1633, The National Archives (hereafter TNA), Colonial Office (hereafter CO) 124/1, 56–8; Jessop to Camock, 11 August 1634, British Library (hereafter BL), Add. Mss. 63854A, #30, 44–6; Jessop to Cap. Ax, 11 August 1634, BL, Add. Mss. 63854A, #29, 41–3; Offen, 'Puritan Bioprospecting', 25–33.
16 Kupperman, *Providence Island*, 310, 167; Newton, *European Nations in the West Indies*, 289; David Watts, *The West Indies: Patterns of Development, Culture and Environmental Change since 1492* (Cambridge: Cambridge University Press, 1987), 170.
17 Order of the Privy Council, 12 January 1638, Calendar of State Papers Colonial, America and West Indies (hereafter CSP), British History Online, http://www.british-history.ac.uk (hereafter BHO), Vol. 1, 80. Michael J. Jarvis, *In the Eye of All Trade: Bermuda, Bermudians, and the Maritime Atlantic World, 1680–1783* (Chapel Hill: University of North Carolina Press, 2010), 219.
18 Offen, 'Puritan Bioprospecting', 21.
19 Derived from the English maroon, itself from the Spanish *cimarron*, a corruption of a Taino word meaning 'flight', English writers in the eighteenth century used 'merooner' to refer to a free person unfit for civilized society, a fugitive, or a masterless man.
20 Lauren Benton, 'Legal Spaces of Empire: Piracy and the Origins of Ocean Regionalism', *Comparative Studies in Society and History* 47, no. 4 (2005): 701. See also Laura Benton, 'Spatial Histories of Empire', *Itinerario* 30, no. 3 (2006): 20–6.
21 Karl Offen, 'English Designs on Central America: Geographic Knowledge and Imaginative Geographies in the Seventeenth Century', *Early American Studies* 18, no. 4 (2020): 399–460. A Jamaican council meeting in 1666 justified letters of marque against the Spaniards on the grounds that privateers furnished the island with coin and bullion, as well as 'cocoa, logwood, hides, tallow, indigo, cochineal' and more which could be traded with merchants from New England for flour and other provisions. Minutes of the Council of Jamaica, 22 February 1666, CSP, BHO, Vol. 5, 1138.
22 Arne Bialuschewski, 'Slaves of the Buccaneers: Mayas in Captivity in the Second Half of the Seventeenth Century', *Ethnohistory* 64, no. 1 (2017): 48.
23 The King to Lord Carlisle, 29 March 1679, CSP, BHO, vol. 10, 950; Thomas Lynch to Lords of Trade and plantations, 29 August 1682, CSP, BHO, Vol. 11, 668.
24 Thomas Modyford to Lord Arlington, 31 October 1670, CSP, BHO, Vol. 7, 310.

25 Murdo J. MacLeod, *Spanish Central America: A Socioeconomic History, 1520–1720* (Berkeley: University of California Press, 1973), 368.
26 Available data on the logwood trade is partial and often contradictory, making estimates difficult. Gov. Lynch to Secretary to the Council for Plantations, 5 November 1672, CSP, BHO, vol. 7, 954; An account of Jamaica, 28 January 1676, CSP, BHO, Vol. 19, 800; Nuala Zahedieh, 'The Merchants of Port Royal, Jamaica, and the Spanish Contraband Trade, 1655–1692', *William and Mary Quarterly* 43, no. 4 (1986): 585; Clarence H. Haring, *The Buccaneers of the West Indies in the XVII Century* (London, 1910), 209; Carl Bridenbaugh and Roberta Bridenbaugh, *No Peace Beyond the Line: The English in the Caribbean 1624–1690* (Oxford: Oxford University Press, 1972), 341–2.
27 John H. Andrews, 'Anglo-American Trade in the Early Eighteenth Century', *Geographical Review* 45, no. 1 (1955): 101; Observations on the present state of Jamaica, M. Cranfield, 14 December 1675, CSP, BHO, Vol. 19, 740.
28 David Eltis, 'New Estimates of Exports from Barbados and Jamaica, 1665–1701', *William and Mary Quarterly* 52, no. 4 (1995): 641, n32. The importance of logwood relative to plunder and trade in the capitalization of Jamaica is discussed in Nuala Zahedieh, 'Trade, Plunder, and Economic Development in Early English Jamaica', *Economic History Review* 39, no. 2 (1986): 205–22.
29 William Dampier, 'Two Voyages to Campeachy, with a Description of the Coasts, Product, Inhabitants, Logwood-Cutting, Trade, &c. of Jucatan, Campeachy, New-Spain, &c', in *Voyages and Descriptions*, Vol. II (London: James Knapton, 1699), 54–7, 80–7. Osnaburg is a plain-weave and coarse fabric made with flax that the English originally acquired from Osnabrück, Germany.
30 McDonald, '"Sailors from the Woods"', 56.
31 Dampier, 'Two Voyages to Campeachy', 54.
32 Bialuschewski, 'Slaves of the Buccaneers', 42–3, 49–51.
33 Modyford to Privy Council, 28 September 1670, CSP, BHO, Vol. 7, 278.
34 Robert Morden and William Berry, 'A New Map of the English Plantations in America Both Continent and Islands, Shewing Their True Situation or Distance from England . . .', London, [1673], John Carter Brown Library, Cabinet Blathwayt 33. See also Offen, 'English Designs on Central America', 438–46.
35 On how imperial powers often used or commissioned maps to reveal their ambitions to rivals, see Ken MacMillan, *Sovereignty and Possession in the English New World: The Legal Foundations of Empire, 1576–1640* (Cambridge: Cambridge University Press, 2006).
36 George Chalmers, *Collection of Treaties Between Great Britain and Other Powers*, 2 vols. (London: John Stockdale, 1790), 2: 37.
37 Gov. Lynch to Lord Arlington, 2 July 1671, CSP, BHO, Vol. 7, 580; Gov. Lynch to the Earl of Sandwich, 14 October 1671, CSP, BHO, Vol. 7, 640; Gov. Lynch to Lynch to Sec. Lord Arlington, 17 December 1671, CSP, BHO, Vol. 7, 697; Gov. Lynch to Joseph Williamson, 13 January 1672, CSP, BHO, Vol. 7, 729; Gov. Lynch to Sir Lyttelton, 28 January 1672, CSP, BHO, Vol. 7, 743; Gov. Lynch to Council for Plantations, 10 March 1672, CSP, BHO, Vol. 7, 777.
38 Gov. Lynch to Secretary to the Council for Plantations, 5 November 1672, CSP, BHO, Vol. 7, 954.
39 Council for Plantations to Thomas Lynch, 8 October 1672, CSP, BHO, Vol. 7, 943; Frank Cundall, *The Governors of Jamaica in the Seventeenth Century* (London: West Indian Committee, 1936), 45.

40 William Godolphin to Earl of Arlington, Madrid, 20 May 1672, TNA, CO 137/48, 41v.
41 On Locke's thinking with regard to logwood and commons, see Eva Botella-Ordinas, 'Debating Empires, Inventing Empires: British Territorial Claims Against the Spaniards in America, 1670–1714', *Journal for Early Modern Cultural Studies* 10, no. 1 (2010): 142–68.
42 Governor Vaughan to Sec. Williamson, 20 September 1675, CSP, BHO, Vol. 9, 673; Garrett Hardin, 'The Tragedy of the Commons', *Science* 162 (1968): 1243–8.
43 Dampier, 'Two Voyages to Campeachy', 131. See also, Offen, 'English Designs on Central America', 455.
44 King to Lord Carlisle, 29 March 1679, CSP, BHO, Vol. 10, 950; Joseph, 'John Coxon and the Role of Buccaneering', 69; Kris E. Lane, *Pillaging the Empire: Piracy in the Americas 1500–1750* (New York: M.E. Sharpe, 1998), 126–8; Offen, 'English Designs on Central America', 448–52.
45 A. P. Thornton, 'The English at Campeachy, 1670–82', *The Jamaican Historical Review* 2, no. 3 (1953): 27–38; Bialuschewski, 'Slaves of the Buccaneers', 52.
46 Governor of New York to Council of Trade and Plantations, 3 May 1699, CSP, BHO, Vol. 17, 343.
47 Dalby Thomas, *An Historical Account of the Rise and Growth of the West India Collonies* [*sic*] . . . (London: Jo Hindmarsh, 1690), 9; Jamaica shipping, 20 April 1700, CSP, BHO, Vol. 18, 347; Council of Trade and Plantations, Whitehall, 5 February 1702, CSP, BHO, Vol. 20, 97; Capt. Haskett to Council of Trade and Plantations, 8 July 1702, CSP, BHO, Vol. 20, 720; Estimates of exports from Jamaica, 20 January 1704, CSP, BHO, Vol. 22, 36.
48 MacLeod, *Spanish Central America*, 384, 370.
49 Governor Hamilton to Council of Trade and Plantations, Jamaica, 19 January 1712, CSP, BHO, Vol. 26, 267; Mr. Cumings to Council of Trade and Plantations, Boston, 22 December 1726, CSP, BHO, Vol. 34, 439. From Curaçao, the Dutch sent home approximately 50 ships a year, most laden with products from British plantations, including logwood. Extracts from three memorials, Whitehall, 19 January 1710, CSP, BHO, Vol. 25, 47.
50 On the removal of the Maya and Miskitu raids, see Grant D. Jones, *Maya resistance to Spanish Rule: Time and History on a Colonial Frontier* (Albuquerque: University of New Mexico Press, 1989); Lawrence Feldman, *Lost Shores, Forgotten Peoples: Spanish Explorations of the South East Maya Lowlands* (Durham: Duke University Press, 2000), 217–20; Mathew Restall, 'Crossing to Safety? Frontier Flight in Eighteenth-Century Belize and Yucatan', *Hispanic American Historical Review* 94, no. 3 (2014): 389; Arne Bialuschewski, *Raiders and Natives: Cross-Cultural Relations in the Age of Buccaneers* (Athens: University of Georgia Press, 2021), 28–38.
51 Daniel R. Finamore, *Sailors and Slaves on the Wood-Cutting Frontier: Archaeology of the British Bay Settlement, Belize*, PhD diss. (Boston University, 1994), 22–3; Lieutenant-Governor Molesworth to William Blathwayt, 17 December 1686, CSP, BHO, Vol. 12, 1066.
52 Jeremiah Dummer to the Council of Trade and Plantations, 25 February 1720, CSP, BHO, Vol. 31, 578; Frank Cundall, *The Governors of Jamaica in the First Half of the Eighteenth Century* (London: Longman Caribbean, 1937), 144.
53 Wilson, 'The Logwood Trade', 5; Galfridus Gray to Council of Trade and Plantations, 9 June 1730, CSP, BHO, Vol. 37, 280.
54 J Tate to Sir John Duddleston, Charlestown, 29 November 1715, CSP, BHO, Vol. 28, 691; Richard Harris to Mr. Popple, with enclosures, London, 27 December 1715, CSP,

BHO, Vol. 28, 129; Galfridus Gray to Council of Trade and Plantations, 9 June 1730, CSP, BHO, Vol. 37, 280.

55 Governor Lawes to Council of Trade and Plantations, Jamaica, 6 December 1719, CSP, BHO, Vol. 31, 479; Governor Lawes to Council of Trade and Plantations, Jamaica, 2 February 1720, CSP, BHO, Vol. 31, 540; Jeremiah Dummer to Council of Trade and Plantations, Jamaica, 25 February 1720, CSP, BHO, Vol. 31, 578; Governor Lawes to Council of Trade and Plantations, Jamaica, 24 August 1720, CSP, BHO, Vol. 32, 213; Duke of Newcastle to the Council of Trade and Plantations, 17 January 1733, CSP, BHO, Vol. 40, 20; Council to Duke of Newcastle, Whitehall, 21 March 1733, CSP, BHO, Vol. 40, 70; Council and Assembly of Jamaica to Council of Trade and Plantations, Spanish Town, 2 May 1735, CSP, BHO, Vol. 41, 564; Martin Bladen to Duke of Newcastle, Albrohatch, 25 April 1739, CSP, BHO, Vol. 45, 148.

56 All quotations and data in this paragraph come from, Copy of Representation from the Board of Trade to His Majesty, relating to the Cutting of Logwood in the Bay of Campeachy, Whitehall, 25 September 1717, TNA, CO 123/3, 265r–276v; Malachy Postlethwayt, *The Universal Dictionary of Trade and Commerce*, Vol. II, 4th ed. (London, 1774).

57 John Atkins, *A Voyage to Guinea, Brasil, and the West Indies* . . . (London: Printed for Ward and Chandler, 1735), 227–8; Nathaniel Uring, *A History of the Voyages and Travels of Capt. Nathaniel Uring. With New Draughts of the Bay of Honduras* . . . (London: W. Wilkins, 1726), 354–7.

58 Extracts from Letters included in one from Capt. Cusack to Admiral Vernon, Port Royal Harbour, Roatán, 25 September 1742, TNA, CO 137/57, 172–3; O. Nigel Bolland, *The Formation of a Colonial Society: Belize, from Conquest to Crown Colony* (Baltimore: The Johns Hopkins University Press, 1977), 50; Finamore, *Sailors and Slaves*, 34.

59 In 1702, Spaniards captured some 20 whitemen and 20 Africans in the Bay of Campeche. Lt. Gov. Beckford report, Jamaica, 2 July 1702, CSP, BHO, Vol. 20, 694. See also Cromwell, 'Life on the Margins', 55–6.

60 Atkins, *A Voyage to Guinea, Brasil, and the West Indies*, 227–8; Uring, *A History of the Voyages*, 354–7.

61 List of entries of ships, Charles Town, 27 May 1700, CSP, BHO, Vol. 18, 476; Thomas Bannister to Council of Trade and Plantations, 7 July 1715, CSP, BHO, Vol. 28, 508; Archibald Cumings to Council of Trade and Plantation, Boston, 17 September 1717, CSP, BHO, Vol. 30, 85.

62 Ross A. Newton, '"Good and Kind Benefactors": British Logwood Merchants and Boston's Christ Church', *Early American Studies* 11, no. 1 (2013): 15–36. For pictures of the plaque on the pew, see Ross A. Newton, 'Logwood Cutters, Merchants, Privateers . . . Religious Gents?', *Commonplace: The Journal of Early American Life* 18, no. 1 (2018). http://commonplace.online/article/logwood-cutters-merchants-privateers/ (accessed 17 April 2022).

63 Historical Manuscripts Commission (ed.), *The Manuscripts of the Marquess Townshend*. Eleventh report, appendix, pt. 4 (London: Her Majesty's Stationary Office, 1887), 148; Vera Lee Brown, 'The South Sea Company and Contraband Trade', *American Historical Review* 31 (1926): 662–78. The war was so-called because a captain of a British merchant ship named Robert Jenkins had his ear cut off by sailors of the Spanish coast guard when they boarded his ship to search for contraband. He was later brought before parliament to generate support for the declaration of war.

64 Although the classic study on Pitt's founding of Black River is Frank Griffith Dawson's 'William Pitt's Settlement at Black River on the Mosquito Shore: A Challenge to Spain in Central America, 1732–1787', *Hispanic American Historical Review* 63, no. 4 (1983): 677–706, it is also notorious for contributing to the erroneous assumption that logwood was available at Black River. See Offen, 'British Logwood Extraction', 113–14.
65 Memorial of Rev. Warren, 1773, BL, Add. Ms. 12413, 4r.
66 Richard Jones to Governor Moore, 3 August 1768, TNA, CO 30/8. See also Barbara Potthast, *Die Mosquito-Küste im Spannungfeld Britischer und Spanischer Poltik 1502–1821* (Köln: Böhlau Verlag, 1988), 223.
67 Gov. Trelawny to Duke of Newcastle, Jamaica, 20 January 1739/40, TNA, CO 137/56, 279v–281v.
68 Robert Hodgson, Reasons for Settling Ratan and Banaca, Black River, 26 February 1742, TNA, CO 137/57, 176r–v; Trelawny to Andrew Stone, Jamaica, October 1742, TNA, CO 157/37, 73; Gov. Trelawny to Duke of Newcastle, 10 December 1743, TNA, CO 137/57, 361v–362r.
69 General Council of War, Spanish Town, 28 June 1742, TNA, CO 137/57, 174; Gov. Trelawny to Duke of Newcastle, [Jamaica], 20 July 1743, TNA, CO 137/57, 351r–352v; Montagu Winchilsea and Gen. Wade to Lords Justices, concerning Roatán, Whitehall, 20 July 1743, TNA, CO 323/11, 37r–43r; [Joseph Smith Speer's Description of Rattan], 1743, TNA, CO 323/11, 42r–43r; Logwood Cutters in the Bay of Honduras to Gov. Trelawny, Honduras, 28 April 1743, TNA, CO 323/11, 72r; Report to the Lords of the Committee proposing that an Independent Company should be kept on the Mosquito Shore, Whitehall, 3 May 1744, TNA, CO 324/12, 284–6.
70 Memorial of Parke Pepper on behalf of Inhabitants of the Bay of Honduras, Rattan, 3 June 1747, TNA, CO 323/11, 134r–140r.
71 Extract of a Letter, Roatán, 25 September 1742, TNA, CO 137/57, 172.
72 Restall, 'Crossing to Safety?', 382–3.
73 Jennifer L. Anderson, *Mahogany: The Costs of Luxury in Early America* (Cambridge, MA: Harvard University Press, 2012), 109–10, 157.
74 Robert Hodgson to Aldworth, Jamaica, 21 April 1751, TNA, CO 137/57, 550v.
75 Lawrence Henry Gipson, 'British Diplomacy in the Light of Anglo-Spanish New World Issues, 1750–1757', *American Historical Review* 51, no. 4 (1946): 632.
76 Admiral Knowles, 1756, Papers relating to the government; to the logwood trade; and to the proposed fort at Belese in the Bay of Honduras, TNA, PRO 30/47/17, n. p.
77 Robert White, *The Case of His Majesty's Subjects Having Property in and Lately Established Upon the Mosquito Shore* . . . (London: T. Cadell, 1789), 4–5.
78 The terms of the treaty can be found at The Avalon Project, Yale Law School, https://avalon.law.yale.edu/18th_century/paris763.asp (accessed 26 May 2022).
79 Joseph Otway to Lords Commissioners, Black River, 25 April 1764, TNA, CO 137/33, 167r–168v; James Lawrie and Edward Despard to Gov. Campbell, Kingston, 16 June 1783, National Army Museum, London, 6807/183-1, 16–17.
80 Lyttelton to Lieut. Robert Hodgson, Jamaica, 18 May 1764, TNA, CO 123/3, 251.
81 See for example, Barbara Bulmer-Thomas and Victor Bulmer-Thomas, *The Economic History of Belize from the 17th Century to Post-Independence* (Belize: Cubola Books, 2012).
82 Bay Settlers to Wm. Trelawny, Bay of Honduras, 19 February 1771, TNA, CO 137/66, 216v. See also Joseph Maud to Gov. Lyttelton, Bay of Honduras, 7 October 1765, TNA, CO 137/34, 5r–6r.

83 Richard Jones to Gov. Elletson, Jamaica, 3 August 1768, TNA, CO 137/35, 24r–27r. See also Joseph Smith Speer, *The West-India Pilot* (London, 1766); Offen, 'British Logwood Extraction', 128–9; Karl Offen, 'Mapping Amerindian Captivity in Colonial Mosquitia', *Journal of Latin American Geography* 14, no. 3 (2015): 40–4.

84 Petition of the Principal Inhabitants of the Mosquito Shore to His Majesty, Great Black River, 31 May 1766, TNA, CO 123/1, 88v; Principal inhabitants to James Lawrie, Black River, 10 October 1776, TNA, CO 137/72, 57.

85 James Lawrie to Rev. William Robertson, Black River, 10 November 1774, National Library of Scotland (hereafter NLS), Robertson-MacDonald Papers, MS 3942, 175v.

86 Robert Hodgson, A View of the Mosquito Shore, London, 12 October 1766, TNA, CO 123/1, 113v; Some Thoughts Relative to Trade lately carried on in the Bay of Honduras and on the Mosquito Shore by the British Merchants, 17 March 1783, BL, Add. Mss. 36806, 213.

87 Letter to Gov. Dalling relative to the Bay of Honduras and the Mosquito Shore, Kingston, 3 September 1779, TNA, CO 137/75, 156v.

88 Bolland, *Formation of a Colonial Society*, 30–1.

89 Robert Liston to Marquis of Carmarthen, Madrid, 7 January 1785, NLS, Liston Papers, MS 5531, 2v–3r. See also Marquis of Carmarthen to Robert Liston, St. James, 29 March 1785, NLS, Liston Papers, MS 5532, 21r–23r; Marquis of Carmarthen to Bernardo del Campo, St. James, 5 October 1785, NLS, Liston Papers, MS 5532, 85r–87r.

90 Between 1799 and 1802, more logwood from Belize was imported to Britain than mahogany, 6,376 tons compared with 6,103 tons, and again in the years 1807–9, the value of logwood imports surpassed mahogany, £84,456 compared to £61,195 per annum. See Nicholas Rogers, *Blood Waters: War, Disease and Race in the Eighteenth-Century British Caribbean* (Rochester: Boydell & Brewer, 2021), 139.

91 Daniel Morris, *The Colony of British Honduras, its Resources and Prospects; with Particular Reference to its Indigenous Plants and Economic Productions* (London: E. Stanford, 1883), 17; C. Hummel, *Report on the Forests of British Honduras with Suggestions for a Far Reaching Forest Policy* (London: Crown Agents for the Colonies, 1925), 14.

7

From abundance to scarcity

Dyewood production and trade in the Colombian Caribbean, 1700–1900

Jorge Enrique Elías-Caro

Studies on Caribbean economic history, including the provinces of colonial Venezuela and Colombia, between the sixteenth and nineteenth centuries, still require new research in several areas. Scholars have primarily concentrated on the examination of slave trade, the commerce of precious metals and sugar production. To a lesser extent, there are works on livestock, tobacco, coffee and mining commodities. However, concerning the former province of Santa Marta in Colombia – a region abundant in history yet lacking sufficient historiographical exploration – there exists a necessity for further investigation into the production and distribution of pearls, beef, sea salt, leather and, notably, natural dyes, which constitute the focal point of this chapter.[1]

Reports of forests full of dyewoods, such as brazilwood or fustic, were registered from an early date by the chroniclers of the sixteenth century, but it was only from the mid-eighteenth century that these trees came to be regarded as of 'economic potential to supply the needs of the Kingdom'. This was the opinion expressed by the Jesuit priest Antonio Julian in his work *La Perla de la América* (1787), who affirmed:

> In the Province of Santa Marta are to be found gold, silver, precious stones, pearls, very fine shells, tortoiseshell, brazilwood, cocoa, tobacco, sugar, wheat, corn, indigo, balsams, oils, and aromatic and medicinal gums, porphyry, stone quarries of jaspers and marble, as well as very tall trees which are valuable for their colors and various qualities; finally, numerous cattle, with abundant pasture, and many horses called *aguilillas*, unique because of their extraordinary (racing) speed.[2]

The economic activities related to dyewoods in the Caribbean were of ancestral tradition. Native Americans used them for their textiles and various artistic expressions. Shipments of brazilwood to Europe were recorded throughout the sixteenth century. In a report of 1587, friar Josef Acosta informed that the Spanish fleet that sailed from Santo Domingo to Spain carried 48 quintals of *cañafistula*, 50 of *sarsaparilla* and 134 of brazilwood and sugar. Altogether, the commodities comprised

898 boxes with an average weight of eight arrobas per box. This thriving commodity trade between Caribbean and Europe would later begin to decline. Even so, by the end of the eighteenth century, the priest and intellectual Antonio Sánchez Valverde, in his classic book *Idea del valor de la Isla Española, y utilidades, que de ella puede sacar su Monarquía*, (1785), observed: 'Foreign nations, especially the Dutch, took advantage. They clandestinely carried their effects, and took our fruits, although the Colony was somehow maintained'.[3] He also noted that 'the very few Indians, who remained, and some blacks, who were introduced, maintained the sugar factories, the plantations of ginger, indigo, *achote*, *cañafístula*, and a little cotton and tobacco'; furthermore, he added, a portion of brazilwood was still being cut (Figure 7.1).[4]

In the Caribbean region of the viceroyalty of Nueva Granada (modern-day Colombia), timber rich in dyes was harvested in different parts of its jurisdiction (see Figure 7.2). Conditions especially suitable for its growth were to be found in the forested areas of the provinces of Santa Marta and Cartagena, in the foothills of the Sierra Nevada, the Serranía de los Motilones, the natural border areas and valleys between the jungles and savannahs of the Ranchería-Cesar and the Sinú and San Jorge rivers, as well as in coastal and inland forested zones of this territory.

In 1802, near the end of the colonial period, the newspaper *El Correo mercantil de España y sus Indias* discussed the environmental requirements for the successful cultivation and production of dye plants and trees. This publication emphasized the importance of the *Tela* tree, which produced a purple dye, noting that 'this tree of America grows in temperate climates; while all the others, such as the *campeche*, the *moralete* and the *brasil*, need warm climates'.[5] In the maps elaborated between the sixteenth and eighteenth centuries it is possible to identify the principal wooded areas in Nueva

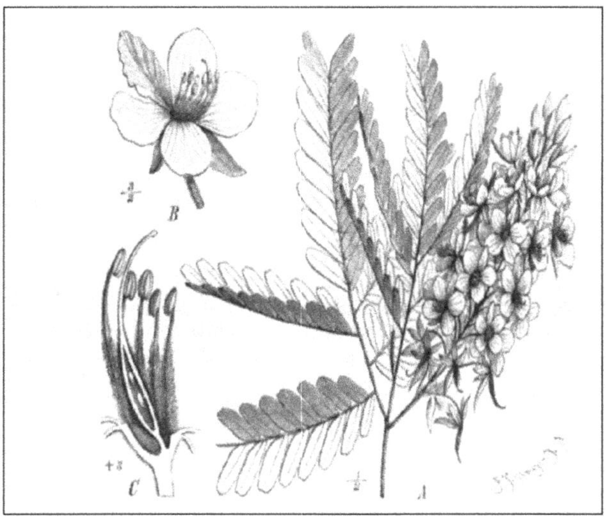

Figure 7.1 Scientific illustration of *Caesalpinia echinate* (brazilwood) (1890). *Source*: Paul Hermann Wilhelm Taubert, '*Leguminosae*', in Engelmann (ed.) *Natürliche Pflanzenfamilien*, 1890.

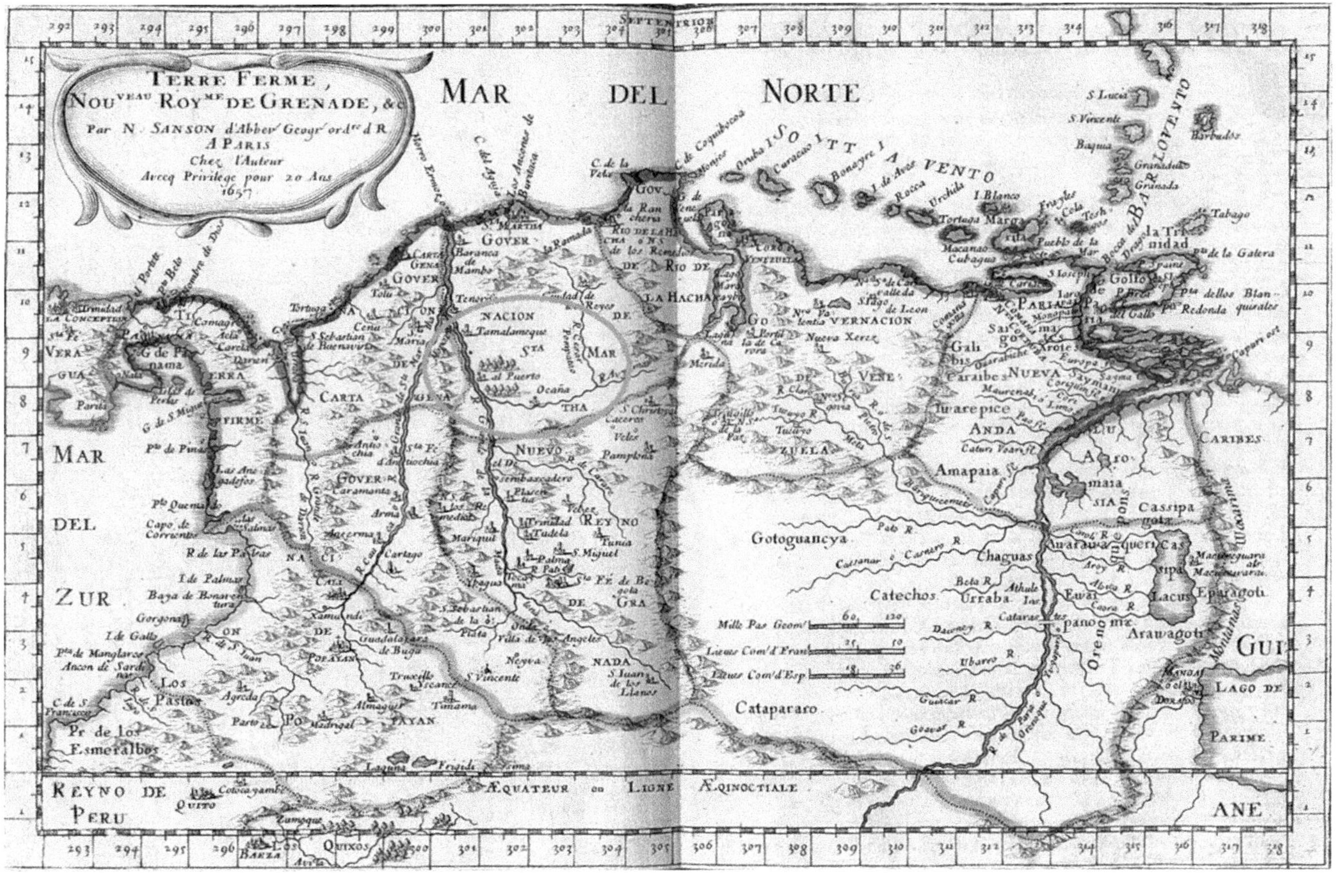

Figure 7.2 Map illustrating the topography and forested areas of Nueva Granada and the southern limits of the province of Santa Marta (1657).
Source: Nicolas Sanson, *Map of Tierra Firme and New Granada* (1657).

Granada (see Figure 7.2), which together with the mountainous regions of the province of Santa Marta (Sierra Nevada and Serranía de los Motilones) were natural emporiums. These forested areas were inhospitable for humans and some British observers described them as places where 'natural wealth (was) in the wrong hands'. However, following the Bourbon reforms and the Decree of Free Trade of 1778, the Spanish authorities began to perceive these forest resources as 'a solution to all problems'.[6]

In any event, Spanish public policies posed more obstacles than solutions for the advancement of a dye industry in Nueva Granada. The lack of official initiatives was underlined in the 1760s by Antonio Julián, then on a religious mission in the province of Santa Marta, before being expelled as a Jesuit priest. He observed the lack of interest in carrying out new economic initiatives, especially regarding brazilwood, pearls, salt, leather and other products that were abundant in the region. In a section of his book *La Perla de América*, he argued that a non-exclusive company should be established in Santa Marta to promote various local commodities, in particular brazilwood.[7] Julián contended that logwood could be harvested on the outskirts of Santa Marta, envisioning the city as a potential hub for gathering precious woods and various other commodities.

Extractive activity

The orography of Nueva Granada is described as early as 1657 in a map made by the French cartographer Nicolas Sanson (see Figure 7.2). This map illustrates the political and administrative jurisdictional boundaries, as well as the mountainous regions and two densely wooded areas. The latter were situated along the banks of the Magdalena River between Tamalameque and Gamarra, to the south of the provincial capital of Santa Marta. Notably, one of these areas is in proximity to Ocaña, a populated centre significantly distant from the Caribbean coast, which we have marked with a circle. The images of trees seem to convey their 'abundance' and hint at the luxuriant expanse of jungles in that location.

Among the most prominent traders of brazilwood residing in Valledupar (a town in the interior of the region of Santa Marta) was Joseph Agustín Parodi, who, by the 1780s, had a network of consignees who sent his cargoes to different cities, especially the ports of the viceroyalty of Nueva Granada and those of the Captaincy General of Venezuela. The dispatches of merchandise from Valledupar to the port of La Guajira were regular as is illustrated by the fact that, between June and July 1787, he made three shipments equivalent to twenty-seven loads of brazilwood. Furthermore, in his capacity as a dyewood merchant, Parodi dispatched goods valued at 145 *pesos* for international distribution of the dyestuff. It should be noted that the local price of brazilwood for that year was 5 *pesos* per load and the total value of the remittance for the two months of June and July 1787 was 556 *pesos* and 775 *reales*, equivalent to 111 loads and 355 pounds of brazilwood.

Another businessman involved in dyewood trade was Manuel Francisco de Blasco, who was based in the port of Cartagena and made several requests to the Governor between 1789 and 1791 for licenses to transport several shipments of brazilwood extracted near Valledupar that he wished to send to the Spanish port of Cádiz.[8] Blasco was one of the few recorded cases of Cádiz merchants who shipped dyewoods directly to Spain. The Blascos, along with the families of Amador, Izquierdo, de Luque, Romero Campo, Mandracha, Rendón and Espinosa, were the most prominent merchants from the Andalusian port of Cádiz, who were residing in the port of Cartagena in the second half of the eighteenth century.[9]

While the cost of transporting brazilwood from Valledupar to Riohacha was generally five *pesos* per load, the sale price in international markets was seven *pesos*, according to reports in the newspapers of Cádiz in the 1780s. Due to the proximity of this kind of timber to the port of Santa Marta, as well as its abundance, local prices were much cheaper and the trees were even harvested by soldiers in their free time, who sold them for between 4 and 6 silver *reales*. A contemporary publication noted on 1 March 1804 that brazilwood had a better price in Europe than fustic and logwood, because of the large supply of the latter dyewoods: 'There is a shortage of Santa Marta and Pernambuco woods, on the contrary, yellow wood and Campeche wood are in such abundance that their prices drop daily'.[10]

Among the entrepreneurs engaged in shipping brazilwood from Valledupar to Riohacha during the late eighteenth century, we have identified notable figures such as Pedro Santiago Molina, Andrés Galván, Julio Antonio Oñate, Ramón Acosta, Francisco Reyna, Felipe Martínez and Joaquín Bernardo Celedón. For their part, among the most prominent commercial houses at Riohacha that received dyewoods from the inland areas of the Neo-Granadian Caribbean were those of Severino Contreras, Pedro Pablo Sarmiento, Tiburcio Reyna, Juan Salvador Daza, Joaquín Pérez and Juan Suárez. Historian Fredy González has affirmed that many of these brazilwood traders belonged to 'the wealthy class of the south of La Guajira, which is made up of landowners and cattle ranchers from San Juan and Villanueva, including sons and grandsons of Europeans like the Daza, Araújo, Socarrás and Celedón families, some Sephardic Jews like the Acostas, as well as Frenchmen such as the Dangond and Lacouture families'.[11] For his part, historian Hugues Sanchez, in a study on cattle *haciendas*, herds and small farms in the district of Santa Marta between 1700 and 1810, provides a list of 21 cattle breeders who also traded in brazilwood; each had at least 1,000 head of cattle, for a total of more than 30,000 cattle in the local herds. These were located in the valleys of the Ranchería-Cesar rivers and the foothills of the Perijá mountain range, mainly in the *haciendas* and farms called *El Patillar, Badillo, El Zorro, La Jagua del Pilar* in the savannas of Maldonado and in those of *El Diluvio, Guarupal, Cimarrón, María Angola and Aguas Blancas*. The merchants Joaquín Bernardo Celedón, Juan Antonio Oñate, Joseph Agustín Parodi and Juan Salvador Daza appear in this list alongside the cattle ranchers with the surnames Pumarejo, Maestre, Bolaños, Ustariz, Baute, de Armas, Pinto and de Rosas, who were also engaged in the brazilwood trade.[12]

In the reports presented by the officials of the royal warehouses between June and July 1787, it is possible to identify a variety of actors and enterprises. Among these individuals, several women were devoted to the extractive economy of the forests, notable among them being Ana Cogollo and Junio Patricia Galván. In the case of Cogollo, she sent cargo to Pedro Pablo Sarmiento (29 loads of 150 pounds) on June 23, being the biggest dispatch of all those registered by the royal officials in this remission, since her consignment alone represented 26 per cent of the total. Junio Patricia Galván, a resident of Riohacha, received shipments from her relative Andrés Galván, underscoring how dyewood cutting and distribution (at the national and international markets) often entailed a family business network.[13] Gregoria Bolaños and Mónica Daza also appear in the reports as timber businesswomen and owners of *haciendas* and herds of more than 1,000 head of cattle.[14]

Another illustrative example of the interconnectedness in the movement of goods through household and familial bonds is demonstrated by the Reyna family. In this case, Francisco dispatched from Valledupar, while his brother Tiburcio received the goods in Riohacha.[15] Certainly, there were entrepreneurs within the timber industry who engaged in direct trade without relying on intermediaries or consignees. They harvested timber and sold it directly to other parts of the country, overseeing the transportation of the cargo themselves. Examples of such individuals include Juan Salvador Daza, Julio Antonio Oñate and Ramón Acosta. The latter, judging by the quantity of loads (15 loads valued at 75 *pesos*), emerged as the third most significant contributor in sending dyewood logs.

In the mid-nineteenth century, production and trading activities in the dyewood industry continued to be dynamic. For example, on 13 April 1848, a contract was signed between Antonio Colinas, a businessman from Valledupar, and the Dutch merchant Nicolás Danies, who lived in Riohacha. The previous year, Danies had exported 180 tons of brazilwood and 413 tons of cowhides to France from La Guajira and Cesar, yielding profits exceeding 1,000,000 guilders. By 1847, Danies was already recognized as one of the most prominent merchants of the region. He had been shipping timber since 1831 (when he moved from Curaçao to Riohacha) to the dyeing industries of France, Holland, England and the United States.[16] This business was continued by his descendants until 1930, when the markets for dyewood declined sharply.

Transportation from forests to ports

Numerous methods existed for transporting the harvested dyewoods from the forested regions of Colombia to the export hubs, facilitating their distribution in foreign markets. From the eighteenth century until the beginning of the twentieth century, the most common method to transport these dyewoods from Valledupar to Riohacha was by means of *recuas* (mule caravans) – of say a hundred mules or oxen – that would take them from the place of harvest in the forest to rivers or streams. When making a comparison between the transport of merchandise by these methods in various economic sectors – for instance in the case of coffee production – the historian Lopera Gutiérrez says the following: 'The role of the muleteers (owners of the *recuas*) and of the *arrieros* (the laborers on foot that accompany the *recuas*), have a considerable significance in all aspects related to the development of the coffee region as an economic and social unit whose differentiation has been achieved over the years'.[17]

It is certainly worth recalling that from the eighteenth century to the first half of the twentieth century the muleteer trade remained highly significant and played a substantial role not only in the dyewood economy but also in the broader Colombian economy. However, as historian Gabriel Poveda indicates, it is difficult to measure the economic impact of the activities of the muleteers.[18] Poveda also goes on to say that the *arriería* (mule driving) was part of a particular context that was composed by groups of free and independent men 'that had their roots in a society in which servile and slave institutions were in crisis' and who lived in a poor land where trade represented the greatest economic opportunity.

Due to the dyewood bonanza, both customs surveillance and control activities were increased in order that the Treasury could obtain taxes. The treasury officials implemented restrictive policies and regulations to increase revenue collection. However, in the province of Santa Marta, the trade in timber exports was exempt from taxes since they were largely free operations carried on by traders from foreign powers and were not intended for the Spanish market since they were catalogued by the royal authorities as illegal. The Spanish Navy carried out numerous seizures of ships, including schooners, sloops or brigantines, laden with dyewood. An example of official controls was the seizure made by two xebecs of the Spanish Navy, called *Liebre* and *Galgo*, that seized an English schooner carrying a large cargo of logwood on 17 April 1755 near the coast of Puerto Cabello, having departed previously from La Guajira. It additionally carried 48,071 silver *pesos* and 535 barrels of flour, as stated in a letter dated 20 April 1755 sent by the head of the coastguard squadron in Cartagena (Pedro de Rada y Aguirre) to the Viceroy of New Granada in Santa Fe de Bogotá (Don Joseph Solís Folch de Cardona). On 10 June 1755, the Viceroy authorized the utilization of the seized funds for a stipend and the consumption of wheat flour to sustain the military convoy.[19]

Seizures of dyewood shipments persisted during the eighteenth and early nineteenth centuries, as evidenced by records from 1805. In these documents, Juan de Sámano, the Governor of Riohacha, reported the apprehension of a British schooner. This vessel had been carrying a load of dyewood logs, having departed from the port and set course for the island of Jamaica.[20] However, not only English, but also Dutch and French ships were involved in illegal dyewood trade, as illustrated by the case of Félix Simón from Curaçao. In 1781, Simon, the captain of the Dutch sloop named *El Delfín*, which had been seized in Riohacha, sought the intervention of José de Osorio, an *oidor* (judge) of the Audiencia of Santa Fe. He aimed to address his case as a situation of 'arbitrary' apprehension and asked for a license to transport dyewood from La Guajira.[21]

As a result of the limitations established for the transportation of dyewoods, there were frequent complaints by merchants. An evident instance of such grievances was articulated by Manuel Joseph de Pumarejo, a cattle rancher and timber merchant from Valledupar. On 23 January 1789, he lodged a formal complaint with the Provincial Governor regarding the situation and requested an exception to transport a shipment of brazilwood from Valledupar to the royal warehouses located in Santa Marta.[22]

The numerous problems involved in transporting timber continued throughout the nineteenth century until the early 1920s, when the dyewood business declined and ceased to be profitable. During the mid-nineteenth century, challenges with transporting dyewoods persisted, given the inherent risks of the journey, substantial freight costs and overcutting. Nicolás Danies Palm, one of the most prominent foreign businessmen of Riohacha, complained and proposed the construction of a railroad between Valledupar and Riohacha with the purpose of loading brazilwood, fustic, hides and livestock, but also to offer a comfortable passenger transportation service. With the construction and commissioning of this train, plus the provision of a multimodal rail transport service (passengers and cargo), this Dutchman suggested that the movement of goods and logistics would be improved as well. In his proclamation, Danies expressed the following:

> Without the slightest exaggeration and based on the knowledge I have of the products of this beautiful country, I can say that today only the thick trunks are used, and the branches are discarded. Today 5,000 tons are exported, but soon 12,000 tons can be shipped out, and offered in the European markets with more comfort than today. Today a load from the 2nd canton and the Valley to this city [Riohacha] costs three *pesos* of freight, but with the iron road it could be lowered to twelve *reales* and even less for the other articles of lower value.[23]

Unfortunately, the project was not executed and remained merely a proposal. Danies expressed his concerns that the transportation of dyewoods was becoming more and more difficult because there was insufficient skilled labour to carry it out. In the opinion of Valencia and Agudelo this was due to: 'Muleteering constituted of a set of habits, techniques, rituals and knowledge which were necessary for the handling of the mules and oxen used for the transportation of goods [. . .], including familiarity with a wide network of bridle paths used, an activity that disappeared after the construction of railroads and the arrival of automobiles'.[24]

The routes most commonly used to transport loads of brazilwood and mulberry (*moralillo*) were those that had been in operation since colonial times, especially between Valledupar and Riohacha. The towns situated along the dyewood trade route served as hubs not only for commercial and economic interactions but also for social and cultural activities.[25] Historiography has labelled the individuals involved in this commodity routes 'businessmen on foot'. Indeed, Lopera Gutiérrez wrote: 'We see how the muleteers' roads, described in many pages of historians and travellers, are a succession of mercantile routes, oral communications, correspondences, cargoes and business of all kinds'.[26]

Writings and other accounts

In 1737 the Royal Spanish Academy published a dictionary of the Spanish language that included a description of *palo de brasil* (brazilwood), referring specifically to a kind of tree with 'very heavy wood' that had the colour of an intensely red 'ember', which – when chopped into 'pieces' and boiled in water – could be used to dye wool, cloth and other things.[27] A few decades later, in 1776, the Spanish monarch issued explicit instructions to colonial authorities, urging them to report any discovery of natural resources within their domains. These resources were to be made available to the Royal Cabinet of Natural History (*Real Gabinete de Historia Natural*) – an institution established by the Royal Court for the advancement of public instruction. The king explicitly referenced brazilwood, along with other plants and medicinal species deemed advantageous for both humans and animals. The text also elucidated the specific plant-based or natural materials necessary for the dyeing process, either as additives or as mordants.[28]

Writings of the time also made clear that the exportation of dyewoods from the coasts of the Santa Marta province during the latter half of the eighteenth century necessitated ongoing vigilance from colonial civil, military and even ecclesiastical authorities. This was expressed by Father Antonio Julian in his book *The Pearl of America* (1787):

> Some will think that this *palo* is called *palo del Brasil* because it is only found in Brazil, and from there it comes to Europe. There is no other reason for this than that it was discovered first in Brazil, rather than in any other part of America, because Amerigo Vespucci discovered the coast of Brazil before Christopher Columbus discovered the Tierra Firme and the province of Santa Marta. For the rest, as it is called of 'Brazil', it can also be called *Palo de Santa Marta*, being found in abundance in this Province. . . . And because, as everyone knows, *Brazil* (which is what all dyers call the wood) is a considerable line of trade . . . I want to expand more undoubtedly on the matter, and show its utility for Spanish trade, and point to the places where it is found, the easy way of transporting it to the ports, as well as the carelessness or inaction of the Spanish merchants in not procuring it for their own interest and that of the (Spanish) Nation, in contrast to the shameless activities of the foreigners who take it to Curaçao, Jamaica, Holland and England, and to other ports in the North.[29]

In an article of 1773, titled *A General History of Travels* in the contemporary Gazette *Diario noticioso, curioso, erudito y comercial público y económico*, published in Madrid, there is a brief description of how brazilwood was manufactured once the wood logs arrived in Europe, especially in the case of the Dutch dye industry.[30] There, from the sixteenth century, beggars and young people who were held in juvenile correctional institutions came to be employed to saw the logs. The resulting sawdust was packed in sacks and subsequently taken to boilers to be cooked at high temperatures. With those boiled logs, the workers obtained the dye water that was used to colour fabrics in red, black or khaki, depending on the wood sawed. These juvenile detention centres employed punishment and compulsory labour to instil discipline and shape character through productive activities, with the most notable one being the 'Rasphuis' of Amsterdam (see Figure 7.3).

In 1767 the gazette *Semanario Económico* pointed out that brazilwood was an excellent dye that 'never spoils'. In this publication, the procedure to obtain the dye from this liquid was detailed as follows: 'Mix one ounce of brazilwood, one pound of common water, half an ounce of fish tail and two ounces of alum rock'. All the ingredients were infused in water for three days, and subsequently boiled in a glazed pot while being kept covered. The boiling process continued until the water portion was reduced by half, resulting in an excellent brazilwood water for dyeing various

Figure 7.3 Illustration of the Amsterdam *Rasphuis* (1662). *Source*: Melchior Fokkens, 'Beschrijvinge der wijdt-vermaarde Koop-stadt Amstelredam' (Amsterdam, 1662).

fabrics and threads.[31] The water infused with dye could be stored for several years, serving not only for dyeing purposes but also for preserving other items, including specific astringents or natural remedies.

Beyond brazilwood: Other dyewood trees

Within the forests of the Colombian Caribbean, a variety of dyewoods extended beyond brazilwood, encompassing *palo de mora* (also known as fustic, *moralillo, or moralete*) and logwood (referred to as *palo de Campeche* or *palo de Guatemala*). The latter resource was present in the Panama region, which was part of the Viceroyalty of Nueva Granada (later known as Colombia) during the colonial period. Certainly, vast expanses of logwood were also discovered within the woodlands of coastal Panama. However, even greater quantities were located in the jungles and tropical dry forests of Yucatan, Chiapas, Guatemala, Nicaragua and, indeed, across Central America and southern Mexico.[32]

Conflicts with the British concerning the exploitation of Central American dyewoods had been ongoing since early times. In 1646, the English requested the *Casa de Contratación* in Seville for permission to engage in dyestuff commerce within the region. This request even encompassed the facilitation of distribution across the Asian

Pacific, including the Philippines. In this request, Panama played a major role, since Panama City, together with Callao, Guayaquil and Acapulco, were the main port cities of the Pacific trades. The enquiry initiated by the merchants of the British East India Company to the Council of the Indies concerning maritime traffic aimed to prompt the body to assess their economic proposition. However, the mercantile Consulate of Seville (*Consulado de Sevilla*) compiled a private report that did not support this business endeavour. The report was influenced by concerns regarding potential conflicts with the interests of the *Casa de Contratación*, particularly pertaining to navigation by English merchants in Spanish dominions.[33]

The response from the Council of the Indies to the British petition was unfavourable, as it contended that the proposal posed a risk to the legal and economic security of the *Casa de Contratación*. The *Casa de Contratación* presented several arguments. The first was that on 22 April 1617, the Basque commander of the Spanish Navy, Sancho de Urdanibia, received two letters signed by Íñigo de Cárdenas, Spanish ambassador in France. These letters informed him of the aim of the French to trade in the West Indies with the British trading house by virtue of the peace treaty as well as to appropriate booty obtained by the pirates, as a result of which the ambassador affirmed that it was necessary to adopt means to stop them, and that Juan de Ciriza y Balanza should inform the Duke of Monteleon.[34] Another negative report came by way of the *Junta de Guerra de Indias* (the high-level military council) that informed the Madrid government on 10 November 1611. The latter report warned of war preparations in Holland against the Philippines, and argued that until this political complication was resolved, it was unwise to proceed with the signing of an agreement that ratified this commerce in dyestuffs because the office of the Governor of the Philippines had not yet pronounced judgement on the matter. Therefore, as these cases were yet to be settled, the report indicated that it was not convenient for the Philippines to trade with the English company. Nonetheless, after years of discussion, the Spanish authorities granted permission in December 1667 for commercial activity by foreigners with the Philippines and the establishment of a state monopoly of logwood. In any case, it was stipulated that a record and control was to be kept in the Secretariat of Peru of all logwoods exported as well as of any commodity of British origin passing through the ports of Acapulco and Panama City.[35]

Historical documentation reveals that state monopolies, particularly for logwood, brazilwood, and *palo de mora*, were instituted in Santa Marta and Riohacha in the year 1786.[36] Curiously enough, in the case of *moralillo*, the increase of its harvest and export came after the establishment of the Republic of Colombia in 1820. Merchants from England sought this dyewood for its beige colour, ideal for dyeing military uniforms and civilian clothing in khaki colour. All the regiments of the British Empire around the world during the nineteenth and early twentieth centuries wore khaki, in various shades. According to Gallego and Sanza: 'in British India the colour khaki for uniforms was adopted in 1848 and was worn by both native and British soldiers; later it was adopted in other parts of the Empire'.[37] Subsequently, it transitioned from the military realm to civilian use, sparking a distinctive and widespread fashion trend.[38]

In nineteenth-century Colombia, the initial mentions of *palo de mora* harvesting emerged following the ratification of international free trade agreements in 1825 with the United States, in 1826 with England and in 1828 with France and the Netherlands.[39] Following the ratification of these agreements, the export of dyewoods increased, especially by British and Dutch merchants. However, in 1827, a year after the government of *Gran Colombia* had signed a Free Trade Treaty with the Netherlands, the Colombian navy seized a ship carrying a large load of *palo de mora* that was illegally leaving the port of Sabanilla. In addition to the logs, the ship carried tobacco, sugar and hides, which were being sent to Curaçao. The customs authorities directed the vessel to Cartagena, requiring the cargo owners and ship's crew to provide explanations regarding the undisclosed merchandise.[40]

Requests to export dyewoods from the different ports of Caribbean Colombia were constant. An example was the request made to the Department of Magdalena by the merchant Juan Glen who sought a permit to export 10,000 quintals of cotton and 5,000 loads of dyewood logs through the port of Sabanilla. More intriguing was the petition by José Izquierdo in 1823, who, as captain of a frigate, implored the head of state, Francisco de Paula Santander, to proclaim that the spoils of private corsairs should not be sold in the public auction house, nor that the trade of brazilwood with foreign powers be encouraged.[41]

Market structure, prices and European demand

The merchants engaged in the trade of American dyewoods faced numerous difficulties, including fiscal levies as well as fluctuations in demand, caused by variations in prices and exchange rates. The merchants in Spain practically never agreed upon all particulars of the goods and conditions of measurement; therefore, they are not constant. The dyestuffs were sold at different prices, and even different measures of volume, which sometimes could be determined by the 'eye' of the buyer rather than by a strict measure of weight. Similar problems arose with the monetary units. In the port of Cádiz merchants dealt in silver *reales*, *maravedíes* and silver *pesos*, while in La Coruña operations were reported in terms of the monetary standard of the *real vellón*. The conversion of the latter was 1 *real* of silver equal to 2.5 reales de *vellón*.

We have compiled a time series illustrating the trends in average annual prices (denominated in silver *reales*) for select American commodities traded in Spain, primarily within Madrid, from January 1793 to December 1808.[42] The creation of this time series involved various monetary conversions aimed at achieving data uniformity. Upon data collection, measurement standardization was implemented. For instance, commodities such as logwood, brazilwood, fustic and *achiote*, initially recorded in quintals, underwent conversion, while cochineal, quantified in *arrobas*, underwent a similar transformation. Notably, based on this data, the highest-priced indigo originated from Guatemala, followed by that from Caracas.

The price statistics show that cochineal as a natural dye was the most expensive product in the market, reaching its highest peak in 1800 at a valuation of 748 silver *reales* per *arroba* (25 pounds of the dyestuff), although the price fell a bit, to 730 *reales* in 1801. In contrast, the years with the lowest values were 1807 and 1797, being registered at 102 and 178 silver *reales*, respectively. The prices of indigo and *palo de mora* were very similar. The cheapest values of the former fluctuated between 22 and 24 silver *reales* per quintal (100 pounds). The average price of the mulberry tree in this period came to 55 *reales de plata* per quintal, while that of indigo was 38. At the same time, the average prices of logwood and brazilwood were respectively 73 and 161 silver *reales* per quintal. Nonetheless, the prices of dyestuffs tended to fluctuate quite notably. In 1806 and 1807 the price of most natural dyes maintained a downward trend and, by 1811, the *Diario de Madrid* reported prices of 7 silver *pesos* for brazilwood and 126 silver *pesos* for cochineal.[43]

Conclusions

One of the most representative economic activities of colonial and republican life in the Colombian Caribbean was the cutting of timber rich in dyes that was exported for subsequent use in processes of textile manufacturing in Europe. As a result, many regional landowners and merchants dedicated themselves to the exploitation of the forests as a profitable economic activity during the seventeenth to the late nineteenth centuries, generating mercantile and business networks and changing mentalities at the local and regional level, as well as playing a significant part in different sectors of domestic economic activity.

Transporting dyewoods across extensive distances over days or even weeks led to the enrichment of numerous individuals engaged in this trade, capitalizing on the ample forest resources. There were dyewood entrepreneurs in the forest production areas and along the trade routes, including Valledupar, La Guajira, the port corridors of Cartagena, Santa Marta and Riohacha, as well as their hinterlands. Frequently enough, the mere ownership of four or five mules or oxen for transporting logs enabled them to accumulate sufficient funds to purchase land, venture into cattle ranching, or amass sizable mule herds. The dyewood merchants endeavoured to expedite their journeys by employing mule caravans on roads, trails and challenging terrains, while also harnessing the utilization of waterways.

The objective of the merchants and muleteers in the region near Riohacha was to load their cargoes on ships navigating the river Ranchería or, in the case of other ports, on ships that travelled on the Magdalena River. From that point, the cargoes embarked on a journey to the sea, where they were exported to Europe through the islands of Curaçao and Jamaica, or via Surinam. Alternatively, they could be shipped directly to the United States, Germany, France, Italy and Holland. These same cargoes of logwood and vegetable dyes were complemented by export of other commodities, including goat and beef hides, wood for the construction of houses, furniture and boats and, of course, shipments of coffee from the Sierra Nevada de Santa Marta and the highlands of southern La Guajira. The same transport system was employed in other timber-producing centres across the Colombian Caribbean (Cauca, Magdalena, Aracataca, Fundación, Don Diego, Tucurinca, etc.). In any case, it was difficult and risky to transport brazilwood, not due to the cargo's quantity or quality, but rather due to the substantial freight expenses and the intricate nature of the journeys themselves. Since the early seventeenth century, dyewood merchants repeatedly voiced their grievances regarding limitations and business obstacles, encompassing inadequate infrastructure, rudimentary logistics and the considerable distances to ports. Despite their pleas for support, success in this regard remained elusive.

Rural communities functioned as the pivotal link in the productive chain of natural dyewoods originating in colonial Colombia, ultimately destined for sale in Europe. These communities undertook the tree cutting and assumed the transportation of loads to regional intermediaries and merchants. Locals were actively involved in the extraction and transportation of dyewoods from La Guajira, the areas around Valledupar and Chriguaná, as well as the Cispatá. These valuable resources would ultimately find their way to the dyestuff and textile industries in Germany, France, Italy, England and notably Holland. But dyewoods were also in demand among painters. For instance, scientific research on the materials used by Van Gogh in his drawings indicates that he used used ink from imported logwood, some which probably came from Curaçao.[44] An example of the frequent shipments of Colombian logwood trade can be found in customs registries. For example, between 1840 and 1860, three ships per week departed from Riohacha, carrying loaded dyewoods bound for Curaçao. Additionally, single vessels set sail for London, Marseille and Rotterdam. Meanwhile, Santa Marta saw two ships depart for Curaçao on a weekly basis.

In the case of the trees harvested in the coastal areas or in the hills of the port of Santa Marta, cutting and loading costs were minimal due to short distances. As a result, transportation expenses remained quite low, greatly benefitting buyers. This meant that in international markets, Santa Marta had a comparative advantage compared to Riohacha, although the latter town also had some advantages since it had fewer administrative and customs controls which could favoured smuggling. In the second half of the eighteenth century, Antonio Julián had noted the diverse difficulties encountered by timber merchants in the neighbouring regions of Valledupar when exporting dyewoods through Riohacha to the coast. He observed that despite the wealth of the forests in the province of Santa Marta, there were only a few individuals that truly benefitted from the dyewood business:

> The transport of the *Brazil* that is cut near the port and city of Santa Marta usually is covered by the buyer, being so close to the points of shipment. But this is not so in Ríohacha where there comes most of the *Brazil* harvested in the mountain town of Moreno and in the valley of Upar. Passing through these mountains I sometimes met some poor men who came with their horses loaded with *Palo del Brasil*, and were taking it down to the Hacha river to sell it to the ships, which regularly await these cargoes. As there is an abundance of horses in most of the province, and the roads are flat and comfortable, transportation is easy, and thus with little work and less care a poor man, who is given to this trade, can support his family. However, despite the abundance of this *Palo* in those lands, there are very few, respectively, who take advantage of it.[45]

This text, which describes so vividly the nature of the local transport of dyewoods during the eighteenth century in a rural province of the Spanish Empire in northern South America, is suggestive of the richness of contemporary chronicles that can prove of interest for future historians who wish to deepen our knowledge of the diverse origins and trajectories of natural American dyes that must be understood in the light of local geographies, societies and economies.

Notes

1 Marcela Escandón Vega, 'Frustración y abandono versus virtud y riquezas. El drama de la "Perla de América" en el siglo XVIII', in *Santa Marta en el siglo XVIII*, ed. Edgar Rey Sinning, Vol. I (Santa Marta: Grupo editorial Cajamag, 2017), 177–210; Joaquín Viloria De la Hoz, 'Economía de la Provincia de Santa Marta en el siglo XVIII: Entre el estancamiento y el contrabando', in *Santa Marta en el siglo XVIII*, ed. Edgar Rey Sinning, Vol. I (Santa Marta: Grupo editorial Cajamag, 2017), 45–74; Jorge Enrique Elías-Caro: 'La Provincia de Santa Marta en el siglo XVIII: Un análisis cuantitativo-cualitativo de su historiografía', in *Santa Marta en el siglo XVIII*, ed. Edgar Rey Sinning, Vol. I (Santa Marta: Grupo editorial Cajamag, 2017), 75–156; And by the same author 'Las perlas de "La perla de América" en la colonia tardía: economía y sociedad', in *Santa Marta en el siglo XVIII*, ed. Edgar Rey Sinning, Vol. II (Santa Marta: Grupo editorial Cajamag, 2018), 209–64.

2 Antonio Julián, *La perla de la América, provincia de Santa Marta reconocida, observada y expuesta en discursos históricos por el sacerdote don Antonio Julián* (Madrid: Don Antonio de Sancha, 1787), 4–5.

3 Antonio Sánchez Valverde, *Idea del valor de la Isla Española, y utilidades, que de ella puede sacar su Monarquía* (Madrid: Imprenta de Don Pedro Marín, 1785), 114.

4 Sánchez Valverde, *Idea del valor*, 114.

5 'La semilla del árbol de la tela', *El Correo mercantil de España y sus Indias* 86 (28 October 1802): 681–8.

6 Camilo Alfonso Torres Barragán, 'Un rojo fugitivo:extracción, circulación y consumo del palo de brasil de la Nueva Granada (1770-1850)' masters thesis, Universidad de la Andes, 2013, 102–119.

7 Antonio Julián, *La Perla de América*, 301.

8 Archivo General de la Nación, (Bogotá, Colombia). 'Solicitud de embarque de palo de tinte por Cartagena entre 1789–1791'. Colonial Section. Legajo Aduanas tomo 15. Title and call sign Aduanas: SC.2,8, D.15, folios 543–614.

9 Adolfo Meisel Roca, 'Entre Cádiz y Cartagena de Indias: la red familiar de los Amador, del comercio a la lucha por la independencia americana', in *Cuadernos de Historia Económica y Empresarial* 12 (Cartagena: Banco de la República, 2004), 62.

10 'Comercio', *El Correo Mercantil de España y sus Indias* 137 (1 March 1804): 137–44.

11 Fredy González Zubiría, *La odisea del acordeón. Historia social de la música vallenata* (Riohacha: Universidad de La Guajira, 2017), 26.

12 Hugues Sánchez Mejía, 'Haciendas de trapiche, hatos, hatillos y "rozas": el mundo rural en la gobernación de Santa Marta (1700–1810)', in *Historia Caribe* 11:28 (Colombia: Universidad del Atlántico, 2018), 264–5.

13 Archivo General de la Nación de Colombia, (AGN). 'Abonos del Oficial Real del Almacén de Riohacha desde el 2 de junio al 28 de julio de 1787'. Relación de cargas de palo de brasil remitida por Gervasio Asterrera a Apolinar de Torres el 17 de septiembre de 1788. Fondo Colonia, tomo 4, 'Estadísticas'.

14 Hugues Sánchez, 'Haciendas de trapiche, hatos', 264–5.

15 See source cited in Table 1, which refers to 'Nine cargoes valued at 45 pesos remitted on June 22, 1787'.

16 Fredy González Zubiría, *Emigrantes holandeses de curazao a riohacha en el siglo xix: historias de vida y genealogía de Danies – Pinedo – Weeber* (Barranquilla, Colombia: Tonos Editorial del Caribe, 2011), 15–16.

17 Jaime Lopera Gutiérrez, *Arriería: los empresarios a pie* (Quindío: Academia de Historia del Quindío, 2017), 1.

18 Gabriel Poveda Ramos, *Minas y mineros de Antioquia* (Bogotá: Banco de la República, 1981), 45.

19 Archivo General de la Nación, Bogotá (Colombia). 'Carta de Pedro de Rada dirigida a Don Joseph Solís Folch de Cardona en Cartagena de Indias indicando el decomiso de goleta inglesa carga de palo de brasil de abril 20 de 1755'. Sección Colonia. Legajo Aduanas tomo 13.,Título y signatura 5 – Aduanas: SC.2, 13, D.5, folios 51–52.

20 Archivo General de la Nación, Bogotá (Colombia). 'Carta de Pedro de Rada dirigida a Don Joseph Solís Folch de Cardona en Cartagena de Indias indicando el decomiso de goleta inglesa carga de palo de brasil de abril 20 de 1755'. Sección Colonia. Legajo Aduanas tomo 13, Título y signatura 5-Aduanas: SC.2, 15, D.15, folios 959–964.

21 Archivo General de la Nación, Bogotá (Colombia). 'Carta de Pedro de Rada dirigida a Don Joseph Solís Folch de Cardona en Cartagena de Indias indicando el decomiso de goleta inglesa carga de palo de brasil de abril 20 de 1755'. Sección Colonia. Legajo Aduanas tomo 13, Título y signatura 5-Aduanas: SC.2, 13, D.42, folios 625–637.

22 Archivo General de la Nación. Bogotá (Colombia). 'Carta de solicitud de auxilio por transporte de palo de brasil de Valledupar a los almacenes reales de Santa Marta de Manuel de Pumarejo al gobernador de la Provincia'. Sección Colonia. Milicias y Marina: SC.37 CO.AGN.SC.37.76. 108, Valledupar, enero 23 de 1789, folios 532–534.

23 Nicolás Danies, *Proclama camino férreo Riohacha-Valledupar* (Cartagena: Imprenta de Eduardo Hernández, 1846), 8–9.

24 Gonzalo Valencia and Fabio Agudelo, *Banca y economía en Armenia y el Quindío, 1929–2002* (Bogotá: Banco de la República, 2002), 37–8. Also see Carrillo Ferreira, *La comunidad mercantil de Riohacha: historias de comerciantes del Caribe en las fronteras de La Guajira, 1840–1870* (Buenos Aires: University of Buenos Aires, 2017). And his essay, '"Una frontera en movimiento". La conformación del puerto de Riohacha en el siglo diecinueve', in *Études caribéennes* (2018).

25 González Zubiría, *La odisea del acordeón*, 28.

26 Lopera Gutiérrez, *Arriería*, 1–2.

27 Real Academia Española, *Diccionario de la lengua castellana en que se explica el verdadero sentido de las voces, su naturaleza y calidad, con las phrases o modos de hablar, los proverbios o rephranes, y otras cosas convenientes al uso de la lengua, dedicado al rey nuestro señor don Phelippe V (que Dios guarde) a cuyas reales expensas se hace esta obra* V (Madrid: Imprenta de la Real Academia Española/ Francisco del Hierro, 1726), 95–6.

28 Biblioteca Nacional de España, Documento de 1776 sobre instrucción del rey, 18.

29 Antonio Julián, *La perla de la América*, 107–10.

30 'Historia General de los Viages', *Diario noticioso, curioso, erudito y comercial público y económico*, 18 January 1773.

31 'Una excelente agua de palo de brasil que nunca se echa a perder', in *Semanario Económico* 33 (29 January 1767).

32 Pascale Villegas and Rosa Torras, 'La extracción y exportación del palo de tinte a manos de colonos extranjeros: El caso de la B. Anizan y Cía', *Secuencia* 90 (2014): 77–93.
33 Archivo General de Indias (AGI), Seville (Spain) 'Consulta sobre propuesta de la Compañía Inglesa de la India Oriental'. Signatura Filipinas, 2, N.73, reference code ES.41091.AGI/23.6.7//Filipinas,2, N. Madrid, 23 August 1646.
34 The Marquis of Montesajo, native of Pamplona (Navarra) was a Knight of Santiago, Secretary of State and advisor to the Council of War. Héctor de Pignatelli y Colonna, was a nobleman and ruler of Italian origin, who held the position of viceroy of Catalonia.
35 AGI, 'Consulta sobre propuesta de la Compañía Inglesa de la India Oriental', Madrid, 23 August 1646. Section Filipinas, 2, N.73, and reference code ES.41091.AGI/23.6.7// Filipinas, 2, N.73.
36 Archivo General de la Nación. Bogotá (Colombia). 'Palo de tinte: se estanca en las provincias de Santa Marta y Riohacha'. Sección Colonia. Milicias y Marina: SC.37: CO.AGN.SC.37.131.30, 1786, folios 236–247.
37 Rosa Gallego and Juan Carlos Sanz, *Diccionario Akal del color* (Madrid: Akal, 2001), 12–13.
38 Jorge Enrique Elías-Caro, 'Los colores de la moda en el mundo occidental. Economía extractiva y comercialización internacional de tintes naturales: del Renacimiento a la Bella Época', in *Memorias del XXIII Coloquio de Historia Canario-Americana* (España: Cabildo Insular de Gran Canaria, 2018). See digital edition at https://dialnet.unirioja .es/servlet/libro?codigo=764904.
39 Jorge Enrique Elías-Caro, 'Las convenciones de comercio, navegación y relaciones internacionales firmadas por Colombia con Estados Unidos y Reino Unido en 1825: al rescate de su memoria, a propósito del TLC', in *Jangwa Pana* 8 (Colombia: Editorial UniMagdalena, 2009), 213–32. See also Jorge Enrique Elías-Caro, 'La Gran Colombia' y los tratados de libre comercio suscritos con Inglaterra, Estados Unidos, Francia y Holanda (1825–1828). una relación centenaria de dependencia económica'. Lecture given at the *XI International Meeting of the Association of Latin American and Caribbean Historians* (Vienna: University of Vienna/ADHILAC, 18 September, 2014).
40 General Archive of the Nation. Bogotá (Colombia). 'Expediente y diligencias que se efectuaron en la aprehensión de un contrabando'. Cartagena 1827. Republic Section: Contrabandos: SR.29, 1, D.19, folios 584–635.
41 General Archive of the Nation. Bogotá (Colombia). 'Solicitudes del Departamento del Magdalena'. Santa Marta 1823. Republic Section: Petitions-Solicit: SR.75, 4, D.31, folios 662–733.
42 Author's calculations based on monthly pricing reports in *El Comercio Mercantil de España y sus Indias*, 1793–1808.
43 *Diario de Madrid* 218 (Spain, August and September 1811). The journal published information related to the purchase and sale of products, negotiations, services and tools in Spain and Spanish America.
44 The REVIGO research project of the Van Gogh Museum and Dutch universities has found that the painter used ink from logwood (imported from Central America or Colombia) for his drawings. See the report titled 'Original Colours of Van Gogh's Drawings' online at the Van Gogh Museum site.
45 Antonio Julián, *La perla de la América*, 107–10.

8

Indigo in eighteenth-century Venezuela

An unfinished history

Frédérique Langue[1]

This chapter provides a comprehensive examination of a relatively understudied aspect of late colonial Venezuela: its indigo economy. The brilliant trajectory of *indigofera tintorea* from its introduction in the Captaincy General by the Basque Antonio de Arbide to its decline on the eve of Independence is the subject of this study. The chapter not only delves into the economic elements of indigo production and trade – such as crop profitability, export quantities and market dynamics – but also examines the social dimension, which involves analysing the functioning of indigo haciendas and other property types.

In the Hispanic New World, indigo (also known as *añil* in Spanish) was highly valued because of the blue dye that could be extracted from this plant. Well known and cultivated on a broad scale in the southern provinces of early colonial Mexico, where dyes were the object of active trade since prehispanic times, the 'pastel' or *xiquilite* continued its expansion in Central America during the seventeenth century and especially in the eighteenth century. Guatemala and the districts of El Salvador soon became centres of the indigo agro-industry, which was based on the *haciendas* and plantations specialized in the cultivation of this plant, next to which there sprung up textile and dye workshops (*obrajes*). In the eighteenth century several texts were dedicated to the description of this dye plant as well as enacting a series of official regulations that were ratified in 1782 and 1784, in the case of Guatemala, aimed to promote indigo production on a large scale.

Colonial Venezuela did not truly experience a commercial boom in indigo until the last third of the eighteenth century. Therefore, a fundamental question arises regarding the chronology of the diffusion of this valuable commercial plant, little mentioned in the area of *Tierra Firme* before this date. The existence of local varieties of indigo in various regions of the Americas was mentioned in numerous travellers' accounts throughout the seventeenth century. Moreover, the official documents of the Spanish colonial regime indicate that the production and use of this plant was regulated in a very precise manner, as confirmed by the *Ordenanzas de Mérida* (1620–1) concerning the productive activities of the *encomenderos* of the Andean region. Subsequent reports from the second half of the eighteenth century indicate that this plant grew 'wild' in the vicinity of the villages in most of the regions that made up the *Capitanía General* (Captaincy General) of Venezuela. However, it was not yet exploited for commercial purposes, in contrast to other regions of Spanish America. In the best of cases, it was used to dye the cotton threads used to make hammocks or *chinchorros*, and also for the clothing of the poor, as was the case in the towns of San Cristóbal and Lobatera in the Andean region of Mérida.[2]

Origins and actors of indigo production in Venezuela

It was not until the 1770s that the commercial cultivation of indigo truly began to gain strength in Venezuela's colonial economy, initially to supply some manufactures and mills, but subsequently for export. It is estimated that already by 1772, the indigo crop was competing with cocoa in the valleys of Aragua, located near the city of Caracas. Two years later, the first exports from La Guaira were recorded. The latter has been the source of confusion regarding the initial introduction of indigo in Venezuela in 1774 (explicitly mentioned in these exact terms in the minutes of the Consulate of Commerce).

After the failed attempt of José de Aponte to introduce it in Venezuela, the Guipuzcoan Antonio de Arbide imported from Guatemala a new indigo variety. Together with the priest and fellow countryman Pablo de Orendaín, he began to plant it in the region of Maracay. It was really from that moment when the referred cultivation began to develop in the province, benefitting from the 'protectionist' measures ratified by the Governor of the Captaincy General of Venezuela, José Carlos de Agüero, and promoted by Spanish emigrants, especially Basque merchants and plantation owners. At this time, the wild indigo was replaced by the new species, which soon prospered. Attempts were also made to acclimatize it in other regions, such as on the shores of Lake Valencia, an attempt coordinated by Alonso Tomás Escobar on a property owned by Tomás de Sandoval. In spite of the success achieved by these individuals, it was in the property 'Agua de Obispos', located in Nirgua (Yaracuy), and belonging to the military official (*maestro de campo*) Blas Ambrosio Jiménez, where the most favourable natural conditions for the cultivation of indigo were found, in particular the appropriate soils and the proximity of rivers to drive the mills in which the dye was processed.[3]

It is of considerable interest to emphasize the role played by the Basque businessman Antonio de Arbide in the commercialization of this crop. Having settled beforehand in New Spain, Arbide arrived in Venezuela around 1768, as an agent of the *Compañía Guipuzcoana*, a Spanish mercantile monopoly which had rights to control commerce in Venezuela and aimed to organize the cultivation and trade of indigo on a larger scale. At the request of Manuel de Clemente y Francia, Arbide sent his brother Manuel to Guatemala so that he could bring the indigo seeds to Venezuela. In his *Viaje a la Parte Oriental de Tierra Firme* (Journey to the Eastern Part of the Mainland) published in Paris in 1806, François Depons points out that it was in 1774 when the priest Don Pablo de Orendaín and Don Antonio de Arbide began to cultivate indigo to the surprise of their neighbours, who had tried in vain to encourage the cultivation of its wild variety on a large scale to dye fabrics, blankets and cotton hammocks.[4]

Arbide established himself on his own, bought land, leased other properties, and became the first producer in the Aragua valleys in the decade of the 1770s. Shortly thereafter, in 1784, General Francisco de Saavedra, who had been appointed military head of Venezuela, decided to regulate the commercialization of this product and to improve the quality of the products destined for export. Arbide was named 'overseer' (*veedor*) for the Maracay region, although he also operated as a merchant, selling imported goods in that city. After having successfully introduced indigo cultivation in Venezuela, Arbide requested 'permission to introduce foreign goods of lawful commerce up to 100,000 pesos and to extract the amount in fruits for the friendly colonies', a request that was denied by the Spanish Crown. He benefitted, however, from the support of the governor-captain general, who proposed that he be granted a pension from the Royal Treasury of 500 pesos per year (to be paid, in the event of the death, to his widow) by Royal Order of 2 October 1805, and guaranteed by the *Fondo de Avería* of the colonial mercantile association of Caracas known as the *Consulado de Comercio*.[5]

The golden age of indigo and its commercialization

The considerable extension of the zones of cultivation of indigo, from Aragua to the southwest (Barinas), as well as the opening up of Atlantic markets, led the harvesters (*cosecheros*) and local authorities to request measures to protect production and thus encourage its export to Europe, where the demand for dyestuffs was expanding. Price increases and tax exemptions were the main measures desired by both parties. An example of such requests was presented by José de Amenabar, a *factor* (an important fiscal agent, in charge of the *Compañía Guipuzcoana*), to Governor José Carlos de Agüero in January 1774. In this document, Amenabar highlighted the possibility of obtaining new income, especially in Europe, through the production and commercialization of indigo, described as 'the most profitable industry of this province'. In February of the same year, the Governor personally informed the Spanish Crown, as he also did the following year. He even enclosed samples of indigo in his official reports (*representaciones*) so that the quality of the plant could be verified. He especially requested that the Crown issue rules for the proper functioning of this new branch of the Venezuelan economy. In a letter dated 16 October 1776, he emphasized the advancements in the cultivation of indigo in the valleys of Aragua. However, he also pointed out growing discontent due to the refusal of the directors of the *Compañía Guipuzcoana* (a privileged mercantile firm) to raise the standard price and their reluctance to buy the harvest at the most convenient time for the plantation owners, which provoked a decrease in interest among the producers in the commercialization of this product.[6]

The creation of the *Intendencia* (intendancy) of Venezuela in 1776, a new administrative structure which might have contributed to the stimulation of local agriculture, did not immediately bring advantages. On the other hand, it did have the more immediate consequence of regulating and then prohibiting the indigo trade with foreign colonies. At the same time, the head of the *Intendencia* of Venezuela, José de Abalos, warned the Spanish Crown, in a letter dated 4 April 1777, that indigo was not being given the importance it deserved. A Royal Order of 26 April in that year, sent to Luis Unzaga y Amézaga (a high-ranking royal functionary who was nominally governor of Venezuela, Cuba and Louisiana), confirmed that the Venezuelan indigo was of a quality comparable to the most famous variety of Guatemala indigo, an assertion that had been defended since 1774 by the Caracas City Council. In the same document, the Crown reiterated its support to the harvesters and pointed out that the factors of the *Compañía Guipuzcoana* were to buy the quantities of indigo brought to them by the *labradores* (small farmers) and plantation owners at a fixed price. Likewise, by Royal Order of 13 April 1777, with the purpose of promoting the cultivation and commercialization of the plant, the extraction rights of the dyestuffs produced in Venezuela were reduced by half, and the sale in foreign countries was authorized. According to these instructions, each year, the price would be fixed according to an agreement between the producers and traders, under the supervision of a board of overseers who would certify the quality of the products. The board soon assembled for this purpose and ruled that there were four kinds of indigo. On 28 June 1777, a Royal Order recommended the following to the governor of Venezuela:

> Efforts shoud be made to advance in this Province, when possible, its planting and harvest, making those dedicated to its cultivation to understand the profits that it promises because of the high price at which the good indigo is sold in these Kingdoms, and the amounts of the surplus that His Majesty will allow for sale to foreign countries.[7]

In the same document, further information was requested regarding the initial cost, the expenses incurred for the production of this crop, and the timing of the duties to be collected on the occasion of their remittance to Spain.

When the ratification of free trade in Spanish America took place in 1778 (although it should be noted that the decree of free trade would not be extended to Venezuela until 1789), the *Compañía Guipuzcoana* – which had a virtual monopoly over large sectors of Venezuelan trade – decided to authorize the sale of indigo in view of the previous success experienced in expanding local indigo production. It also eliminated the mediation of the supervisory board in charge of fixing prices in spite of the royal decree issued the previous year in this regard. This provision was confirmed by another Royal Decree of 20 October 1778. Henceforth, indigo prices would be regulated on the basis of the existing supply and the corresponding demand. Furthermore, in order to encourage production this natural dyestuff was exempted from taxes when leaving Venezuela and entering Spain, as long as it met a minimum quality requirement. As of 1786, indigo remained free from export duties.

As early as 1780, Intendant Abalos had expressed concern about the level reached by prices in Venezuela. Consequently, he granted permission to export indigo to Curaçao, opening up the potential for trading it in exchange for Dutch commodities. This initiative extended beyond mere local trade, involving commercial exchanges with mainland Spain, even during periods of intricate diplomatic relations. Indeed, trade with Curaçao was not limited to the simple transit of indigo since Abalos also authorized the barter of cocoa, indigo, tobacco and mules for flour and other Dutch commodities.[8] Intendente Saavedra estimated that the price of a pound of primer indigo leaf had fallen from 13 to 7 *reales*, despite the fact that the cost of production was close to 10 *reales* and that it was sold for 21 *reales* in the Iberian Peninsula.

The promotion of indigo cultivation in the commercial centre of Maracay, as well as in Turmero and La Victoria, not only contributed to the transformation of these localities but also incited conflicts with sugarcane producers and cattle ranchers (*ganaderos*). Antonio de Arbide himself had to bear the consequences of these rivalries in 1780, when he was demanded to pay some 50,000 pesos because of the damage that would be caused to public health, livestock and other crops by dumping the water used for indigo processing in the irrigation ditches. Although his opponents did not manage to win this lawsuit, the number of claims in this regard began to multiply thereafter. Landowners such as Andres Fuenmayor, Pedro Francisco Jiménez, Miguel Alvarez and Manuel Macero led the objections. On the other hand, indigo producers complained about the lack of labour when other crops needed it or when it was requisitioned for public works in La Guaira or for the local Veteran's battalion. The landowners unanimously denounced this forced hiring, invoking the decline of their indigo plantations and requesting the intervention of the Spanish Crown. In 1788, they asked the Intendant to receive commodities (cocoa and indigo) in payment for slaves (*negros*) at the established prices at that time (12 *reales* for a pound of indigo of the best quality).[9]

Similar to tobacco – an important Venezuelan crop – indigo predominantly relied on free labour, although not exclusively. If we consider Humboldt's estimates, from 1789 to 1795, it took an annual average of between 4,000 and 5.000 day labourers in the valleys of Aragua to carry out the harvest of indigo and prepare the dyestuff. When the overproduction crisis of 1787 occurred, Intendant Francisco de Saavedra decided to investigate the reasons for such a situation. Indigo producers in the Maracay region confirmed that most, if not all, of the farms specializing in the indigo harvest employed day labourers who were paid six or seven *pesos* a month for their work, besides food. Approximately 20,000 day labourers were already employed in this branch of the Venezuelan colonial economy.[10]

Despite these difficulties, indigo became the second largest export product of Venezuela at the end of the colonial period. As a result, the Spanish royal administration attempted to more directly control local institutions in matters related to indigo, particularly judicial functionaries who exercised jurisdiction in administrative and military courts. From the 1780s, the landowner Antonio Fernández de León began to greatly influence the appointments in the region of Maracay, so that, by the eve of independence, virtually all those who held these critical positions were close to him. Among the officials that required the careful scrutiny of the interested parties were those of inspectors (*veedores*).[11]

In 1794, the Consulate of Commerce considered the need to appoint indigo inspectors who were capable and trustworthy, recommending that:[12]

> They be white and honest people, of probity and have knowledge of how to obtain the most perfect of the qualities of the dyestuff, to avoid the serious damage

> suffered in trading activities when there is lack of intelligence or lack of legality in the case of blacks and mulattoes who practiced this operation that has resulted in the detriment or production of the said dye.

Prominent members of the local elite were chosen under these circumstances, including figures like Francisco Baez, Francisco Arámburu, Martín Belaunzarán, Antonio Arriezurrieta, Josef Vicente Galguera and Felix Martinez. It was also decided that each supervisor would receive respectively 4 *reales* for each sack of 100 pounds of the dyestuff inspected, a sum to be paid half and half by the seller and the buyer. If the sacks weighed 200 pounds or more, the payment to be made would be of 8 *reales* respectively by the buyer and seller.

Indigo landscapes: The production zones in Venezuela

From 1784 onwards, several areas specialized in indigo cultivation: San Carlos, Tinajos and San Francisco Javier de Agua de Culebra for the wild variety; Sabana de Ocumare, Cúa and Guanare, recently created; as well as Cojedes, Sarare, El Pao and Guacara. In the case of the *haciendas* of La Victoria (Aragua valley), it was estimated that they yielded a production of 150 *arrobas* (the *arroba* was approximately equal to 25 pounds) per year. As early as the 1720s, Pedro José de Olavarriaga provided the first information on the facilities that existed for the planting and harvest of indigo in Venezuela, even if it was wild. However, it was much later that cultivation intensified. Indigo cultivation was also recorded in the valleys of Tuy, Cagua and Turmero, but it was the lands near Maracay that proved most productive, with an annual production estimated at 4,000 *arrobas* for the sixty *haciendas* installed in the vicinity of the city. Of the 220 indigo *haciendas* registered in Venezuela in the years 1785–7, more than half were located in the jurisdiction of La Victoria, that is, in the Aragua Valley, the other main place of concentration being the Tuy Valley.

One historical study estimates that the physical size of large indigo *haciendas* in Aragua ranged between 188 and 600 hectares, an area smaller than that of the *haciendas* that specialized in cocoa production, which reached an average extension of 6,595 hectares. The indigo properties included offices – although rather rudimentary –, hydraulic mills which helped activate paddles and wheels to process the indigo leaf (carried out by means of a process of maceration), water tanks, and containers for the conservation of the liquid indigo dye before being dried. In the most productive valleys, the large indigo *haciendas* used slave labour, although this was not a guarantee of higher productivity. In 1793, there were recorded to be 320 slaves on 15 of these farms, or an average of 21 slaves per establishment. In the jurisdiction of Maracay, the indigo and cocoa plantation belonging to a certain Andrés Díaz had a total of 39 slaves; in that of Juan de Castellanos, which was also a sugarcane plantation, 47 slaves were identified, but only 18, or barely 40 per cent of the total, were actually working on the plantation (a high percentage were disabled due to age or health problems). It must be emphasized, however, that the *haciendas* in this region that combined slave labour and day labourers had smaller areas, between 24 and 160 hectares approximately. They were predominantly owned by Basque, Canary Island, Catalan and Spanish immigrant 'entrepreneurs', but the way they operated did not differ much from that of the large slave-owning estates.[13]

The expansion of the indigo trade resulted in the association of small producers, especially the harvesters, in sharecropping schemes or by means of leases, which made it easier to establish indigo, coffee and other local crop farms on land that had hitherto been unoccupied but were sometimes owned by them. Other planters requested loans in cash and various effects from merchants or winemakers, offering the harvest and their belongings as collateral. Such was the case of the third category of indigo exploitation that has been identified in the Aragua Valley. These were medium-sized *haciendas* (with an average of 60 hectares) and were in fact similar to commercial enterprises, which used sophisticated accounting methods and sometimes employed significant capital. They were often managed by large merchants such as Felipe de Llaguno, Juan Joseph de Mintegui and Pablo Orendaín, who were associated in one of those companies.

Pablo Orendaín had worked with Juan Joseph de Mintegui (buyer of two of the aforementioned properties and, who was, apparently, financial agent of the group) in up to three *haciendas* producing indigo dye. The labourers came from the plains of Guárico, Cojedes and from the Aragua valley itself, with salaries estimated at between 1 and 3 *reales* per day paid in cash, or partly in goods and specie. This remuneration was much lower than the salaries of the hacienda supervisors (*mayordomos*) who could receive up to 286 pesos per year. Nonetheless, the scarcity of free labourers, and the nature of the work, which was considered unhealthy, forced many of the entrepreneurs to pay day labourers even at times when they were not needed, or in advance, in order to have labour available at the right time. Hence, the rivalries that arose in the Aragua Valley between several of these landowners to retain this floating labour force sometimes proved to the detriment of the immediate profitability of their enterprises. [14]

Indigo cultivation extended under these conditions to the whole of the *Intendencia* of Venezuela, as far as Maracaibo, the island of Margarita (San Juan del Valle), Barcelona and Cumaná. The preparation of indigo was carried out in Cumanacoa, San Fernando and Arenas whose indigo factories became famous at the end of the eighteenth century. Their products, which were comparable in quality to those of Guatemala, became more renowned than those of Caracas, because the former were frequently adulterated in order to increase their weight. Other factories were also established in the vicinity of Angostura in the years 1787–8. In the province of Barinas, 17 indigo farms (Barinas, Barinitas, Nutrias, Obispos) yielded a production of 588 *arrobas*.

The case of Barinas is particularly illustrative of the strict policies put in place by the local authorities, especially with regard to indigo cultivation and processing. According to the evaluations made by Governor Fernando Miyares, at the end of 1787 the cultivation of indigo, as well as that of cotton, coffee, cocoa, sugar cane and even *achote* (*onoto*), was 'very much at the beginning', in spite of the existing natural advantages, including the fertility of the soil and the transportation facilities provided by the rivers of Apure and Orinoco. When Miyares set out to transform the economic development of Barinas, he took several measures to promote agriculture and stimulate commerce, placing special emphasis on the cultivation of indigo.

In 1783, the value of the total production in this area was already estimated at 40,000 *pesos*, but the lack of labour led many indigo farm owners to abandon their plantations. Furthermore, as the Governor indicated in a letter of 10 June 1788, addressed to Antonio Valdés, there were few natural commodities that could be exported profitably to Spain, with the exception of dry-cured tobacco. A similar state of mind was expressed by the wealthy Don José Ignacio del Pumar, ensign (*alférez*) of the city of Barinas, when Miyares assumed the position of Governor of the province. He then owned three indigo *haciendas*, which he considered to be 'perfectly established' but nonetheless he was thinking of abandoning them. Encouraged by the Governor to go ahead with indigo cultivation, Pumar pointed out that, in 1787, many plantations had indeed been launched regionally at the Governor's behest. There were then 39 indigo *haciendas*, and they produced 1,356 *arrobas* a year. He himself had requested 150 slaves in order to address the lack of labourers. In any case, the shortage of slave labour was resolved by the Governor arranging for Indian crews to work for wages equivalent to those paid on the Spanish ranches. On 23 May 1788, the town council of Barinas praised Governor's work, underlining that:

> The providence of favouring agriculture by employing Indians has greatly improved the state of this Province, the proof of which cannot be more visible, since the indigo has been sufficient over the past year and a half to cover the payment in Caracas and Maracaibo of the large quantities of goods imported from Castile in this short time, of which there was no prior example until now.[15]

In July of the same year, Governor Miyares ratified to Antonio Valdés the success of his measures, and pointed out the following:

> The cocoa and indigo crops, which are easy to cultivate, are increasing significantly, and will make up three strong branches of commerce, if the provision of blacks that I have requested from Your Excellency is granted.[16]

In fact, the figures speak for themselves: in 1789, Barinas sent 1,863 pounds of indigo from San Fernando de Apure to the province of Guayana by river. Likewise, this commodity began to be shipped to Maracaibo and Caracas quite frequently, in exchange for Castilian goods, tools and wine. The previous year, the indigo sent to Spain had allowed the purchase of 100,000 *pesos* of various goods and the commercial relations of Barinas with the other Venezuelan provinces had intensified, especially with Guayana, as indicated by Alexander von Humboldt, who did not hesitate to describe the small city of San Fernando de Apure as a 'warehouse of river commerce'.

Between 1775 and 1783–5 dyestuff production increased from an annual average of 8,710 pounds to 163,221 pounds. In the years following, the relevant figures were 294,000 pounds produced in 1786–90, 718,074 pounds in 1791–5, and 697,451 pounds in 1809. The prices of the pound of indigo fluctuated between 11 and 13 *reales* from 1783 to 1786, between 12 and 14 *reales* from 1793 to 1796, and were quoted at an average of 12 *reales* in 1808-1809. In fact, prices fluctuated significantly, as can be seen from the variations in average prices between 8.5 and 13 *reales* per pound in the Caracas market, depending on the quality of the three types of indigo (*flor, sobre y corte*). The expansion of these exports led to the diversification of the Venezuelan economy, which had clearly advanced by the end of the eighteenth century. The Atlantic trade indeed contributed to highlight the complementary role of indigo in the regional and costal economies, thus in the Capitanía General de Venezuela. During this period, the proportion of indigo shipments relative to the total export value surged from 1 per cent in 1775 to 15 per cent in 1784–5. It should be noted that the latter was on the basis of trade registered exclusively between the port of La Guaira and Spain and that at that time 95 per cent of the Venezuelan trade transited through La Guardia. Moreover, the value of indigo shipments as a portion of the total reached 40 per cent in 1791–5 and 30 per cent in 1809.[17]

Circumstantial interventionism

The large variations in indigo prices justified that the Spanish State, through its representatives, repeatedly intervened on behalf of the producers. Many plantations, at least in the case of the Aragua Valley, were not owned by large landowners (the *mantuanos*, in other words, the creole aristocracy) but by small growers who repeatedly complained about the cost of leasing the land and the difficulty in obtaining free labour. As part of the measures aimed at promoting indigo production, the *Intendencia* not only authorized the export of indigo to foreign colonies in exchange for slaves and money, but it also proceeded to buy the crops corresponding to the year 1786, investing in them the surplus realized from the sale of tobacco at the precise moment when prices had reached a minimum, that is 9 *reales* per pound, 2 *reales* less than the previous year.

These official purchases were never completely suspended. In December 1786, 17,000 pounds of indigo dyestuffs were sent to Spain. In 1787, 8,163 pounds arrived in Cádiz, at a cost of 14,326 pesos. In 1788, in spite of receiving an order to suspend purchases, 137,093 pounds were sent and at the beginning of the following year, another 13,712 pounds were shipped. A Royal Order of 28 February 1787, and another of 26 November of the same year, ordered the use of 125,000 strong *pesos* from the tobacco rent to 'give them value (to the remittances of cocoa and indigo: the latter cost 10 *reales* per pound at that time) and avoid the ruin of the laborers'.[18] In 1788, the Cabildo of Caracas requested the Intendant to buy indigo – but also cocoa and cotton – from the harvesters and to reinvest the tobacco monopoly's liquid assets in commodities of the province. In this way, the members of the Cabildo aimed to avoid a drastic decrease in prices, already caused in part by the lack of outlets, both for Veracruz and for Spain:[19]

> The neighbours are at the mercy of a single buyer in all the cities, resulting in a decline in their prices due to the lack of alternative markets, to the detriment of the State.

Arbide himself came to sign *representaciones* (documents of protest) written by the producers, in which he denounced the monopoly of the merchants – despite having ratified the freedom of trade – and the role played in this respect by several trading houses of the Iberian Peninsula (especially in Cádiz), and asked for concessions similar to those that had been granted in favour of those companies. Along with the dominance of a few merchants such as the brothers Iriarte and José Escorihuela, there was also another risk of monopoly, namely that of the *Compañía de Filipinas*, a new mercantile establishment that to some extent replaced the *Compañía Guipuzcoana*.[20]

The situation improved a few months later, however, and in 1792 the pound of indigo was quoted at 13 *reales*. This interventionism of the state was clearly manifested on another occasion: the Royal Order of 12 September 1792 authorized ships coming from Veracruz to call at Venezuelan ports and to ship local products from there, among them indigo, which was still exempted from taxes on arrival to Spain. The wars between Spain and France (1793) and against England (1796) made trade relations difficult and offered the Caracas merchants the opportunity to impose minimum prices on the harvesters. In 1793, the pound of indigo was worth 13 *reales*, but in 1796, its price was in practice only 10 *reales*, or even less if we recall the provisions taken in 1793 to guarantee these prices by Intendente Fernández de León which motivated another intervention by the Intendencia to raise the price level again and attempt to reach 14 *reales* per pound of dyestuff, common before the war.

The economic crisis at the end of 1797, a consequence of the Spanish Navy's defeats by British forces, led to a decline in both prices and the volume of exports. This situation endured beyond 1800, with the pound of indigo being valued at 9 *reales*, and persisted until 1828, when the pound of indigo barely reached 8 *reales*. Exports dropped to from 2,955,973 pounds in the period 1793–6 to only 793,210 pounds between 1797 and 1800, with an estimated average price for the whole period of 13 *reales* per pound of indigo dye. The increase in cotton and coffee exports did not manage to compensate for the deterioration of indigo and cocoa volumes and income, although trade was diversified. Indeed, between 1797 and 1800, only 151,749 pounds of indigo were exported to Spain and 641,461 to foreign nations and colonies.[21]

The small producers were the most affected by these downward trends. It was estimated that two-thirds of indigo producers ceased exploiting their plantations during the crisis that lasted from 1797 to 1801. But other circumstances also contributed to the destabilization of indigo production in Venezuela including the drought of 1804, which made all types of agricultural products more expensive, as well as robberies and thefts (even of seeds) on the *haciendas* by labourers or free slaves, who then sold them illegally. Consequently, the Lieutenant of Justice at Maracay went so far as to decree the prohibition of clandestine commerce, as ratified by the document known as the *Bando* of 1 October 1794.[22]

Conclusions

At the end of the eighteenth century, indigo stood as one of Venezuela's principal export commodities, with an annual shipment average of 544,608 pounds. This was a result of the substantial demand prevailing in the Iberian Peninsula during that era, with Venezuela effectively supplying nearly half of the Spanish dyestuff market. It is worth noting that while Venezuelan indigo successfully rivalled the leading producer of its time, Guatemala, it also reaped benefits from interventionist public policies enacted by the local state administration, with the backing of the Spanish Crown. However, the decline of the crop was as surprising as its initial success. It was not until 1828 that Venezuela witnessed the true downturn in indigo production and trade in the face of competition from other production areas. In 1805, the Consulate of Commerce of Caracas, in a session on 27 November, pointed out the following regarding the increase in the consumption of coveted indigo and the consequent eagerness of the nations to produce it in their colonies:

> because today it is harvested in abundance in the East Indies, as well as in the West Indies: in the former it is much cheaper because of the low wages, due to its immense population, and although it used to be of inferior quality, it is noticed by the public papers that the superior variety is already in the same rank in the European markets as ours and that of Guatemala, because of the care and efficiency of the English in the sowing and harvest of this product in their extensive possessions of Hindustan [*sic*], from which the greater part of the factories are already supplied, and that nation, natural and eternal enemy of the prosperity of all the others, aspires nothing less than to appropriate to itself exclusively the benefit and provision of this precious and necessary article.[23]

The scarcity of labour denounced by indigo producers in Venezuela – together with the profit motive and the monopolies of the Spanish merchants – raises many questions. This situation perhaps finds partial explanation in the status of the producers, who were predominantly small growers. On the other hand, it should be noted that the available slave labour force was working mainly in the large *haciendas*, producing yet another essential commodity for the local economy: cocoa. While there were haciendas solely devoted to cocoa production, indigo typically held a secondary place in the production of these large estates. Of a sample of twelve *haciendas* that produced indigo, in four of them it was grown as a secondary product, and only one was on leased land. The land rents amounted to ten *pesos* or more for each bushel of the dyestuff produced during the year, not counting the workers' wages and the cost of machinery. This resulted in very limited possibilities for profits, depending on whether or not the producer owned his land (and slave labour), since the market price was low, barely fluctuating between 10 and 12 *reales* per pound of indigo.[24]

The increasing disaffection for indigo cultivation, however, was not based on local circumstances or because of the impoverishment of the soils, which was inevitably a consequence of repeated cultivation over time, and of the consequent decrease in yields. It was mainly related to the development of the chemical industry and, above all, to the invention in 1841 by the German Fritzoche of the dye that would later be known as aniline, as well as to the spread of iron prussiate. At present, the two indigo species mentioned in the introduction to this chapter continue to grow wild in the hot lands of Venezuela and their leaves are frequently used to cure scabies.[21] We could even say that indigo has experienced a relative resurgence in the global textile industry, capitalizing on modern techniques. This is evident from the abundance of fabric items, including popular products like blue jeans, which are currently marketed and celebrated for being crafted with this vibrant natural blue dye that holds significant historical importance.[25]

Notes

1 A previous version of this chapter was originally published in Spanish as Frédérique Langue, 'El añil en la Venezuela ilustrada. una historia inconclusa', *Revista de Indias* 58, no. 214 (1998): 637–53. The current text has been updated with more recent bibliographical references. The acronyms for primary sources used in this chapter are as follows: ACM: Municipal Council Archive, Caracas; AGI: General Archive of the Indies, Seville; and AGN: Archivo General de la Nación, Caracas. The editors of the aforementioned journal have given permission for the translation and publication of this article in the present volume.

2 For several studies on the history of colourants in the Mesoamerican area see Jean-Pierre Berthe, 'El cultivo del pastel en Nueva España', *Historia Mexicana* IX, no. 3 (1960): 1–28; María Justina Sarabia Viejo, 'Evolución de los colorantes mesoamericanos en el siglo XVIII', in *Europa e lberoamérica. Cinco siglos de intercambios* I (Sevilla: AHILA/Junta de Anadalucía, 1992), 321–42. See also, by the same author, María Justina Sarabia Viejo, *La grana y el añil. Técnicas tintóreas en México y América Central* (Sevilla: EEHA, 1994); José Manuel Santos Pérez, 'Los comerciantes de Guatemala y la economía de Centroamérica en la primera mitad del siglo XVIII', *Anuario de Estudios Americanos* 56 (1999): 463–84; Meritxell Tous Mata, 'El Añil y los pueblos de indios en la Provincia de San Salvador, siglos XVI y XVII', in *Poder local, poder global en América Latina: Ponències presentades al XI Encuentro-Debate América Latina ayer y hoy* (Barcelona: Universitat de Barcelona, 2008), 53–68. On indigo in Mexico, see Gerardo Sánchez Diaz, 'Cultivo, producción y mercado del añil en Michoacán en el siglo XIX', *Nuestra Historia* 1 (1991): 35–44. On Venezuela see Astrid Avendaño Vera (eds), *Diccionario de Historia de Venezuela* I (Caracas: Fundación Polar, 1988), 147. And Eduardo Arcila Farías, *Economía Colonial de Venezuela* I, 2 vols. (Caracas: Italgráfica, 1973), 234. María Teresa. Zubiri Marín, 'Algunos productos básicos de la agricultura venezolana en el siglo XVIII: añil, cacao y tabaco', *Boletín Americanista* 47(1997): 249–60.

3 Ana Zabalza Seguín, 'Añil, grana y vainillas. Los comerciantes navarros en América central (XVII-XVIII)', in *Identidad y estructura de la emigración vasca y navarra hacia Iberoamérica siglos (XVI-XXI). Redes sociales y desarrollo socioeconómico*, ed. José Manuel Azcona Pastor (España: Thomson Reuters Aranzadi, 2015); Arcila Farías, *Economía Colonial de Venezuela* I, *idem*. For a brief review of these early years in the local cultivation of indigo, see Adela Sánchez Nario, 'La introducción del añil como planta comercializable en Venezuela', in *Venezuela en el siglo de las Luces*, coord. María Carmen Mena García (Sevilla-Bogotá: Muñoz Moya Editores, 1995), 76–7.

4 Francisco Depons, *Viaje a la parte Oriental de Tierra Firme en la América meridional* II (Caracas: Banco Central de Venezuela, 1960), 25–6.

5 Depons, *Viaje*, 167. See also *Diccionario de Historia de Venezuela*, 167. AGN, Actas del Consulado, 2529: Junta ordinaria, 26 September 1804 (instruction from the Captain

General). Eduardo Arcila Farías, *Documentos del Real Consulado de Caracas* (Caracas: UCV, 1964), 109–13.

6 About this commercial monopoly, see Montserrat Gárate Ojanguren, *La Real Compañía Guipuzcoana de Caracas* (San Sebastián: Sociedad Guipuzcoana de Ediciones y Publicaciones, 1990).

7 Arcila Farías, *Economía colonial de Venezuela*, I: 337. *Diccionario de Historia de Venezuela*, I.

8 Arcila Farías, *Economía colonial de Venezuela,* I: 352. *Diccionario de Venezuela*, I: 147–9.

9 ACM, Minutes of the Cabildo, 1788: session of 28 July 1788.

10 ACM, Minutes of the Cabildo, 1788: 148. Alexander Humboldt, *Viaje a las regiones equinocciales del Nuevo Continente* II (Caracas: Ed. del Ministerio de Educación, 1956), 70–1. AGÍ, Caracas, 501: *Memorial de los cultivadores de añil*, 26 May and 2 July 1787. See also '*Informe de Francisco Saavedra*', 23 November 1787.

11 Michael McKinley, *Caracas antes de la Independencia* (Caracas: Monte Ávila, 1987), 175.

12 AGN, Actas del Consulado, 2526: Junta Ordinaria del 26 de agosto de 1794.

13 Germán Pacheco Troconis, 'Las haciendas de añil en los valles de Aragua en las últimas décadas del período colonial (1767–1830)', *Estudios de Historia Social y Ecónomica de América* 11 (1994): 349–64. PhD Thesis of this autor: 'El Añil: historia de un cultivo olvidado en Venezuela 1767–1870' (Universitat Autònoma de Barcelona. Departament d'Economia i d'Història Econòmica, 2001). See also Pedro José de Olavarriaga, *Instrucción general y particular del estado presente de la Provincia de Venezuela en los años de 1720 y 1721* (Caracas: Fundación Cadafe, 1981), 59–60.

14 Pacheco Troconis, 'Las haciendas', 362–3.

15 Virgilio Tosta, *Historia de Barinas 1577–1800* (Caracas: Academia Nacional de la Historia, col. 'Fuentes para la Historia Colonial de Venezuela' no. 183, 1986), 386–8.

16 Tosta, *Historia de Barinas.*

17 McKinley, *Caracas antes de la Independencia*, 56, 155 (mentions a letter from Saavedra to Gálvez with this information, dated 28 June 1784). *Gaceta de Caracas* facsimile edition, 1809–1810 (Caracas: Academia Nacional de la Historia, 1960); Arcila Farías, *Economía colonial de Venezuela*, I: 335–7 y Diccionario de Historia de Venezuela, I: 147–9. Germán Pacheco Troconis, 'El I: comercio colonial americano del añil: Guatemala y Venezuela, 1774–1810', *Boletín americanista* 61 (2010): 151–75.

18 'Reales Órdenes' (28 February and 26 November 1787).

19 Ibid.

20 ACM, Actas del Cabildo, 1788: representación del síndico procurador al Ayuntamiento, 18 February 1788; A.G.I., Caracas, 104: mentioned in the report of the Governor Guevara Vasconcelos, 10 March 1804. See also Arcila Farías, *Historia colonial de Venezuela*, 2nd ed. (Caracas, FCE, 1973), II: 42; Germán Pacheco Troconis, 'Hacienda y pequeña producción de añil en la Provincia de Caracas: complementariedad y contradicciones, ss. XVIII-XIX', *Historia Agraria* 31(2003): 87–110.

21 ACM, Agriculture: 'Informe sobre la providencia para sostener el precio de los frutos que a instancia del Ayuntamiento tomó el Intendente Fenández de León' (julio de 1793); ACM, Actas del Cabildo: Junta extraordinaria del 10 de abril de 1793; Humberto Tandron, *El Real Consulado de Caracas y el Comercio exterior de Venezuela* (Caracas: UCV, 1976), 160; Arcila Farías, *Economía colonial de Venezuela*, II: 75–7.

22 A.G.I., Caracas, 501: Memorial de los productores de añil, 2 de julio 1878; Tandron, *El Real Consulado*, 125. AGN, Actas del Consulado, 2529: Junta Extraordinaria del 1 de octubre de 1804.

23 Arcila Farías, *Documentos del Real Consulado*, 17–23.

24 AGN, Actas del Consulado, 2528: Junta Ordinaria de 28 de junio de 1802; McKinley, *Caracas antes*, 78.

[21] McKinley, *Caracas antes de la Independencia*, 149. Rafael Cartay, *Historia económica de Venezuela, 1830–1900* (Valencia, Venezuela: Vadell Hermanos Editores, 1988), 51, 54; *Diccionario de Historia de Venezuela*, I: 149. Claude Viel, 'Colorants naturels et teintures du XVIIe siècle à la naissance des colorants de synthèse', *Revue d'histoire de la pharmacie* 347 (2005): 327–48.

25 Michel Pastoureau, *Bleu. Histoire d'une couleur* (Paris: Le Seuil, 2000); Georges Roque, *La cochenille, de la teinture à la peinture. Une histoire matérielle de la couleur* (Paris: Gallimard, 2021).

9

Above and beyond Eliza Lucas Pinckney

Slave expertise and South Carolina indigo

Andrea Feeser

From the mid-1740s until the Revolutionary War in what is now the United States, indigo was a major crop in South Carolina, making almost as many fortunes for plantation owners as rice, the most successful commodity grown in the colony in the eighteenth century. Eliza Lucas Pinckney, revered today not only for having helped establish indigo in South Carolina, but also because she did so as a young woman in a man's world, had her success because her Wappoo Plantation stood on rich land formerly inhabited by Wappoo Indians[1] and because her twenty Black, Indian and mixed-race slaves kept her planting concern and household running. Like other South Carolina indigo planters, Lucas Pinckney established a fortune on Native territory with human chattel. Publications, archival information and an extant eighteenth-century indigo vat demonstrate the crucial labour and expertise slaves in general, as well as slaves at Lucas and Pinckney properties, brought to the making of indigo in South Carolina.

Producing indigo: Written documents

Publications on why and how indigo should be produced were part of a wide body of knowledge with multiple contributors that enabled planters' success with the dye plant in South Carolina.[2] Written information on the crop was furthered by imported dye makers' expertise, and especially by slaves' physical experience with indigo, which involved intensive field labour and dye-making practices. Moreover, the efforts of all of these individuals were made possible by slave activities that enabled plantations to run: the craftsmanship that went into building and repairing homes and outbuildings; the domestic work that kept spaces clean as well as people fed, clothed, healthy and tidy; the fishing, hunting, animal husbandry and gardening that put food on the table; along with the packing, hauling, and boating that got goods to market. White planters, overseers and dye makers rarely engaged such physicality in the sustained, lengthy and repetitive manner that slaves did. Britons certainly learned from doing, but their expertise came chiefly from observation rather than from the repeated and strenuous bodily efforts that indigo production demanded. This realm was occupied by slaves and involved all the hard work that made an indigo plantation function, from the housekeeping that assured a planter's standing with a comfortable and attractive home to packing dye cakes for shipment overseas.

White authors' written knowledge obtained from observing indigo making, and by gathering others' understanding and experience with the plant, is no more intrinsically valuable than slaves' know-how obtained from immediate interaction with the material conditions of cultivating and processing indigo. Nevertheless, the former – texts about planters' involvement with the dye plant – dominate today's literature on the staple. This is in part because these documents, which include ledgers, letters, periodical articles, pamphlets and books, have been archived and published. Further, they provide detailed information about conditions and processes that is easy to obtain merely from reading. In this respect planters' writings offer accessible, useful data that describe how all individuals involved with indigo worked.

Indigo was often a challenge to grow in South Carolina (it fared better in more tropical climes such as the Caribbean and Florida where warmer weather and richer soil were consistently available), and very tricky to make dye from,[3] so information about and experience with indigo production was an invaluable resource. Planters who tried growing the crop drew on existing knowledge about it from publications (predominantly those by French producers in the Caribbean but including one by an early South Carolina grower), as well testing their own hypotheses about indigo's growth and manufacture. For example, Eliza Lucas Pinckney made indigo through trial-and-error experimentation; with shared experience from a neighbour planter; and with know-how sold to her father by dye makers from Montserrat, a British Caribbean possession that had exported the blue dye to England in the early eighteenth century. Further, texts read by planters who tried indigo contained information gleaned from Chelsea Physic Garden head Philip Miller, who in turn had relied on another author, the French Dominican Jean Baptiste Labat, who carefully described early-eighteenth-century indigo manufacturing on Martinique.[4] Labat's observations replicated in Britons' publications provided specific information about when and how to plant, nurture, harvest and make dye from indigo.

A treatise written by Elias Monnereau translated from French and published in London in 1769 demonstrates just how complete such information could be. In *The Complete Indigo-Maker. Containing, An Accurate Account of the Indigo Plant; Its Description, Culture, Preparation, and Manufacture with Œconomical Rules and Necessary Directions for a Planter How to manage a Plantation, and employ his Negroes to the best Advantage*, Monnereau carefully expounded upon all the topics noted in his publication's title. Specific to Caribbean conditions, the text nevertheless addresses realities that concerned the South Carolina planter: how to supervise the overseer and slaves who worked an indigo plantation in a warm climate. With great specificity, Monnereau described how the overseer should manage the planters' property: not only the land on which indigo grows but also the slaves who labour with it. The writer also minutely detailed when and how to have slaves prepare soil for planting; drop seed into tilled ground; weed fields and rid plants of pests; harvest indigo and put it through the many processes necessary to make blue dye. The thoroughness of Monnereau's instructions is especially evident in the pages he devoted to overseeing slaves: in them he related in full what he took to be slaves' indolent and devious natures and further suggested how to address their wrongs:

> I have just hinted at the rigour that it is necessary to observe with regard to the negroes; but care should be taken that it does not degenerate into cruelty, which is already but too frequent in our islands, where, upon a simple suspicion or a slight fault, unwarrantable cruelties are exercised. The chastisement should always be proportioned to the crime; if atrocious, it should not be spared; but small faults should often be winked at, otherwise punishments would be incessant.[5]

Although Monnereau's text conveys to some degree the mental and physical energies slaves brought to indigo growing and dye making – and certainly communicates the bodily pain slaves experienced if punished for misbehaviour – his treatise offers no insight into how indigo manufacture was indebted to indigenous African expertise or know-how slaves injected into indigo making based on their practical, ongoing labour. Some additional writings point towards the former, while several images and an artefact index the latter.

With respect to African knowledge, published accounts establish that in the Caribbean during the seventeenth century, European and British labour organization practices combined with indigenous African indigo-making skills. In Barbados, British colonizers adopted the Spanish indigo-workshop model run by slaves, some of whom were observed employing dye-making procedures adapted from African practices. These slaves used hollowed out logs to soak plants that were subsequently pounded with mortars in pestles, a technique of African origin.[6] Further, when British and European merchants introduced South Indian methods of indigo production in the Caribbean – which entailed the used of multiple vats and an extensive beating process – the resulting larger economies of scale required wood paddles and, on some plantations, wood vats. Fabricating this dye-making equipment required a great deal of wood working, skills that many slaves had obtained in their homelands.[7] Thus indigenous African knowledge found its way into Caribbean indigo manufacture processes that also included South Indian and European as well as British techniques and procedures. These amalgamated methods travelled to South Carolina with the first colonists from Barbados, who brought instructions and seed with them to their new home.[8] Although their attempts to establish indigo were not a great success, the later indigo experiments of the 1740s in the South Carolina were, for Eliza Lucas Pinckney's famous project was indebted to a Caribbean model of dye manufacture: dye makers from the island of Montserrat taught her and her slaves how to produce indigo dye.[9] Further, it is entirely possible that African indigo expertise was directly and literally imported to South Carolina: many slaves brought to the colony during the indigo boom (from the later 1740s until the Revolution) came from regions in Africa where indigo was cultivated and turned into dye.[10] Indeed, depressions called indigo pits on the grounds of a former indigo plantation on Edisto Island[11] suggest this likelihood.[12] For centuries, Africans such as the Yoruba of Benin, the Manding of Mali and the Hausa of Kano were renowned for indigo making and dyeing, with the latter employing pit vats to colour cloth.[13] Although African blue dye is typically made by burning indigo plant material and pressing the ash into small balls, vat remnants can be an important ingredient for dye.[14] It is therefore conceivable that African slaves used their native expertise with dye making and dyeing at an Edisto indigo concern.

Producing indigo: Information from illustrations and an artefact

Whereas only a small clue concerning indigenous African knowledge exists on Edisto island, and mere echoes of such expertise appear in written documentation, illustrations from eighteenth-century publications on indigo along with an extant South Carolina artefact provide clear insight into the sustained, demanding and expert work slaves contributed to indigo culture. Whether informed by knowledge from their homelands or based on skills acquired while responding to directives, it was this labour that insured indigo's success in South Carolina. For without this know-how, planters and merchants could not have implemented and extended their knowledge to achieve wealth and power. Careful study of indigo-making equipment, and pictures that show slaves engaged with the plant, enrich planters' how-to-make-indigo documents and their statements about slaves toiling with the dye plant. A large set of dye works from botanist and physician Alexander Garden's Otranto plantation (rare extant eighteenth-century indigo vats), along with several illustrations from indigo-making guides that show slaves at work with indigo, provide a clear and more human sense of how very material and physical slaves' indigo expertise was, as well as how indispensable their efforts with the dye plant were.

Although the illustrations of indigo labour from M. de Beauvais Raseau's 1770 *L'art de l'indigotier*, as well as Henry Mouzon's roughly contemporary map of Saint Stephen Parish were not a source for South Carolina planters, they amplify written instructions with images of how slaves carried out the work prescribed. They therefore emphasize who did the hard labour of indigo making while suggesting just how demanding and skillful this work was.[15] De Beauvais Raseau's illustration of planting indigo shows that it was arduous work performed by both male and female field hands: it involved either hoeing or man-powered ploughing of spaces for dropping seed, which was done by women in low crouching positions.[16] In the picture, slaves of both sexes are portrayed with visibly engaged muscles and bare feet, and the men pulling and pushing the plough, as well as the women planting seed, are shown with bent backs. Small hills and trees dot the horizon, suggesting that the slaves are tilling and seeding a very large field, and because the slaves' bodies are arranged symmetrically and in synch with one another, the overall idea conveyed is that strenuous, repetitious labour was carried out over a lot of ground. Planter records note that more repeated small- and large-motor movement characterized crop maintenance, for slaves were responsible for plucking grasshoppers and other pests from indigo plants and weeding the ground around them.[17] The slaves' minimal attire in the de Beauvais Raseau illustration remind viewers that indigo growing was largely conducted in warm and hot weather, so heat, sometimes extreme, made the field workers' straining, crouching, pinching and pulling even more difficult.

Although very physical labour such as this is often not considered skilled, and was not designated as such by eighteenth-century planters, it certainly required real understanding to produce strong and healthy plants as well as calculation to economize on effort and reduce discomfort. The symmetrical arrangement of figures and ground markings in de Beauvais Raseau's image not only produce a pleasing sense of order and mark the appropriate distance between plants for optimal growth, but also suggest that the slaves worked deftly in unison in their planting efforts. In the picture, men and women are shown to have coordinated their moves so that tasks follow smoothly upon one another; the established rhythm suggests that gestures are not repeated unnecessarily, and that energy is thereby conserved. Although the illustration's symmetry is most certainly in part the artist's conceit – a way of communicating order, control, and harmony at a plantation where forced labour could conceivably incite insurrection – it equally communicates slaves' ability to produce skillful working methods that made demanding work easier to bear.

Mouzon's image of indigo harvesting and processing also emphasizes physical activity by showing slaves with bowed backs: it contrasts bent bodies carrying cut indigo to dye-making vats with the upright backs of a planter and overseer watching slaves toil.[18] Here the distinction between the work of White and Black people is made explicit: the former observe (and perhaps guide) varied processes the other two conduct. The labour is characterized as both taxing – the human-size bundles of indigo stalks transported by slaves weighs them down – as well as less demanding but deliberate and repetitive. Both hard and lighter work is visually described in vignettes by the indigo vats. Three slaves man the equipment by maintaining the water supply that runs in and out of the vats, and insuring that the indigo is kept under water by weights. Although the slaves are not shown beating the fermented liquid that results from submerging plant matter, the extremely long and thick paddles for this endeavour are propped nearby. The major effort their wielding entails is suggested effectively: the huge paddles are set at a dynamic diagonal that points to the bowed slave backs burdened with indigo brush. Beneath the paddles sit the trays for the so-called indigo mud – the material that falls from solution after oxygen is introduced from beating and an agent such as lime has been added – and to their upper right slaves cut dye cakes from the dried matter, stack them in a shed for further drying, and pack them in barrels for shipping. The individual producing the cut squares of dye looks female, and planter documents reveal that women did do this work, which like managing indigo in its liquid state, required a great deal of finesse. Indeed, planter Henry Laurens greatly prized his slave woman Hagar for how well she tended his indigo paste and set great store in her expertise, praising her for 'her great care of Indigo in the Mud'.[19]

Some of the equipment used in the final stages of indigo processing is documented in de Beauvais Raseau's illustration on the dye's manufacture: hanging bags for straining mud, trays and trowels for storing and smoothing it prior to cutting, and a large mortar and pestle – which resembles those used to clean chaff from rice – for breaking indigo pods to collect seed. Two male slaves are shown crushing pods with the latter equipment, and their tensed muscles along with braced stances inject an impression of indigo production's physicality into a largely didactic representation of tools. The necessary coordination of their efforts suggests a regular rhythm, and the sense of repeated, deliberate movement is echoed in the tray of cut indigo mud with its numerous, equal partitions. Precision and physical processes are thereby conjoined to communicate dexterity and know-how. Taken together, de Beauvais Raseau's and Mouzon's illustrations resolutely put labouring slave bodies into White writers' texts that describe indigo making, and these images reveal that the work was demanding as well as skillful.

These conditions can be readily inferred from studying a set of remaining colonial South Carolina indigo vats at Alexander Garden's country residence. Although it is not clear whether they were Garden's vats (preservation consultants date the vats to sometime between 1760-1800),[20] it is evident from their composition and scale that they were a challenge to build and to use. Long, tall and built from brick, set and sealed with a type of plaster, the vats are the product of a master mason's design and several construction workers' efforts. Fitted with sluices to drain water from one tank to the next and subsequently outside the works, the vats required not only a clear understanding of dye-making procedures, but sophisticated engineering. White planters, overseers and dye makers in South Carolina originally would have acquired the knowledge necessary to create such vats from experienced indigo makers in the Caribbean, although major White South Carolinian planters would not have built the vats themselves, but instead tasked Black bondsmen with the labour. Since the Otranto indigo works were produced after the staple was well established in the colony, slaves had enough experience with indigo, and certainly with building, to have designed and created the vats on their own.

The large scale of the Otranto dye works not only necessitated slave skill and strength to build but also to run. Because the vats are big, slaves managing them would have expended a great deal of physical and mental energy. They would have had to lift many heavy weights into the upper vat to submerge countless mounds of indigo stalks under water and would have had to evaluate a big expanse of fermenting liquid to gauge how much lime the indigo required and how much oxygen had to be added through beating. Slaves also would have had to work paddles hard and long to agitate the indigo pool sufficiently. During this difficult work, they would have needed to examine carefully all dye residues that floated to the top of the liquid so as to decide when to drain the murky water: first off of plant material and then from the last vat. Finally, removing the indigo mud from the bottom vat would have required time and careful handling.

The scale of the Otranto indigo works makes concrete another dimension of slave labour with the dye plant, an aspect not visible in the vats themselves or in any illustrations of slave efforts. Clear instead from planters' records and published complaints about indigo making in colonial periodicals, this dimension entailed a different sort of strength and ability: slaves' capacity to excel in noisome circumstances. The chemical that produces indigo dye is the product of fermentation, and the quantity of dye produced at any given indigo concern was in direct proportion to the number of plants harvested and submerged in vats of water. The fermentation produces sickening odour and large-scale works such as those from Otranto were therefore putrid places to labour. Slaves were not in a position to complain about the stink (White colonists who lived near their neighbours' vats were) and must have inured themselves to the stench and swarming flies attracted to decomposing matter in order to focus on the mental and bodily demands of the procedures that went into dye making.[21]

Even though written descriptions of work with indigo rarely provide human detail about slaves, and although the bodies pictured in the illustrations of indigo making in de Beauvais Raseau's and Mouzon's texts are Black, Indians were without doubt tied to indigo manufacture. An illustration from a 1773 map of South Carolina commissioned by parliament makes this visible by demonstrating that colonists needed both Africans and Indians to support indigo production. The map represents an Indian alongside two White colonists with an African slave hauling a container of indigo to a dock for shipping. The image's composition and the way the figures are depicted suggest that the wealth and power of planters and merchants stands at the centre of and is propped up by the work of Africans and Indians. The two colonists, positioned in the middle of the image, are shown in the fine clothing that wealthy planters and merchants could afford, and one is pointing towards the scantily clad Black slave off to the right carting indigo on his bent back. The Indian is pictured at the far left, leaning slightly inward while seated on a rock amidst plants that suggest South Carolina's natural terrain. He is shown with a quiver of arrows on his back and holding a bow and arrow pointing towards the other figures. The implied menace is contained by a tree stump that abuts the bow, and the Indian's hand appears slightly upturned as if making an offering. Wearing a headdress and a mere blanket draped around his waist, he appears as 'uncivilized' as the barely dressed slave: both are almost naked compared to the elaborately attired colonists. The Indian and African turn in towards the White men as if they are framing the colonists. Indeed, these postures replete with curved shoulders, communicate the message that eighteenth-century South Carolina trade in staples such as indigo was built on the backs of hostile Indians either kept at bay or enlisted as warrior allies, and of Africans forced into slave labour. Although the Indian is not here represented as a slave, the two are made almost equivalent as props of White indigo planting and trade because of their shared positioning, near naked state, and colour.

Most South Carolina plantation narratives fail to account for the Indian slaves among the thousands of enslaved Africans who toiled hard to produce staples like indigo. Native slaves comprised one-quarter of the colony's slave population in 1708,[22] and although their numbers paled beside those of African slaves in later decades – imported in large volumes as rice planting came to dominate South Carolina's economy – they are still present in late colonial and early federal records.[23] That Native slaves performed the same types of labour as their Black fellows is evident in wills, inventories, runaway notices, and work advertisements that name slaves and describe the work they performed: William Robert Snell, who has produced an extensive analysis of documents pertaining to Indian slaves in South Carolina, maintains that these individuals 'were employed in most ways that Negro slaves were utilized'.[24] Typically, more valuable slaves did specialized tasks that not only improved their owners' material circumstances, but that enabled masters to earn extra money by hiring out these skilled slaves.[25] For women, such work consisted of all the activities that went into making and maintaining clothing as well as nursing (both the care of people and the feeding of babies); for men it included all types of construction with wood, metal, stone, and brick; and for women as well as men it might entail cooking. Often slaves who were listed in documents without a noted skill set were field hands. Slaves labelled Indian, described as 'mustee' or as having Indian features, and with names prefaced by 'Indian' appear in documents of all types throughout the colonial period and certainly during the later 1740s to mid-1770s when indigo flourished.[26] Indeed, the names of Indian and mustee slaves are embedded in the Lucas and Pinckney papers, family records which also contain enough information about a Lucas and Pinckney mulatto slave to demonstrate just how profoundly his labour guaranteed the success his owners had with indigo.

Quash's/John Williams's expertise at Lucas and Pinckney properties

Wappoo is where Eliza Lucas (Eliza Lucas Pinckney when she married Charles Pinckney in 1744) watched to see which Caribbean staples might flourish in the South Carolina Low Country, and she drew not only on expertise she acquired as a self-taught student of all things botanical, but also on the knowledge and work of other people. Lucas wrote in a letter to a close friend that she spent some time in her fields every day,[27] but certainly not to supervise all the work performed there: she would have required a driver to manage the particulars of her field hands' labours. Although no one person is named such in her letters, it was most likely her mulatto slave carpenter Quash. As Joyce E. Chaplin states in her book on eighteenth-century agriculture in the Lower South, period travellers observed that Black drivers frequently 'managed much of the day-to-day direction of agricultural activities'.[28] In his work on eighteenth-century slave life, Philip D. Morgan further elaborates that 'one identifiable type' of driver was a slave tradesman, a person who would far exceed the appraised value of other slaves because he played two critical roles on a plantation.[29] That Quash may have served as Eliza Lucas's right hand man at Wappoo is further suggested by an entry in the young woman's letterbook of 1742, a document that points to her close ties to the man. In this entry she noted a letter she sent to her father concerning slaves charged with trying to run off to St. Augustine in Florida where the Spanish offered them sanctuary to antagonize their British enemies. Lucas wrote 'They accused Mol[att]o Quash. I was at his trial when he proved himself quite Innocent. The ring leader is to be hanged and one Whyped'.[30] This alarming event took place three years after South Carolina's Stono Rebellion, which occurred not far from Wappoo and which was the largest slave revolt in British North American during the colonial period. Eliza Lucas's advocacy for Quash, implicitly at his trial and explicitly in her letter to her father, suggests that he was a trusted and valued member of her household and business concerns: a slave with whom she closely worked.

While it is not certain that Quash served as Lucas's Wappoo Plantation driver, it is certain that he had an active role in the ultimate success of the indigo project she began there with other slaves' and some free people's assistance. These experiments with the dye plant are described in Lucas's early 1740s letterbook entries and in a missive she wrote to her son Charles in 1785,[31] and can be pieced together further from additional period records. Eliza Lucas first had indigo seed planted in summer 1740; the resulting crop was largely taken by frost, but her slaves harvested some seed which they grew the following summer. Although her 1741 yield was low, she had enough plants to discuss with her neighbour Andrew Deveaux, who was also experimenting with indigo, and with which the dye maker Nicholas Cromwell could work. Cromwell was a contract labourer who hired himself out to profit from his expertise and was sent from Montserrat by George Lucas to teach his daughter and two family slaves how to make dye cakes from indigo plants.[32]

Although Cromwell fulfilled his contractual obligations to the Lucas family, he also appears to have sabotaged his own efforts and the Lucas's ambitions, the latter of which were ultimately realized not only through Eliza Lucas's continued efforts but also through those of her husband Charles Pinckney and especially through Quash's labours. Eliza Lucas reported that Cromwell 'made some brick Vats on my Fathers plantation on Wappoo Creek' but also made 'a great mistery' of the dye-making process, which she said that Cromwell compromised by adding too much lime at a crucial mixing stage.[33] Despite this setback, perhaps the result of the dye maker's remorse over sharing indigo techniques that kept Montserrat in dye production,[34] the young woman had her slaves undertake subsequent plantings. In both 1742 and 1743 Wappoo produced little indigo, but in 1744, with the aid of Nicholas Cromwell's more reliable brother Patrick, Eliza Lucas's slaves made a good quantity of dye. At this point she had her overseer at the Lucas Garden Hill Plantation on the Combahee cultivating indigo and had much success with this project as well.[35]

In a 1744 letter to her father, written several months after her marriage to Charles Pinckney, Eliza Lucas Pinckney wrote that her husband had begun the necessary steps to get the family's indigo into the English market by sending six pounds of her dye to London to be evaluated.[36] It was received there with favour, for a 3 December 1744 letter to Charles Pinckney published in the 1 April 1745 issue of the *South Carolina Gazette* stated the following:

> I have shown your indigo to one of our most noted brokers in the way, who tried it against some of the best French, and his opinion it is as good. . . . When you can in some measure supply the British demand, we are persuaded, that on proper application to Parliament, a duty will be laid on foreign growth, for I am informed, that we pay for indigo to the French £200,000 per annum.[37]

George Lucas, while surely pleased with his daughter's accomplishments, remained concerned about the quality of dye produced on his plantations, finally turning to Quash to insure the excellence of the family's indigo. George Lucas retained the services of Nicholas Cromwell (his value as a dye maker must have exceeded his disreputable nature, for in addition to ruining the first batch of indigo that he made for the Lucases, he was accused of thievery),[38] but he worried that the man would remain untrustworthy and that his brick vats in South Carolina were unreliable. George Lucas wrote to Charles Pinckney that he had made additional but failed attempts to find a 'Negro dye maker' from Montserrat to send to South Carolina,[39] an effort that shows the planter knew such workers were very skilled, and also that he expected a slave but not Cromwell to be compliant. Another Lucas letter to Pinckney suggests that the former was willing to consider Cromwell's claim that his South Carolina dye failed because his brick dye works had not been built in time to season properly.[40] However, Lucas's ultimate distrust of Nicholas Cromwell and of his brick vats is evident in a 24 December 1744 letter he penned to his son-in-law. It is upon receipt of this letter that Quash became an indisputable player in the family's indigo venture and of exceptional value to them. Lucas wrote:

> I have since recollected upon my wife telling me the indigo Nicholas Cromwell made gave linen a red cast, that it may be possibly continue some time, if not always, as the work is made of bricks. I therefore desire Quashy may be put immediately, to make an indigo work of the same dimensions, with the brick one at Wappoo on Garden Hill Plantation, with plank and timber, the plank to be well joined, and pin'd [?], with wood to the frame of the inside, that it may be tried there with a small quantity of ground, planted there, and if a difference appears, another such work may be made at Wappoo.[41]

Although Nicholas Cromwell's brother Patrick had produced good dye from brick vats, George Lucas feared that subsequent dye made in these vats could not be counted upon. Convinced that brick could compromise dye colour but that timber likely would not, Lucas charged Quash with improving the family's indigo concern.

Indeed, it is very possible that Garden Hill's first successful crop was indigo produced in 1745 after Quash made wooden dye works at the plantation. Garden Hill overseer William Murray's account sheet of 1745 shows that in this year he shipped to George Lucas's London agent indigo valued at more than £225.[42] Although this entry may refer to the plantation's 1744 crop, Charles Pinckney's accounts from the later 1740s show that by 1746 the Lucas-Pinckney clan preferred wood to brick in their indigo operations, for one record from this period states that Quash spent thirty-three days making wooden indigo vats at Garden Hill and elsewhere.[43] Charles Pinckney's records establish without a doubt that Quash's carpentry skills made a key contribution to the indigo experiments Eliza Lucas had begun with her father and the Cromwells and that she continued with her husband.

Charles Pinckney's mid-to-late 1740s letters and accounts also show the extraordinary skills that Quash possessed and how they were used by the Lucas and Pinckney families to further their standing. Letters written between Pinckney and his father-in-law demonstrate that the two men quietly battled for the use of Quash's services. Sometime between late 1744 and spring 1745, George Lucas instructed Charles Pinckney to insure Quash and send him by a 'prime sailer' to Antigua.[44] Pinckney demurred, arguing that he had great use for the slave, and on 22 May 1745, Lucas relented, nevertheless maintaining that he could 'make considerably more of [Quash's] labor. . . .'[45] Some time that year, Pinckney began purchasing materials for a city residence in Charles Town, and in 1746, he put Quash in charge of the entire construction project. In a long list of building specifications dated 4 November 1746 that Pinckney wrote out for Quash (which demonstrates that the latter read, wrote and kept accounts), Pinckney concluded: 'In your calculation you are desired to distinguish and set down what the Carpenters work comes to; what the Joiners, including the Dining room and what without the dining room, and what the stairs & Venetian window separately come to'.[46]

Pinckney's instructions describe a stately home with a great deal of woodwork and Quash was paid well for overseeing the ambitious building project. The planter's accounts book features the following debit next to the carpenter's name: for '3 years & 3 months allowance at £200 pr. annum from 1st Jan. 1746 to 1st April 1750, – £650'.[47] In addition to the handsome allowance he paid Quash, Pinckney also paid his building supervisor on three separate occasions 'to encourage him in his carving work'.[48] This accounting note suggests that besides managing a large construction project, Quash applied his own woodworking skills in the building of the Pinckney home. Although the house burned down in Charles Town's great fire of 1861, a family member who knew the house well published a lengthy description of its features in 1896. The portion of her narrative that details the home's woodwork reads as follows:

> The window on . . . [the] staircase (one of the most remarkable features of the house) was very beautiful, of three arches with heavily carved frames, and a deep window-seat extending the whole length of the landing-place. . . . The whole house was wainscoted in the heaviest paneling, the windows and doors with deep projecting pediments and mouldings in the style of Chamberlayne. The mantelpieces were very high and narrow, with fronts carved in processions of shepherds and shepherdesses, cupids, etc., and had square frames in the paneling above, to be filled with pictures.[49]

Given the authority and high pay accorded to Quash, it is entirely possible that he produced or contributed to the production of this sophisticated carving.

One additional notation in Pinckney's list of building expenditures demonstrates how highly the planter valued his carpenter: a debit that reads 'to his freedom £750'.[50] Pinckney purchased Quash and a few other Wappoo Plantation slaves in January 1749 or 1750 after George Lucas's death and upon the mortgage foreclosure of Lucas's South Carolina properties. In the deed that records the sale, Quash is also identified as John Williams,[51] and it is by this name that Pinckney refers to him in the 12 May 1750 document by which the planter manumits the slave. Whatever personal services John Williams rendered Charles and Eliza Lucas Pinckney that persuaded them to free their slave, Williams's professional skills as an overseer and carpenter were obviously crucial to the Lucas and Pinckney families' accomplishments. Williams very possibly worked as Eliza Lucas's driver at Wappoo Plantation while she conducted her initial indigo experiments, and he built many of the wood indigo vats that produced the dye which brought fortune and prestige to the Lucas-Pinckney clan. His skills as a supervisor and carpenter also guaranteed that Charles and Eliza Lucas Pinckney had a Charles Town residence that visibly demonstrated their standing.

More Black as well as native contributors help make blue

If John Williams's final circumstances remain mysterious, the entire lives of the other slaves that worked Lucas and Pinckney plantations at first appear wholly so. Snippets of information appear when the slaves are listed as property in financial records or when the enslaved persons' skills were desired or marketable. For example, each slave is recorded by name in appendices to plantation deeds, and sometimes such lists include whether a person was a field hand or an artisan.[52] Further, the letters between George Lucas and Charles Pinckney that feature a polite battle over Quash for his carpentry skills, include a tussle over a cook named Dick, who must have been equally admired for his abilities to warrant potentially strained relations among father- and son-in-law. Moreover, Pinckney kept careful records when he hired out slaves, especially for his house slave Bettina and 'Nurse Molly' who did considerable work for others while he and his family lived in England. All of these brief notices, which mark people as chattel, at times give a sense of how and how hard they worked: Dick must have delighted palettes, Bettina kept a very clean house, Molly healed the sick, and from other records, it becomes clear that along with Quash, Barbuda, C. Quaca, N. Quaca, Pompey, Quamina, Say and Sogo were also woodworkers, and that Pompey was Quash's apprentice.[53]

Additional family records provide racial designations for a few of these enslaved people. Indian Peter must have been named thus because he was entirely or part Native American, and the latter would also have been true for Mary Ann and her two children Prince and Beck, since Mary Ann was labelled a 'mustee', the period term for a person both African and Indian.[54] Further, Dick and carpenter Quash were both described as a mulattos.[55] These Indian and/or part Indian, as well as part White slaves' origins, like those of the other enslaved persons at Wappoo, are undocumented. However, given that the Wappoo slaves were originally owned by John Lucas (an absentee landowner whose South Carolina property was administered by another person until George Lucas installed his daughter there), they were likely acquired in South Carolina and very possibly born there. Although some Native American slaves were sent to the Caribbean to rid the colony of actual or perceived enemies, others were retained, for a significant number appear as chattel in South Carolina property records of the time.[56] And although mulattos born in the Caribbean might be sold at the Charles Town slave market, the Lucas and later Pinckney mulattos were more likely the product of a South Carolina White planter's or overseer's sexual relations with a slave woman, for such dealings were then very common.[57] It is even possible that Quash and any other mulatto slaves at Wappoo were the offspring of the man who oversaw the plantation on behalf of John Lucas, for if they were born shortly after John Lucas acquired the property in 1714, they would have been in their early twenties when Eliza Lucas took over the plantation in 1739.

In a certain sense, these family records that only feature slaves' names and roles relay key information. In the first instance, and with respect to Wappoo, that it took twenty slaves to run a successful indigo plantation: Quash, Dick, Mary Ann, Dick, Lynn, Sawney, Indian Peter, Isaac, Pompey, Sarah, Mo (left blank), Beck, Molly, Nanny, Mary, Peter, Pompey, Douglas, Betty, and little Gulla.[58] That alongside Eliza Lucas and John Williams, some of these slaves brought indigo seed to plants that yielded dye from which dye cakes were made. And further, that others must have worked as house slaves, helping Eliza Lucas run her household. In the second instance, descriptive additions to names – Indian placed in front of Peter and 'mustee' placed after Mary Ann – show that in the Lucas and later the Pinckney household Native Americans were enslaved alongside Africans. Although it is not clear whether these Indian individuals worked in the house or in the fields, like all of the other family slaves they contributed labour to keep Lucas and Pinckney plantations running and thereby if not directly, certainly indirectly aided the planters' indigo concerns.

Indeed, on land named for a Native tribe – Wappoo – slaves who were part Indian, and in Peter's case perhaps wholly so, worked with Black and mulatto slaves to make the Lucas-Pinckney clan famous for bringing a major staple into South Carolina's and Britain's trade networks. Although the Lucases and Pinckneys did not consider these slaves' contributions as valuable as John Williams's vats and later his extensive work on a Charles Town Pinckney mansion, for references to these persons in family documents are limited, they most certainly contributed to the planters' success with indigo. Black and Native expert labour and Indian land made colonial South Carolina indigo culture flourish at Wappoo and every other colony plantation that grew indigo. May the lives of other figures like John Williams emerge from archives.[59]

Notes

1 For an overview of the Wappoo, see 'South Carolina–Indians, Native Americans–Wappoo', South Carolina Information Highway, 2022, accessed 5 December 2022, www.sciway.net/hist/indians/wappoo.html. Any surviving Wappoo were probably absorbed into the mixed population of so-called Settlement Indians, peoples who maintained a livelihood by living alongside Britons established in Charles Town and its environs (for a brief discussion of Settlement Indians, see Chapman J. Milling, *Red Carolinians* [Columbia: University of South Carolina Press, 1969], 62–4).

2 David L. Coon, 'Eliza Lucas Pinckney and the Reintroduction of Indigo Culture in South Carolina', *Journal of Southern History* XLII, no. 1 (1976): 69–75.

3 Max S. Edelson, *Plantation Enterprise in Colonial South Carolina* (Cambridge, MA and London: Harvard University Press, 2006), 184.

4 Coon, 'Eliza Lucas', 71–2.

5 Elias Monnereau, *The Complete Indigo-Maker. Containing, An Accurate Account of the Indigo Plant; Its Description, Culture, Preparation, and Manufacture with Œconomical Rules and Necessary Directions for a Planter how to manage a Plantation, and employ his Negroes to the best Advantage. To which is added, A Treatise on the Culture of Coffee* (London: P. Elmsly, 1769), 50.

6 Frederick C. Knight, *Working the Diaspora: The Impact of African Labor on the Anglo-American World, 1650–1850* (New York and London: New York University Press, 2010), 93–4.

7 Knight, *Working the Diaspora*, 97.

8 Knight, *Working the Diaspora*, 100.

9 For the mixed blessings Eliza Lucas Pinckney received from the first Montserrat dye maker with whom she worked, see Coon, 'Eliza Lucas', 66.

10 Virginia Gail Jelatis maintains that Africans enslaved in South Carolina were frequently from African regions where indigo was grown and used in dyeing cloth, and she provides detailed data for the dates during which and the regions from which such slaves were imported. For her this suggests that slaves contributed their indigenous knowledge to indigo making in South Carolina. See her *Tangled Up in Blue: Indigo Culture and Economy in South Carolina, 1747–1800*, PhD diss. (University of Minnesota, 1999), 152–3. Knight notes that mid-eighteenth-century South Carolina indigo planters preferred slaves from Senegambia, which has a long history of indigo production. He thereby suggests that Africans with direct experience making indigo used their skills in South Carolina (Knight, *Working the Diaspora*, 104–8).

11 Charles Spencer, *Edisto Island, 1663 to 1860: Wild Eden to Cotton Aristocracy* (Charleston: The History Press, 2008), 69.

12 This possibility is evocatively portrayed in Julie Dash's 1991 film *Daughters of the Dust*: one scene shows slaves dyeing material in such pits.

13 Kate P. Kent, review of *Into Indigo: African Textiles and Dyeing Techniques*, by Claire Polakoff, *African Arts* 14, no. 1 (November 1980): 28.

14 Duncan Clarke, 'West African Indigo: An Introduction', *Adire African Textiles*, 13 February 2010. http://adireafricantextiles.blogspot.com/2010/02/indigo-in-west-africa-introduction.html (accessed 5 December 2022).

15 For an evocative description of this labour, gleaned in part from planter's accounts, see Edelson, *Plantation Enterprise in Colonial South Carolina*, 159–60.

16 Beauvais Raseau's 1770 *L'Art de l'Ingotier* digitized by the Biodiversity Heritage Library: https://www.biodiversitylibrary.org/item/125174#page/1/mode/1up

17 Philip D. Morgan, *Slave Counterpoint: Black Culture in the Eighteenth-Century Chesapeake and Lowcountry* (Chapel Hill and London: The University of North Carolina Press for the Omohundro Institute of Early American History and Culture, Williamsburg, Virginia, 1998), 160.

18 Henry Mouzon's 1773 *A Map of the Parish of St. Stephen* at the South Carolina Historical Society: https://schistory.org/exhibit/a-marriage-of-science-artistry/

19 Morgan, *Slave Counterpoint*, 164.

20 David B. Schneider, 'Otranto Plantation Indigo Vats', United States Department of the Interior, National Parks Service, National Register of Historic Places Registration Form, 25 March 1988. http://www.nationalregister.sc.gov/berkeley/S10817708012/S10817708012.pdf (accessed 5 December 2022).

21 Morgan, *Slave Counterpoint*, 163. For indigo as a pollutant, see William S. Coker, 'Spanish Regulation of the Natchez Indigo Industry, 1793–1794: The South's First Antipollution Laws?', *Technology and Culture* 13, no. 1 (1972): 55–8.

22 Morgan, *Slave Counterpoint*, 481.

23 See the Wesley D. White papers, 1521–1993 (1311.00), South Carolina Historical Society, Charleston for powerful documentation of this fact.

24 William Robert Snell, *Indian Slavery in Colonial South Carolina, 1671–1795*, PhD diss. (University of Alabama, 1972), 150. His Appendix V (150–60) provides details of Indian and Black slaves' employ, gleaned from a variety of colonial records.

25 Morgan discusses this matter as well as some slave artisans' ability to hire themselves out. Morgan, *Slave Counterpoint*, 348–53.

26 See the White papers for relevant and extensive documentation.

27 Eliza Lucas to Mary Bartlett, [1742], *The Letterbook of Eliza Lucas Pinckney*, 34.

28 Joyce E. Chaplin, *An Anxious Pursuit: Agricultural Innovation and Modernity in the Lower South, 1730–1815* (Chapel Hill and London: University of North Carolina

Press, for the Institute of Early American History and Culture, Williamsburg, Virginia, 1993), 85.

29 Morgan, *Slave Counterpoint*, 222.

30 7 January 1742/3 entry, *The Letterbook of Eliza Lucas Pinckney*, 57.

31 See letterbook entries between 1740 and 1744 in *The Letterbook of Eliza Lucas Pinckney*, 5–71 and Eliza Pinckney to Charles Cotesworth Pinckney, 10 September 1785, *Manuscript Collection*, Charles Town Library Society, Charles Town, SC (hereafter MC/CLS).

32 George Lucas to Charles Pinckney, 30 January 1744, *Pinckney Family Papers*, Manuscripts Division, Library of Congress (hereafter PP/LC).

33 Eliza Lucas Pinckney to Charles Cotesworth Pinckney, 10 September 1785, MC/CLS.

34 Eliza Lucas Pinckney to Charles Cotesworth Pinckney, 10 September 1785, MC/CLS.

35 *The Letterbook of Eliza Lucas Pinckney*, xix.

36 Harriott Horry Ravenel, *Eliza Pinckney*, Women of Colonial and Revolutionary Times, South Carolina Heritage Series 10 (1896) (Spartanburg: The Reprint Company, 1967), 104–5.

37 *South Carolina Gazette*, 1 April 1745.

38 H. Hyrnes to Charles Pinckney, 13 May 1747, PP/LC.

39 George Lucas to Charles Pinckney, 12 July 1745, PP/LC.

40 George Lucas to Charles Pinckney [sometime between November 1744 and 22 May 1745], PP/LC.

41 George Lucas to Charles Pinckney, 24 December 1744, PP/LC.

42 *The Letterbook of Eliza Lucas Pinckney*, xix.

43 Alice R. and D. E. Huger Smith (eds), *The Dwelling Houses of Charles Town, South Carolina* (Philadelphia and London: J.B. Lippincott Company, 1917), 366.

44 George Lucas to Charles Pinckney [sometime between November 1744 and 22 May 1745], PP/LC.

45 George Lucas to Charles Pinckney, 22 May 1745, PP/LC.

46 *Dwelling Houses of Charles Town*, 371. That John Williams was extremely knowledgeable about building is further evident from a later entry in Charles Pinckney's account book: the former bought an architecture book from the latter sometime around 1754 (see fold out in the back of the ledger, PP/LC).

47 Smith, *The Dwelling Houses*, 367.

48 Smith, *The Dwelling Houses*, 366.

49 Smith, *The Dwelling Houses*, 371–2.

50 Smith, *The Dwelling Houses*, 366.

51 Andrew Rutledge, and William Boone, attorneys for Charles Alexander to Charles Pinckney, bill of sale for five slaves, 5 January 1750, South Carolina Department of Archives and History (hereafter SCDAH).

52 Charles Pinckney to George Lucas, acceptance of Wappoo Plantation and twenty accompanying slaves as dowry, 1 May 1744, Charleston Archive, Charleston County Public Library (hereafter CA/CCPL), Miscellaneous Records, Charleston County, 1746–49, 375; George Lucas to Charles Alexander, release from mortgages on five plantations, 25 August 1746, Miscellaneous Records (Main Series) [Selected Volumes], SCDAH.

53 George Lucas to Charles Alexander, release from mortgages on five plantations, 25 August 1746, Miscellaneous Records (Main Series) [Selected Volumes], SCDAH.

54 Andrew Rutledge, and William Boone, attorneys for Charles Alexander to Charles Pinckney, bill of sale for five slaves, 5 January 1750, SCDAH.

55 Charles Pinckney, certificate that he received from George Lucas upon his marriage to Elizabeth Lucas, the plantation Wappoo and twenty accompanying working slaves, 1 May 1744, Miscellaneous Records (Main Series) [Selected Volumes], SCDAH. In this document, Mary Ann (spelled Mariann) was also described as a mulatto.

56 For a record of Indian slaves owned by South Carolinians from 1683 to 1795, see Snell, *Indian Slavery in Colonial South Carolina*, 184–240.

57 See Winthrop D. Jordan, 'American Chiaroscuro: The Status and Definition of Mulattoes in the British Colonies', *The William and Mary Quarterly* 19, no. 2 (1962): 183–200 and Donald L. Horowitz, 'Color Differentiation in the American Systems of Slavery', *Journal of Interdisciplinary History* 3, no. 3 (1973): 509–41.

58 Charles Pinckney to George Lucas, acceptance of Wappoo Plantation and twenty accompanying slaves as dowry, 1 May 1744, CA/CCPL, Miscellaneous Records, Charleston County, 1746–49, 375.

59 For a detailed account of Quash's many achievements, see Tiffany Momon, 'John "Quash" Williams', *Journal of Early Southern Decorative Arts* 42 (2021). https://www.mesdajournal.org/2020/john-quash-williams-charleston-builder/.

10

Local production, Atlantic trade

The logwood economy in Laguna de Términos during the nineteenth century

Pascale Villegas and Rosa Torras Canangla[1]

During the nineteenth century, sailing ships entering the ports of El Carmen and Campeche – located on the Yucatan Peninsula on the southern coast of the Gulf of Mexico – arrived loaded with a wide variety of European and American manufactured goods and returned home with logwood colourant, the most valuable commodity produced in this peripheral region of Southeast Mexico. Divided into two species by today's biologists according to their geographical distribution – *Haematoxylum campechianum L.* and *Haematoxylum calakmulense* – logwood trees were well known for their colouring properties by royal officials and merchants on the Yucatan shore since at least the last third of the sixteenth century during the era of the Viceroyalty of New Spain.[2] The production and trade of this colourant, already in existence since colonial times, took on unprecedented proportions during the second half of the nineteenth century, defining the socio-economic development of this region articulated by a system of rivers that flowed into the *Laguna de Términos* (Terminos Lagoon) and crossed the jungle and mangrove swamps where the logwood trees grew (Figure 10.1).

This chapter studies logwood production and commercial structure in the Términos Lagoon during the nineteenth century. Throughout this century, this extractive enclave was inserted in the transnational networks of the Atlantic World economy. It was known for its large community of wealthy merchants and landowners – both locals and European and American emigrants – and for being a production area characterized by different degrees of forced labour under a regime of debt peonage. The chapter draws upon information from legal and civil archives, along with press sources, to reveal the complex production dynamics and commercial and social networks that emerged within the logwood industry in El Carmen, shedding light on how this commodity was sourced in the area and how this region was integrated into the broader Atlantic trade network.

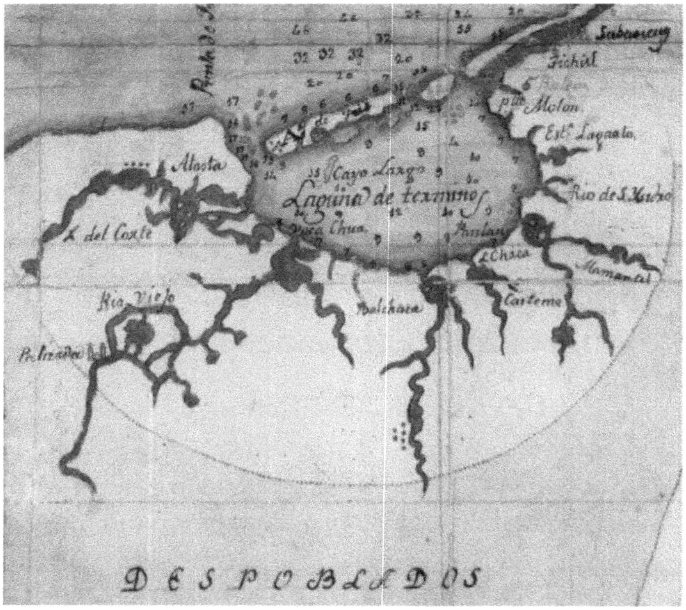

Figure 10.1 Map of Laguna de Términos (1814). Source: 'Mapa corográfico de la Provincia de Yucatan que comprende desde la Laguna de Términos en el seno mexicano hasta Los Zapotillos en el Golfo de Honduras', 1814, Mapoteca Digital Orozco y Berra, 1784- OYB-7264-B.

Logwood trade in the Laguna de Términos

Throughout the colonial era, the extraction and export of logwood in the Yucatan Peninsula – whose dyeing uses were part of an ancestral Maya tradition – were of great importance in the regional economy, which was devoid of other natural wealth such as minerals and with a land unsuitable to the sowing of wheat. Logwood production joined the hitherto most important commodities such as salt, cotton blankets, wax, honey, indigo and scarlet.[3] In essence, there was a continuity with pre-Hispanic and colonial trade patterns, but on a much larger scale.[4] In Europe, soon after the introduction of this dyewood, it became one of the favourite dyes imported from the Americas, along with brazilwood, cochineal, indigo and annatto (*achiote*), as evidenced by the numerous treatises and dyers' regulations published over the following centuries.[5]

Due to the inhospitable nature of this tropical area and its low population density, the Spanish Crown gradually lost control over the production and trade of logwood during the earlier period of colonial rule. The Crown's authority further diminished as English pirates settled in the area, engaging in illicit logwood extraction until their definitive expulsion.[6] Initially, the English favoured logwood production in the extreme northeast part of the Yucatan Peninsula, in an area called Cabo Catoche. This zone was their favourite place for loggings until they began to locate other coastal lands where this tropical tree was even more abundant. As the Englishman William Dampier –

sailor, traveller, pirate and logwood cutter – pointed out, since 1665, the forests around the Laguna de Términos had several advantages: the swift transportation of logs facilitated by the junction of multiple rivers; floodable lands with optimal conditions for the rapid growth of this tree; sparse settlement due to low population density; and cooperation with English smugglers until their definitive expulsion in 1717.[7]

In order to protect the entry or exit of ships from the Laguna de Términos, the Spanish Crown established the Presidio del Carmen on the island of the same name in 1724. This fort had a permanent garrison of soldiers and was also populated by their relatives and some inhabitants of the surrounding area, most of them *pardos* (brown, mixed race). Years later, in 1772, the governor of the Presidio, Pedro Dufau, founded the town of San Joaquin de la Palizada on the banks of the river of the same name. This river was the main route for trading logwood towards Laguna de Términos and, therefore, to the port of El Carmen, where it was shipped for export to Europe and the United States.[8]

The commercial circuits of logwood between 1750 and 1807 have been the subject of thorough work, among others, by historian Alicia Contreras Sánchez.[9] During this period, logwood legitimate and illegal trade routes connected Campeche with the Spanish and Spanish-dominated ports of Cádiz, Santa Cruz de Tenerife (Canary Islands), Santo Domingo, Havana and Veracruz. However, it was only in the nineteenth century, amidst the insertion of the Yucatan Peninsula into the Atlantic World capitalist system, that logwood production and trade would reach a significant boom in Laguna de Términos. During the nineteenth century, virtually all ships departing from the ports of Campeche and El Carmen were loaded with logwood, which had emerged as the primary commodity of the region. The situation was described in the mid-nineteenth century by a publication edited by Mexico's Ministry of Development (*Ministerio de Fomento*) as follows:

> El Carmen, a port established to the Southwest of the island of the same name, in front of the point of Jicalango, is the gateway to Laguna de Términos. This port, today belonging to the State of Campeche, is the deposit point, the centre of the logwood trade, called Campeche [. . .] All the ranchers take it in small boats without keels through the Laguna, as if to a common centre, and from there, they go to Carmen, where foreign ships flock to load such a precious article to take it to the main markets of Europe and even America.[10]

Between 1847 and 1874, more than 500,000 tons of cut logwood logs left the port of El Carmen (Table 10.1). Although annual exports remained around 20,000 tons per year, the drop in the biennium 1850–1 is explained by the fact that a presidential decree had forced the closure of the said port. As a result, only foreign ships previously unloaded in another Mexican harbour were allowed to load at the port of El Carmen, thus impeding direct trade. Moreover, the sudden decline in logwood trade in 1863 and 1867 coincided with the onset and conclusion of the French Intervention in Mexico, during which Emperor Maximilian of Habsburg was installed on the Mexican throne. These events disrupted maritime traffic but did not affect woodcutting in the haciendas, where logwood continued to accumulate in warehouses.

Table 10.1 Logwood Exported from El Carmen, 1847–74 (Selected Years)

Year	Logwood Trade (in Tons)
1847	13,909
1850	17,036
1855	26,114
1860	21,514
1865	14,475
1870	19,022
1874	15,676
Total (1847-1874)	517,917

Source: Belgian Diplomatic Records, Ministère des Affaires Etrangères. Direction du Commerce et des Consulats. N° Perso 897. Consulat de Belgique a Laguna de Terminos (Mexique). The dearth of primary sources posed a significant challenge in reconstructing logwood exports from Campeche, El Carmen and Sisal ports. This difficulty arose from various factors, including inadequate preservation practices, the loss of records over time and deliberate secrecy maintained by merchants and customs authorities. Their control over the trade and involvement in smuggling activities complicates a comprehensive and accurate account of logwood trade from these ports.

Before the outbreak of war between Mexico and the United States (1846–8), North American ships led the list of foreign ship arrivals to Puerto del Carmen, mainly from New Orleans, Mobile (Alabama) and New York. From the second half of the nineteenth century, French ships from Marseille, Bordeaux and Havre dominated the logwood trade, followed by British ports. During this period, the European merchant marine brought a diverse array of commodities to the Yucatan Peninsula, including items such as sewing needles, bronze nails and bottle labels. They also introduced grains like barley and flaxseed, along with products like bone fans, corset whalebones, wooden buttons, whips, perfumery, umbrellas and ivory combs. Furthermore, the imports included various consumables such as olive oil, cognac, champagne and mustard, as well as Dutch cheese and wines. Additionally, the merchants brought cotton fabrics and clothing, canvas, silk, wool and musical instruments like accordions, harmonicas and rosewood pianos.[11] The commercial novelty of that time was the massive introduction throughout the west coast of the Yucatan Peninsula of tiles and bricks that arrived in ballast from France – and to a lesser extent from Belgium – to replace the thatch of the roofs that caused devastating and repeated fires.Ultimately, logwood trade set the stage for a continual rise in maritime traffic between the European ports of Antwerp and Marseille and the Mexican port of El Carmen.[12]

As a consequence of Mexico's independence in 1821, foreign legations gradually opened in Campeche. Given the port's long-standing operation on the Yucatan Peninsula since colonial times, it was expected that maritime activity would notably expand, particularly with Mexico entering into peace and navigation treaties with friendly nations. The consulates of the United States (1819), France (1832), Spain (1841), Bremen (1841) and Belgium (1847) settled in Campeche. However, they witnessed this city maritime decline as it was supplanted by the port city of El Carmen when it was opened to deep-sea trade in 1822. All these diplomatic legations – including the British, absent in Campeche – commissioned diplomats to represent their countries' interests on Mexican soil.[13]

El Carmen, initially a presidio and later evolving into a town and city from the mid-nineteenth century onwards, enjoyed several noteworthy advantages over Campeche. First, logwood was available at a low price, as it was harvested around Laguna de Términos during the dry season and transported by river to El Carmen during the rainy season. Second, the port boasted an excellent natural infrastructure, making it easy to manoeuvre and anchor large ships of more than 300 tons. Despite the favourable hydrological structure that facilitated the extraction and shipment of logwood from El Carmen to European and American ports, overexploitation of this resource led to a need to venture deeper into the tropical forest for harvesting. As a result, throughout the nineteenth century, the ownership of forest land fell into the hands of a select few individuals – namely, the merchants settled in El Carmen who held complete control over the logwood business.

Logwood production and local elite

In the eighteenth century, following the expulsion of the English from the region, the Spanish Crown implemented cutting licenses to regulate the extraction of logwood. The new context resulted in jurisdictional matters and disputes over the lands around Laguna de Términos, where residents of Spanish origin at Presidio del Carmen were given privileges over other neighbours.[14]

In early republican Mexico, the new liberal legislation made available public vacant lands (*terrenos baldios*) and granted licenses that gave rights to exploit formerly communal lands that would soon become private property. Moreover, the insatiable demand for land coincided with widespread logging in the region. As deforestation took its toll on more accessible areas, a need to explore new territories arose, leading to the expansion from the coast towards the peninsula's interior, following the course of the main rivers.

The Mexican policies promoting the colonization of the forest were intended for both nationals and foreigners, with a preference for Europeans. In a decree issued in February 1856, the federal government granted authorization to foreigners residing in the Mexican Republic, allowing them to own and acquire property under the same conditions as any Mexican national. However, in practice, land ownership rights were repeatedly prohibited to foreigners from neighbouring countries. In these later cases, the argument was that internal colonization policies had to privilege the 'Mexicanization' of the territory. In 1861, the governor of El Carmen's district (*partido del Carmen*), Perfecto Vadillo, took a step forward by publishing a project that presented an opportunity for individuals interested in dedicating themselves to agricultural work, such as cereal cultivation. The project offered so-called vacant and unproductive lands for this purpose.[15] Nevertheless, the decisive push for the privatization of extensive land holdings came with the legislation enacted during the presidency of Benito Juárez in 1863. This law echoed General Santa Anna's declaration in 1853 that vacant lands belonged to the Mexican nation, setting forth the mechanisms for their privatization. Under this law, landowners were granted a three-month window to initiate the process of obtaining their land titles following the new procedures established by the federal authorities. Mexican internal colonization of the forest frontier, from Laguna de Términos into Quintana Roo and El Petén, came to an end with the Herrera-Mariscal Treaty of 1895, in which Mexico imposed itself on Guatemala.[16]

The wealthiest and most powerful merchants in the Laguna de Términos area were foreigners or descendants of migrants from Europe and the United States. They held a virtual monopoly over the logwood economy. They not only controlled its trade but bought large extensions of land in the basin where the logwood tree grew and obtained licenses for its exploitation. During the republican period, the majority of families who acquired licenses for vacant lands had previously possessed extensive forested areas dating back to colonial times before Mexican independence. Additionally, these very families held significant local and regional political positions, and, as the nineteenth century progressed, they were joined by new migrants from Europe and the United States. Out of the entire foreign resident population in El Carmen in 1853, there were thirty-one individuals from Spain, eleven individuals of mixed Spanish heritage, nine from France and eight from the United States.[17]

The commercial houses of Spanish origin included Gutierrez and Cie., which had its roots in Santander, Spain; Juan Ferrer y Otero, who was Mexican but had Catalan heritage through his father; Paullada and Cie., where Esteban Paullada, the founder, was born in El Carmen but had Spanish and French ancestry; Preciat and Gual, both born on the Catalan coast in Torredembarra and Valls; and Francisco Pallás and Cie., also of Catalan origin. Among the French, notable names included Benito Anizan and Cie.; the Paoli brothers, Casimiro and Pablo, born in Corsica; and Genoese merchant Juan Repetto Simonet, whose son, Manuel, married Ana Niéves, the daughter of Victoriano, nicknamed 'the Peninsular Croesus' due to his immense wealth, making him the primary creditor in the Laguna de Términos region.

As the nineteenth century progressed, these commercial houses extended their presence throughout all sectors of the local and regional economy. In addition to logwood production and trade, they were intermediary agents, provided credit, and were owners of ships and large extensions of valuable forest lands. Furthermore, these very families held the monopoly on the logwood exportation from El Carmen, with most of the timber being harvested from their own land (Figure 10.2). Gutierrez and Cie., based in Campeche, stood as the foremost commercial house in the logwood trade and partnered with Belgian consul, Pedro Requena, based in El Carmen. Other significant participants included the merchant Juan M. Roura, who held the position of consul for the United States and B. Anizan and Cie., directed by Benito Anizan and whose brother, François, was appointed vice-consul of France in 1853. These foreign merchant houses established a widespread network of political connections, with family members occupying diplomatic positions. As a result, they not only held control over forest lands and the logwood trade but also exerted influence over local political decisions, which facilitated their business operations.

Particularly noteworthy is the case of the B. Anizan Company and the Anizan brothers.[18] The Anizan brothers, Francisco and Benito, French immigrants originating from Sainte-Marie d'Oloron in the French Pyrenees, arrived in Mexico in the 1830s. They arrived through the port of Veracruz and promptly began setting up clothing shops. Over

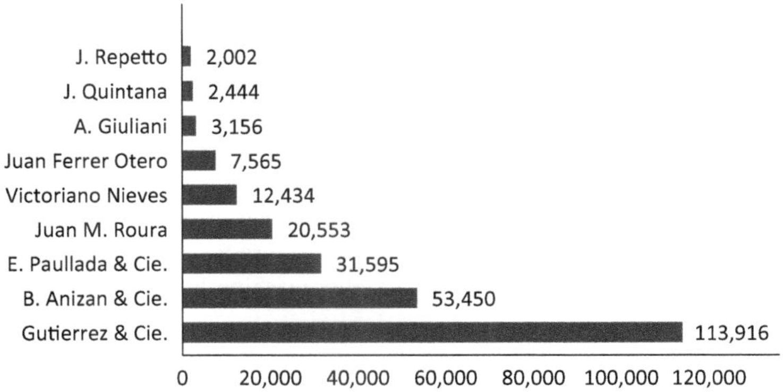

Figure 10.2 Logwood export consignees from El Carmen, 1849–69 (in tons). *Source*: Diplomatic Archives of Belgium, Ministère des Affaires Etrangères. Direction du Commerce et des Consulats. N° Perso 897. Consulat de Belgique a Laguna de Terminos (Mexique).

time, they diversified their ventures, expanding into the sugar trade, engaging in lending activities, and venturing into the production of leather goods and clay tiles. However, from its inception, the company's primary goal was to exploit and commercialize campeche wood, and to a lesser degree, other forest products such as mahogany and cedar. This family business acquired substantial land in the region, leveraging it for the extraction and transportation of logwood logs to Europe through the deployment of its own chartered boats. The expansion of Benito and François Anizan's businesses was only possible thanks to the new Mexican laws that allowed the commodification of public vacant lands in the Yucatan Peninsula. In addition to acquiring vacant lands, the company also functioned as a creditor, leading several haciendas and ranches to become the property of the Anizans due to non-repayment of debts.

Taking advantage of the aggressive land property policies implemented by Juarista and Porfirian governments, the Anizan family came to own 165,123 hectares of land between 1850 and 1896. Following the death of Benito Anizan in 1895 (who had taken over the family business after the passing of his brother François several years earlier in 1862), his successors decided to sell all of his properties to Ana Niévez de Repetto for a sum of 580,000 pesos. It is worth noting that Ana Niévez de Repetto was also a prominent landowner herself. She inherited the San Pedro estate from her father-in-law, Juan Repetto, which covered an area of 878 hectares. The southern boundary of this hacienda was marked as the demarcation line separating Mexico from Guatemala. By 1893, Ana Niévez de Repetto had significantly expanded her ranch's size to a vast 52,110 hectares.

Another telling example is the case of the San José de Aguada Seca estate, located adjacent to the San Pedro estate. In 1887, the Mexican Ministry of Development (*Secretaría de Fomento*) awarded the San José de Aguada Seca property to Phoebe Apperson Hearst, the wife of American magnate George Hearst. This estate encompassed an extensive area of 142,315 hectares of tropical forest. This land concession became possible due to a personal exemption granted by Mexican president Porfirio Díaz from the law that prohibited foreigners from owning land in border areas.

In summary, the territorial occupation of the Laguna de Términos and the surrounding forests during the nineteenth century allowed the conformation of a regional elite – white and Catholic – with property rights over large land extensions. This territory had one of the most precious dyewoods in international markets and the ideal conditions for logistics and trade. This merchant elite producing and trading logwood lived in urban areas, especially in El Carmen, and, to a lesser extent, in the thriving town of Palizada. From these municipalities, they exercised control over local logging and processing activities within the tropical forests of the Yucatan Peninsula and overseed logwood trade across the Atlantic, reaching European and American ports.

Labour dynamics

Throughout the logwood boom of the nineteenth century, the population inhabiting the tropical forests of the Yucatan Peninsula found themselves confined to progressively smaller communal lands (*ejidos*). The great majority were Indigenous Mayan people and *pardos* living in the towns formed during the sixteenth century's first wave of colonization. The population working on logwood production was mobile and suffered harsh and insecure labour conditions. As a scarcely populated area deemed a refuge during the Spanish colonial era and the regional conflict known as the Caste War (1847-1901), the primary concern voiced by forest landowners was the shortage of labour (*escasez de brazos).*

The Mexican legislation of the nineteenth century persisted in enabling the exploitation of forest workers. A relatively new debt peonage system ensured that workers could not leave forest camps, effectively harnessing labour control and restricting direct resistance. In 1859, the 812 men exclusively employed in logwood cutting in the Laguna de Términos area owed a collective debt of $155,546 to their respective owners, who were mainly members of the urban merchant elite mentioned earlier in this chapter.

Historical records clearly evidence the deplorable living and labour conditions of logwood cutters and other workers in the logwood economy. Working conditions in logwood camps of the Laguna de Términos were so appalling that, in 1842, the Government of the State of Campeche sent a commission to investigate 'the criminal traffic that they carried out in the robbery and sale of indigenous people and the mistreatment of their servants'.[19] The report of this official commission exposed a wide range of labour control mechanisms. It included not only personal indebtedness but the coerced transfer of debts among family members or from victims to inmates; forced marriages; payment of wages in non-negotiable cash tokens only redeemable at the local store; absolute prohibition of leaving the estate; and various forms of mistreatment such as women's haircuts, whipping and shackles for prisoners.

European travellers visiting the haciendas of the Yucatan Peninsula also left testimonies of labour conditions in the forests. For instance, the French naturalist Arthur Morelet observed in 1857 that:

> It would be impossible to exploit the forest without the help of the indigenous people. But, to have their labour at a vile price, it is helpful to bind them by employing a pecuniary obligation. It is a principle that the contractor never loses sight of them. The workers that he employs almost always depend on him; they settle in the place of exploitation with their wives and children, when they are married; they are given a poor hut and an axe, and then everything they need to subsist is sold to them because the hacienda is ordinarily situated far from markets. Believe it or not, the retail sale of these provisions, bought in bulk and at a discount, is often the most positive benefit of logwood cutting. The shameful balance of this activity sometimes rises to 150 or 300 percent in favour of the merchant, depending on his rapacity or that of his agents.[20]

In the second half of the nineteenth century, approximately 40 per cent of the land surrounding the Laguna de Términos was exclusively devoted to logwood cutting, 19 per cent to cattle raising, 12 per cent to sugar cultivation, and the remaining per cent to corn and vegetable crops.[21] Moreover, the number of workers employed in logwood cutting and transport increased in the area between 1859 to 1865.[22] In less than six years, the number of workers in the haciendas and ranches that cut logwood in the Laguna de Términos basin doubled, which shows the intensification of this activity during the second part of the nineteenth century (see Table 10.2). For instance, the Limonar hacienda went from 51 workers to 236, the San Juan Las Piñas ranch from 3 to 48, San Geronimo from 33 to 210 and the Anizan ranch from 42 to 113.

The full-time work and residence on haciendas and ranches entailed a reliance on food supplies, hand tools, and clothing provided by landowners and their foremen. However, the low salary of logwood workers did not sufficiently meet these needs. In addition to these goods, workers received regular payment in the form of coins minted with the name of the property and its owner. Nonetheless, this currency was only valid within the confines of the productive unit, as there was a store where workers could use these coins. Yet, the high prices at the store meant that workers' debts always seemed to remain the same. Hence, these stores played a dual role of indebting and isolating the workers from the wider society. The coercive system imposed in logwood production units aimed to maintain a cheap, hard-working, submissive, and loyal workforce by perpetuating the servants' indebtedness and keeping them bound to their obligations.

In theory, the workers had the freedom to chose their employers and switch occupations as they desired, on the condition that the new employer would pay off their accumulated debt. However, this so-called freedom was severely restricted by the weight of their debts, significantly limiting their actual choices and mobility. In this context, the most common means of acquiring workers was through the transfer of debt from one owner to another. This is evident from the records of the sale and purchase of ranches and haciendas, which included the names of the workers and the corresponding amounts of their debts.

Even after the French Intervention (1862–7), and despite the Law of 3 November 1868, working conditions in logwood production remained unchanged.[23] Judicial complaints from workers did not begin until some servants had access to federal

Table 10.2 Number of Employees in Selected Haciendas and Ranches in the Laguna de Términos for the Years 1859 and 1865 (Working in Logwood Production)

Name of the Hacienda / Ranche	1859 *	1865 **
Municipio de Palizada		
Hacienda Limonar	51	236
Hacienda San Geronimito (Anizan)	33	210
Rancho Anizan o Río Blanco (Anizan)	42	113
Municipio del Carmen		
Hacienda San Miguel del Pom antes Sitio Trinidad (Victoriano Góngora)	68	95
Hacienda Dolores (Antonio Gutiérrez Barrera)	17	35
Rancho Santa Gertrudis (Victoriano Nieves)	22	41
Rancho Quemado (Joaquín Jiménez)	24	43
Rancho San Juan las Piñas (Juan M. Roura)	3	48
Rancho La Montaña de Montecristo (Esteban Paullada)	34	123
Rancho San José Marentes (José Leyva)	40	96
Rancho San José del Este (Victoriano Nieves)	41	133
Rancho San Joaquín (Andrés Pérez Laforet)	31	122
Rancho San Antonio de la Sabana (Juan M. Roura)	27	58
Rancho San José de la Victoria (Antonio Gutiérrez Barrera)	60	107
Rancho Balchakah (Pilar Sánchez)	16	30
Rancho Encantada (Victoriano Nieves)	46	128
Rancho Salsipuedes (Andrés Aznar)	52	190
Municipio de Mamantel		
Hacienda Polvoxal	28	85
Hacienda Candelaria (H. Pauling)	61	101

Sources:
**Estadísticas del Estado de Campeche Agricultura e Industrias Anexas*. Bajo de la Administración de la Agencia del ministerio de fomento de Campeche, 10 volúmenes, 1859.
** Villegas, Pascale and Rosa Torras Conangla, *Naturalezas peninsulares, visiones imperiales. El Departamento de La Laguna-El Carmen en 1865-1866* (Mexico: Universidad Nacional Autónoma de México, Instituto de Investigaciones Filológicas, Centro de Estudios Mayas, 2018).

justice at the end of the nineteenth century. These court records show how the practice of the owners not accepting payment of the debt was maintained as a mechanism of subjection in logwood estates.

Conclusion

Laguna de Términos in the Yucatan Peninsula was the main centre of the logwood economic boom of the nineteenth century. A modernizing and transnational elite of wealthy merchants reinvented an Atlantic business exploting subaltern workers and accelerating the territory's environmental degradation. This oligarchy, largely composed of Europeans and Americans, resided in cities such as El Carmen and wielded a monopoly on logwood cutting, pricing and global exportation. For their part, local workers were a mobile community that remained hooked in a subordinate way to an extractivist economic logic. Despite the importance of logwood dyes in the colony and the nineteenth century, European textile factories and dye manufacturing workshops were never interested in importing this commodity already processed. It was more convenient for European interests and maritime merchants to use the weight of the raw logs for stability during the voyage of sailing ships. This is the reason why attempts to establish processing factories of dyewood in the Yucatan Peninsula, like 'La Estrellita' in El Carmen, ultimately failed.

After the decline of the Logwood economy, chicle production emerged as the predominant extractive activity that characterized Laguna de Términos in the first half of the twentieth century. As the value of logwood for international markets diminished, local merchants in the area transferred their lands to large American companies and capitalists. This tendency solidified land grabbing and reinforced a logic of economic dependence on commodity exports that proved extremely challenging to reverse in the twentieth century.

Notes

1 This chapter has received funding from the project 'La construcción histórica de una región depredada: la cuenca transfronteriza de la Laguna de Términos (siglo XVIII-XX)', PAPIIT-UNAM no. IN-401022.
2 R. Cruz and M. Sousa, 'Haematoxylum calakmulense (Leguminosae, Caesalpinoideae), una Nueva Especie Mesoamericana', *Novon: A Journal for Botanical Nomenclature* 23 (2014): 31–6; Joseph M. Gilbert, 'British Loggers and Spanish Governors: The Logwood Trade and its Settlements in the Yucatan Peninsula: Part II', *Caribbean Studies* 15, no. 4 (1976): 43–52.
3 Manuela Cristina García Bernal, *Campeche y el comercio atlántico yucateco, 1561–1625* (Campeche: Gobierno del Estado de Campeche, CONACULTA, 2006).
4 Pascale Villegas, 'Del comercio prehispánico al comercio colonial en la Nueva España. Entre trastorno y continuidad', in *Ideología y cultura en América colonial: ¿Puentes entre dos mundos?*, eds. M. V. Guevara Erra and M. C. Cordero Huertas (YBRIS, Universidad de Sevilla, 2011), 179–97; Pascale Villegas, *Las relaciones entre los aztecas y los mayas. Comercio, conquista e intercambio cultural* (España: Editorial Académica Española, 2012).
5 Charles François Du Fay, *Observations physiques sur le mélange de quelques couleurs dans la teinture* (Paris: Académie royale des sciences, 1737), 253–68; Juan Pablo Canals y Martí, *Instrucción sobre el conocimiento de prueba de la tintura de lanas y de sus texidos* (Madrid, 1764); Miguel Jerónimo Suarez y Núñez, *Arte de teñir las lanas, sedas, hilo y algodón* (Madrid, 1779); Eugenio Larruga, *Memorias políticas y económicas sobre los frutos, comercio, fábricas y ordenanzas expedidas par su gobierno y fomento* IX, 1790 (64); vol. XII, 1791 (67); vol. XIII, 1791 (68); XXVI (81).
6 Carlos Justo Sierra, *Breve historia de Campeche* (Mexico: Fondo de Cultura Económica, El Colegio de México, 1998), 68.
7 Víctor Alfonso Medina Lugo, 'Fear and Trepidation: The Socio-Cultural Impact of Maritime Piracy and illicit Smuggling in San Francisco de Campeche, 1630–1705', Master Dissertation, Missouri State University, Graduate Theses no. 3431 (2019); William Dampier, *Dos viajes a Campeche* (México: Miguel Ángel Porrúa, 2004), 53.

8 Rosa Torras Conangla, *La tierra firme de enfrente. La colonización campechana sobre la Región de Los Ríos (siglo XIX)* (Mexico: UNAM, 2012).
9 Alicia del Carmen Contreras Sánchez, 'El palo de tinte, motivo de un conflicto entre dos naciones, 1670–1802', *Historia Mexicana* 37, no. 1 (1987): 49–74; Alicia del Carmen Contreras Sánchez, *Historia de una tintórea olvidada. El proceso de explotación y circulación del palo de tinte, 1750–1807* (Mérida: Universidad Autónoma de Yucatán, 1990).
10 *Las mejoras materiales* (Campeche: Imprenta de la Sociedad Tipográfica por J.M. Peralta, 1859), 373–4.
11 For insights regarding the volume of commodities unloaded by European ships see *Las Mejoras y Materiales* (Campeche: Imprenta de la Sociedad Tipográfica por J. M. Peralta, 1859).
12 Pascale Villegas, *Del puerto de Marsella a las casas de Campeche. El comercio de las tejas de barro, 1852–1932* (Campeche: Universidad Autónoma de Campeche, Gobierno del Estado, 2014).
13 Claudio Vadillo López, *La región del palo de tinte: El partido del Carmen, Campeche: 1821–1857* (Campeche: Fondo Estatal para la Cultura y Las Artes, 1994), 100.
14 Rosa Torras Conangla, 'Palo de tinte y territorialidad en la península de Yucatán, en las postrimerías del periodo colonial novohispano', *Temas Americanistas* 43 (2019): 231–48.
15 Carlos Justo Sierra, *Acción gubernamental en Campeche* (Mexico: Imprenta de Estampillas y Valores, 1972), 33.
16 Rosa Torras Conangla, '¿Colonos o montaraces? Territorialidades en las selvas fronterizas de Guatemala y México (siglos XVIII-XX)', *Revista de Historia* 83 (2021): 5–23. See also, J. Fenner, *Los deslindes de terrenos baldíos en Chiapas, México, en el contexto internacional y nacional, 1881–1917*, Doctoral dissertation (Michoacán: El Colegio de Michoacán, 2010).
17 Padrón de los extranjeros avecindados en el partido del Carmen en Archivo General del Estado de Yucatán (AGEY), Fondo Poder Ejecutivo, Serie Censos y Padrones, Jefatura política del Partido de Campeche, caja 65, vol. 15, exp.18.
18 Pascale Villegas and Rosa Torras Conangla, 'La extracción y exportación del palo de tinte a manos de colonos extranjeros. El caso de la B. Anizan y Cía', *Secuencia* 90 (2014): 77–93.
19 Archivo General del Estado de Yucatán (AGEY), Poder Ejecutivo, Ramo Gobernación, CD 15, caja 50, vol. 3, exp.54. Año 1842 f. 1.
20 Arthur Morelet, *Viaje a América Central, Isla de Cuba y Yucatán*, ed. Carolina Depetris and Rosa Torras Conangla, Vol. II (Mérida: UNAM, 2018, original: 1857), 107–9.
21 Vadillo López, *La región del palo de tinte*, 56.
22 Estadísticas del Estado de Campeche Agricultura e Industrias Anexas. Bajo de la Administración de la Agencia del ministerio de fomento de Campeche (10 volumes, 1859). See also Pascale Villegas and Rosa Torras Conangla, *Naturalezas peninsulares, visiones imperiales. El Departamento de La Laguna-El Carmen en 1865–1866* (Mexico: UNAM, 2018).
23 Law published in *El Espíritu Público, Periódico semioficial del Gobierno del Estado de Campeche*, No. 140 (6 November 1868).

11

Costa Rican neotropical dyewoods in global context, 1885–1940

Anthony Goebel McDermott and Ronny J. Viales Hurtado[1]

Introduction

The history of the exploitation and exports of precious woods and dyewoods in Central America has been overshadowed by that of other tropical commodities that dominated the history of globalization in the region from the late nineteenth century. In Costa Rica, coffee and bananas generated a regional specialization in production, export, transportation and marketing of a transnational nature from the 1880s onwards. The actors and networks engaged in the production and trade of these plantation commodities extended beyond the boundaries of the national state. Moreover, export markets were situated in distant states and regions.[2] In recent decades considerable progress has been made in the analysis of the Latin American 'agro-export model' (or export-led growth), characterized, in general terms, by dependence on the international markets, foreign capital and credit concentration.[3] In the case of Costa Rica and Colombia, for example, much emphasis has been placed on the so-called 'coffee oligarchy' and transnational companies and their partners, mainly the United Fruit Company.[4] These studies have emphasized mono-export (not monoculture), the tendency to increase imports and the fiscal dependence on indirect taxes on imports and, to a small extent, on regressive exports, at least until 1950.[5] However, more studies are still needed to shed light on regional differences in this agro-export model, in line with the work of Pérez Brignoli on the dynamics of the banana export economy in Central America.[6]

This chapter situates the Costa Rican economy in its broader international framework, aligning with recent trends in global economic history.[7] More precisely, our purpose is to analyse the process of transformation of Costa Rican forest resources into export commodities from the origins of agrarian capitalism to the present, a subject which has been studied in some detail but that needs to be placed in the context of the broader historiography dealing with global commodity chains.[8] We focus our attention on the economic and socio-environmental dynamics generated from the exploitation of the dyewoods in the neotropics, in particular *Palo de Mora* (mulberry tree), for both artisanal uses and conspicuous consumption of wealthy elites in industrialized nations in Europe and North America. We illustrate the transition from the occasional exploitation of some precious woods of the North Pacific area of Costa Rica to new extractive forms of commercial exploitation of forests in new zones, among them the peripheral regions of the Central Valley and the Atlantic and Caribbean regions. Our study starts from the conceptualization of forest exploitation as an activity endowed with its own dynamics and not as a simple prerequisite for the reorganization of the natural biophysical environment with a view to introducing higher value-added activities. To this end, we look at timber exports from Costa Rica (by species) and the specific demand by the main importing countries, as well as their evolution over time, to account for the rise of global resource trade and its socio-environmental consequences in this country.

The dye commerce in Central America has a long-standing tradition. It is worth recalling that the product that exerted the greatest weight in the regional economy towards the end of the colonial period was indigo, whose profits were reinvested in highly profitable activities of the time such as cattle raising. The indigo boom was spurred by its high demand in the market of natural dyes that occurred during the first stages of the European industrialization, particularly in Great Britain, that had turned into 'the workshop of the world'. The dye business grew dramatically from the middle of the eighteenth century until 1799 when it began to decrease.[9]

The increase in the indigo production as the lead export product of the *Reino de Guatemala* – which at the time comprised what today are the republics of Guatemala, El Salvador, Honduras, Nicaragua, and Costa Rica, as well as the Mexican State of Chiapas – encouraged the inter- and intra-regional productive specialization, becoming one of the driving forces in environmental change in the region.[10] On the other hand, dyewoods played a less significant role in the regional economy than indigo. Nonetheless, it cannot be overlooked that since the sixteenth century dyewoods were extracted in small quantities from the tropical coast of Belize. Although the commercial dynamics were relatively limited, in a few decades the wood from easily accessible places disappeared almost completely from the coasts of Belice and were the first tropical species of Central America to become severely exhausted. However, the importance of the exploitation of the mulberry trees in the Nicoya Peninsula (in the North Pacific coast of present-day Costa Rica) should not be ignored. It was spurred by the lifting of the ban in 1774 of the commercial exchange between the viceroyalties of New Spain and Peru, as a part of the new commercial policies adopted by the Bourbon monarchy. The brazilwood of Nicoya was exported to Peru in exchange for a variety of goods, and was considered 'one of the goods that played a main role in the commercial exchange in the first years of independence of the Central American provinces and one of the oldest in all its economic history'.[11]

What seems unusual at first glance is the exploitation of dyewoods during the full industrial apogee, when according to the conventional narrative the synthetic dyes had replaced those of natural origin. However, this is precisely the case of the mulberry trees. The commerce of this dyewood between the end of the nineteenth century and the first decades of the twentieth has been relatively neglected, as has occurred with other dyewoods. Such is the case of the logwood exploited massively in the Yucatan Peninsula and recently rescued from the historiographical oblivion through a deep analysis of its exploitation and interrelations with the technological changes from a global perspective.[12] The same situation occurred with other natural dyes of extractive origin such as cochineal (*grana cochinilla*), whose production, consumption and marketing have received more attention in the historiography.[13]

Our analysis is based on the premise that the processes described above generated constant pressure on the country's forest cover in Central America, due to the dynamic and changing nature of the timber business, in constant adjustment to the demands of the world market, as well as to the ecological possibilities of exploiting forest resources.

Timber species and economic exploitation (1883–1955)

The analysis of the volume and types of timber exported from Costa Rica between 1883 and 1955 indicates two clear trends. On the one hand, we underline the variety of wood exports, since a total of 53 timber species were traded abroad. On the other hand, we emphasize the visibly differentiated character of the quantities exported, and therefore, of the economic pressure on the species considered as timber, something that was changing throughout the period studied.[14]

The species that received the greatest pressure from the world timber market was Cedar (*Cedrela* spp.), exploited in an extractive manner mainly in the northern Pacific region of the country until the remnants of primary and secondary forest in Guanacaste were almost wiped out, at least in exportable quantities. We consider that these exports were primarily of Bitter Cedar (*Cedrela odorata L.*) and Sweet Cedar (*Cedrela tonduzii* C.DC) given their relative abundance and varied uses and, by the end of the study period, they were no longer being exported.[15] During the entire period, timber valued at US$4,899,386.91 was exported.

In the second place, as the most exported species from Costa Rica and, very distant from Cedar, was the 'palo de mora' (*mulberry tree*), another precious wood of extractive exploitation of which US$1,250,783 was exported. The mulberry tree (*Chlorophora tinctoria*) is a shrub or tree, between 2 and 30 metres high, sometimes with axillary thorns and glabrous stems with whitish latex. Its geographical distribution ranges from Mexico and the Antilles to the north of Argentina. It is a plant of humid or dry climates, rare in primary forests and generally grows on the banks of rivers. It is deciduous during the dry season. In Costa Rica, it is a very uncommon species in primary forests while more abundant in secondary forests, but its regeneration is very scarce.[16]

Palo de mora produces a wood that although moderately difficult to work, and whose immunization is virtually impossible, has the advantages of its high natural durability, due to its remarkable resistance to fungi and insects, which allows it to have a duration in exterior use of more than fifteen years. Among its many uses are joinery, heavy construction, shipbuilding, carpentry, railroad ties, bridges, and others.[17] As can be seen, this wood was of particular use in vital infrastructure elements in the processes of capitalist modernization that were taking place throughout the nineteenth century and beyond, both in the centre and on the periphery of the world economy. It was also the only dyewood exported in considerable quantities from Costa Rica towards the end of the nineteenth century. In addition to the uses mentioned above, this wood species was a commercial source of a yellow dye (*maclurin*) used for colouring handicrafts and other textiles.[18]

Palo de mora: A commodity or a specialty?

According to Tarek Ismail Kakhia, since the eighteenth century there were two natural yellow dyes that were important in Europe and were extracted from American trees.[19] They were quercitron, from the quercitron oak (*Quercus velutina*) native to North America and the 'yellow mulberry' (also known as *fustic*, *palo de mora* or *Maclura tinctoria*) that was located in the Caribbean (West Indies, Mexico and Cuba) and that was highly valued.[20] It is appropriate to point out that, in addition, the variety of *palo de mora* (*Chlorophora tinctoria*), from which the *old fustic* yellow was extracted, had been imported from Cuba, Jamaica, Puerto Rico and other Caribbean islands since the sixteenth century.[21]

Historically, one of the most intensive applications of palo de mora (*Chlorophora tinctoria*) has been as an industrial dye. The yellow dye extracted from this species acts quickly on textiles when used with mordants that allow the fixation of colours in textiles, whether they be metallic salts, acids or organic substances. This means that it can be mixed with other dyes and mordants to produce a variety of colours ranging from yellowish to greenish. Thus, the *khaki* colour (*kakhi*) typical of military uniforms can be obtained with *palo de mora*, which could explain, as a first approximation, the prominent place of this wood in Costa Rican exports in the period studied, as well as its current scarcity.

Khaki military uniforms were introduced by Sir Harry Burnett Lumsden and William Stephen Raikes Hodson in 1848 for British colonial troops in India and proved effective for field service and battle. *Khaki* was widely used at the time of the Indian Rebellion (1857–8) and subsequently served as the official colour for the uniforms of British, native and colonial armies. It was later adopted in other parts of the British Empire spreading gradually to other nations around the globe.[22] In addition, military uniforms of this colour constituted a camouflage technique that was used during the First and Second World Wars, as well as in other conflicts located outside Europe and America.[23] Therefore, *palo de mora*, more than a commodity, was a specialty, being a type of product with a selective demand by a particular market niche, especially as a natural dye for textiles.

The dye industry suffered an important impact between 1775 and 1860, when W. H. Perkin innovated by manufacturing the first coal tar dyes and, in 1869, when alizarin was synthesized as a red dye for the uniform of the English parliamentary army, as the first natural pigment to be synthetically duplicated.[24] This resulted in a decrease in global demand for all natural dyes.[25] Although this may have been a general trend, it is clear that the demand for *palo de mora* had other important cycles, even after the innovation of synthetic dyes, because some of these presented problems to fix the colour in some fabrics. For example, other varieties of *palo de mora* were an important inputs for silk production in Utah, United States, in the 1860s.[26]

Costa Rican timber exports and environmental impacts (1883–1955)

The cycle of Costa Rican exports of the ten timber species most in demand during our study period shows a clear transition. First, there was a phase clearly dominated by the export of a few slow-growing, extractively harvested precious timber species, especially in the North Pacific, where cedar and *palo de mora* paid, for the most part, the high ecological price of the world demand for Costa Rican timbers. Second, there came a phase marked by the predominance of fast-growing 'cultivated' woods, especially *balsa*, and other extractive but mass-consumed woods, such as *cativo*, obtained from new logging regions for export in Costa Rica, in the context of a diversified and increasingly mass-produced supply (see Figure 11.1).

The first phase occupies most of the period analysed here. It is located in the subperiod 1883–1941, where the final year was the last to record cedar exports of some importance. The second phase begins in 1942, a year that registered a clear predominance of balsa exports over any other species, until the end of our period, when male cedar, bitter bedar, cativo, and balsa dominated the international market for Costa Rican timber in similar proportions. At the same time, we could consider the subperiod between 1920 – the first year in which balsawood exports were recorded – and 1941, as the transition period between one type of commercial logging and the other.

It is worth noting that, although Cedar was the most exported species in the first phase, which we will call the 'extractive phase', when it accounted for 93.21 per cent of Costa Rica's total timber exports (1907), the fact was that *palo de mora* represented 74.32 per cent of the exportable supply in 1894. On the other hand, during the second phase, which we will call the 'agro-industrial phase' due to the dominant type of exploitation, balsa exports reached 86.26 per cent of Costa Rica's timber supply. This historical maximum was recorded in 1944, which seems to leave no doubt of the high demand for this wood, especially by the United States in the context of the Second World War, while at the same time it seems to account for the success of state policies aimed at stimulating the development of balsa 'plantations', reflected in projects such as the contract signed between the government and the International Balsa Company S. A. As a result, balsa could be considered in this phase as the prototype of cultivated export timber, even though it lost ground in the last years of our range of analysis to other timbers with regard weight, such as cativo, which accounted for almost 50 per cent of timber exports in 1952. The ecological bill of both phases was undoubtedly high, although its features were clearly differentiated in a manner analogous to the type of exploitation.

In the case of the 'extractive phase', it seems clear that both the destruction of the forest and the expansion of agro-livestock activities associated with this change in land use that took place in Guanacaste[27] had a decisive impact on the irreversible alteration of the trophic webs and the hydrological cycle – altering the climate of the region, the intensification of the erosion processes, especially visible in the rainy season, and the fragmentation of the forests, which implied the reduction of the area and connectivity

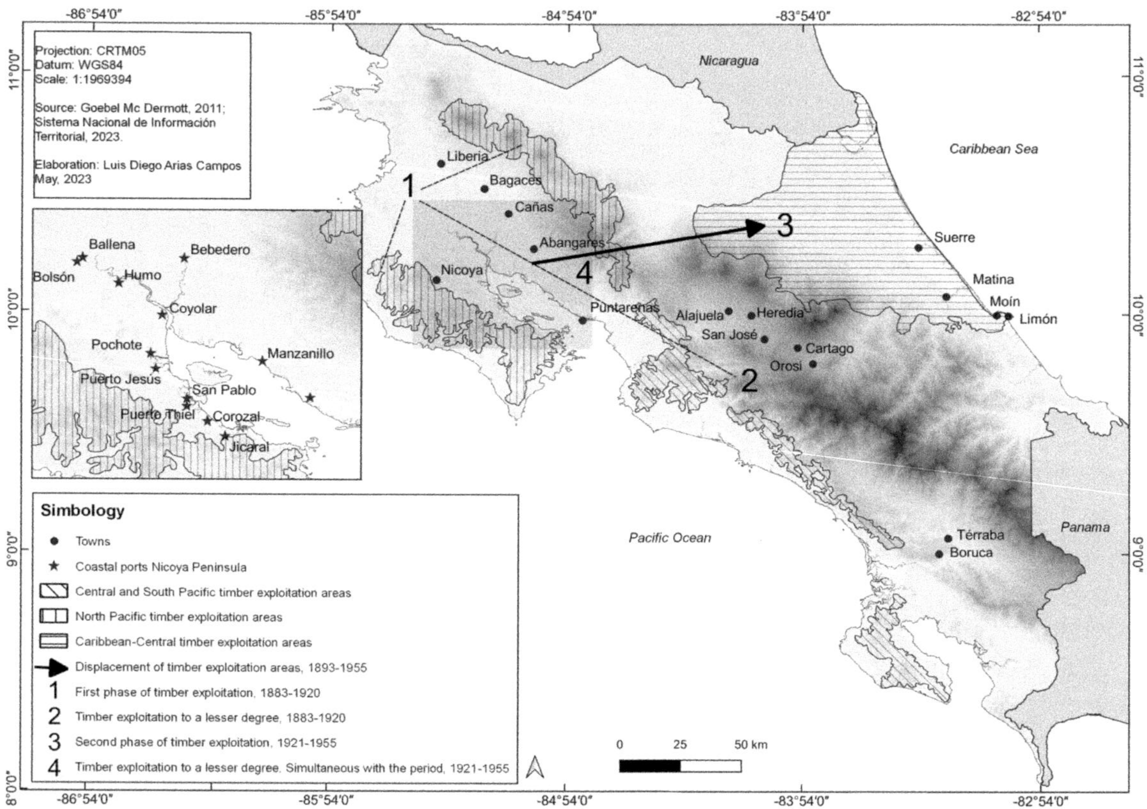

Figure 11.1 Phases of the economic exploitation of the forests in Costa Rica (1893–1955). *Sources*: Anthony Goebel Mc Dermott, 'El mercado mundial de productos forestales y la presión económica sobre los bosques de la Costa Rica "liberal": hacia una tipología básica (1883-1955)', in Ronny Viales Hurtado and Anthony Goebel Mc Dermott (eds.) *Costa Rica: Cuatro Ensayos de Historia Ambiental* (San José, C.R.: Sociedad Editora Alquimia 2000, 2011): 125–96; Sistema Nacional de Información Territorial. 'Visores temáticos'. Accessed 20 May 2023. Elaboration: Luis Diego Arias Campos, May 2023.

between fragments, just to mention some of the consequences of this intense pressure of the international market on Costa Rican forests, especially those of Guanacaste.[28] As a result of the depredation of Guanacaste's forests given the desirability of their timber on the world market, by 1910, 50 per cent of this type of timber had already been felled.[29]

From the point of view of ecological economics, we can say that this depredatory process of extractive exploitation squandered a good part of the natural capital of the Costa Rican North Pacific region. Given the virtually free access to exploitation of the forests and the non-internalization of the present and future environmental cost, the destruction proceeded under the strict observance of the profitability criteria characteristic of the capitalist economy. During our period of study, the ecologically unequal exchange between Costa Rica and the natural resource importing countries – in which timber and the exploitation of forests played an unfortunately prominent role – was consolidated. As a result of the constant expansion of the ecological footprint, such exploitation absorbed more and more of the environment from the natural capital supplying countries, generating an ecological debt that has not been sufficiently recognized to this day.[30]

Given the long history of the extractive exploitation of precious wood species, and the slow or sometimes impossible replacement of felled trees, it is worth noting that most of the species whose wood was in great demand in the international market at this stage were now threatened or endangered due to their diverse uses and special qualities for processing.[31] Some authors, such as Jiménez, consider that the process of incessant deforestation, that the planet has undergone and is undergoing, will inevitably lead to 'the extinction of hundreds of thousands of plants, which will undoubtedly result in genetic isolation that will produce inbreeding and, subsequently, the loss of their genetic variability'.[32] This profound impact of deforestation is a long-term historical process that sunk its roots in the material and symbolic consolidation of the environmental premises of Western modernity. Indeed, we would like to point out here that deforestation in Costa Rica is not only a product of the socio-economic, cultural or socio-environmental changes of the last twenty years. This had its corollary in nineteenth-century economic liberalism, where nature, and in this case forests – as one of its central components – had no value before its insertion into the capitalist world market, where exchange value irreversibly replaced use value.

The ecological and environmental transformations generated in the 'agroindustrial phase' were not minor. As mentioned earlier, the deforestation of the biodiverse forest which led to the planting of uniform forests of a single species, following the nineteenth-century management standard of maximum sustainable timber production, caused the loss of many of the ecological functions of forests, as well as their essential production for human life and sustenance.[33] The simplification of ecosystems, land degradation, loss of fertility, water retention – since plantations of fast-growing exotic species are generally considered more water consumers than producers[34] – and the loss of grass for grazing are, as we have already said, only some of the ecological functions lost in the substitution of the 'real' forest by commercial plantations.[35] In fact, as recent studies have shown, maintaining the forest adjacent to the cloud forest is an effective way to maintain ecological functions.[36]

We have so far tried to explain how factors such as changes in consumption patterns by the main importers of Costa Rican timber and the relative depletion of various species – whose fine timber was widely sought after on the international market – had a decisive impact on the market pressure on certain species of the country's forest cover. But it is no less true that the *forest footprint* of the different importing countries must have been differentiated in terms of species. This is because, as we know, consumption is not only related to price and market conditions but is also a cultural and historical construct with a certain degree of autonomy.[37] Thus, the material and symbolic needs for forest products were not identical among all countries importing Costa Rican timber, and this must have been reflected in the differentiated pressure that each of them exerted on the different exportable species. The analysis of these specificities of timber consumption patterns, to which we will turn next, help us to describe the way in which certain countries historically conditioned the intensity with which certain species were exploited and inserted into the world market.

Timber and markets: Main importers of Costa Rican timber by species (1883–1955)

As already mentioned, timber consumption has played a significant role in driving deforestation. In the global context, the first thing that allows us to corroborate the available information is that the United States was the largest ecological debtor of Costa Rica considering only its forest cover.[38] This country led the imports of four of the six species that we have selected in order to study the relationships between markets and types of wood exported. Of all the mahogany (*Swietenia humilis* Zucc., and *Swietenia macrophilia* King)[39] exported by Costa Rica in the years for which we have information, 57.79 per cent went to the United States. The United States was also the leading importer of cocobolo (*Dalbergia retusa* Hemsl.) (77.78 per cent) and palo de mora (45.13 per cent), followed closely by France (40.18 per cent), and accounted for virtually all of Costa Rica's balsa exports (96.68 per cent) (see Figure 11.2). Three of the timbers of which the United States was the leading importer are high-value precious woods not available in its vast, though decreasing, forest cover. Mahogany and Cocobolo, it is worth noting, were very fine woods with multiple uses, many of them for the luxury consumption of the wealthiest sectors of society. Thus, these fine woods were used in the manufacture of furniture, handicrafts, sports objects (such as chess pieces), tool handles, musical and scientific instruments, piano cases, veneers and plywood, among others,[40] which also gives us a clear idea of the industrial activities, both in the United States and in other centres of world capitalism, that generated the greatest destruction of forests during an extensive period of time, especially in Guanacaste.

Balsa, likewise, being grown in tropical America and not in US soil, was widely sought after for its versatility and wide possibilities in the construction of military equipment, as well as in the manufacture of toys, models and many other uses in which its intrinsic features, such as its low specific weight, resulted in an undeniable advantage over other materials, as mentioned above. The United States, in this way, led

Country	CEDAR	CAOBA	COCOBOLO	MULBERRY TREE	BALSA	ESPAVEL
United States	7.30	57.79	77.78	45.13	96.68	0.00
France	0.32	3.10	1.92	40.18	0.01	0.00
Germany	63.11	7.17	12.74	7.00	2.69	0.00
United Kingdom	3.09	18.53	0.67	5.40	0.01	0.00
Spain	0.04	0.08	0.00	0.00	0.00	0.00
Italy	0.42	6.48	0.53	0.09	0.00	0.00
Holland	0.30	0.01	0.06	0.00	0.27	0.00
Denmark	0.02	0.00	0.00	0.00	0.00	0.00
El Salvador	0.04	0.08	0.00	0.00	0.00	0.00
Peru	24.06	4.13	0.36	0.33	0.00	100.00
Panama	1.27	2.36	4.02	1.87	0.34	0.00
Honduras	0.00	0.00	0.00	0.00	0.00	0.00
Colombia	0.01	0.18	0.00	0.00	0.00	0.00
Nicaragua	0.02	0.00	1.93	0.00	0.00	0.00
Chile	0.01	0.07	0.00	0.00	0.00	0.00
TOTALS	100.00	100.00	100.00	100.00	100.00	100.00

Figure 11.2 Costa Rica: Export of wood species by country (percentage of total exports) (1883–1955). *Source*: Dirección General de Estadística y Censos (DGEC). *Anuarios Estadísticos*, years 1883–1955 and *Resúmenes Estadísticos*, years 1883–1910. Authors' calculation.

the changes in logging by species, as it was at the forefront of imports of both precious woods for the conspicuous consumption of the wealthier sectors, and 'cheap' woods for mass consumption, whose use became widespread, among other reasons, due to the creation and expansion of the suburbs characteristic of the middle class in that country from the 1950s onwards, although the boom of the war economy had already stimulated this consumption in the first half of the 1940s. This demand was closely linked to the increase and diversification of the consumption of forest products in the United States, which had led it to become a net importer of such products after having historically been a net exporter, especially of timber. Indeed, by the mid-eighteenth century, for example, every port on the Eastern coast from Falmouth in Maine to Sunbury in Georgia was exporting lumber, mainly to the Caribbean.[41]

This panorama contrasts, ostensibly, with that presented by the United States in the first decades of the twentieth century when, according to the meticulous work of Zon and of the Research Section of the United States Forest Service, being the first to make an estimate of the planet's forest resources, the production-consumption ratio of forest products showed a massive deficit in the developed world. While production was 845.4 million cubic metres, consumption reached 1312.4 million cubic metres.[42] This generated a deficit of 467.0 million cubic metres, of which an impressive 638.7 cubic metres (above the overall deficit of the developed world[43]) corresponded to the United States, which, as can be seen, expanded its *forest footprint* globally. This is also evidenced by the fact that, at the same time, the United States accounted for 15 per cent of world timber imports.[44]

The US market, however, was not alone in leading the economic exploitation of certain species. In the case of cedar, Germany was the country that overwhelmingly concentrated the demand for this wood, with 63.11 per cent of the total imports of this species, which made it undoubtedly the largest importer in the entire period under analysis. Peru was the largest importer of Costa Rican timber in Latin America, accounting for all exports of espavel, and was also the second largest importer of cedar (24.06 per cent) (see Figure 11.2). The high consumption of cedar by Germany can be explained by Europe's traditional import of precious woods from the tropics, especially for the manufacture of fine furniture, musical instruments and other luxury goods, although, as Michael Williams points out, in a context of growing scarcity of timber that dominated the timber trade since the late nineteenth and early twentieth centuries, tropical forests were beginning to be conceived simply as a source of timber, regardless of its type,[45] which seems to confirm the growing process of commodification of forest products in general, as mentioned above.

The high Peruvian imports of espavel seem more difficult to explain, although in this case the picture is not entirely clear. Michael Williams, in one of his works, gives an account of the existence of vast timber marketing networks on both sides of the Pacific Ocean where, for example, the 'cheap' timber massively exploited in Oregon, met the construction needs of cities and countries that by the end of the nineteenth century were not self-sufficient in timber for this type of activity. This was also the case of Victoria (Australia), particularly after the excesses that followed the gold rush, which was supplied with the hardy redwood used to produce railroad ties.[46] Within these Pacific countries with a deficit in timber production, Williams highlights the fact that Ecuador, and more specifically Peru, proved to be lucrative markets for timber entrepreneurs settled along the Pacific Ocean coast.[47] According to Williams, this deficit consequently led to dependence on foreign timber, given the accelerated deforestation caused by the introduction and generalization of diverse economic activities such as mining, cattle raising and export agriculture that were present in the other countries of the Pacific basin, with the notable exception of New Zealand, where a flourishing timber industry developed.[48] It is not surprising, then, that low-value but resistant and useful woods, especially for construction work, such as espavel, of which Costa Rica still had important remnants in the first decades of the twentieth century, were exported to countries that, like Peru, had long required enormous quantities of this type of forest products. To this we must add the possibility that Peru was a centre for the collection and redistribution of timber to other South American countries.

The UK, on the other hand, exerted most pressure on mahogany, by becoming the US escort in the import of this wood, accounting for 18.53 per cent of total exports, which seems to be related to the export of fine woods by European countries, as mentioned above. In fact, England stands out for its constant consumption of this type of wood.

Although many countries imported Costa Rican timber at one time or another (a total of fifteen), it was the five countries mentioned earlier that imported constant and considerable quantities and thus exerted real pressure on the country's forest cover. This pressure, as can be clearly seen, was differentiated according to species, although the United States, it should be noted, exerted the greatest on the country's timber species, something it shared to a very limited extent with its importing counterparts.

The United States seems to have required a high and constant quantity of Costa Rican timber for various uses, while the other countries seem to have required huge quantities of a single species for specific uses and marginal quantities of other timber to supplement their regular supply. These were obtained from other countries and regions of the planet, as was the case, we presume, of the European powers which, for much of our period of study, had colonies or nations with preferential treatment that supplied them with the timber required for their economies under favourable conditions.[49]

The US predominance in the exploitation of Costa Rican timber by species could also be explained by other factors such as the proximity and size of the Costa Rican economy. However, this should be taken with due caution, as it is worth remembering that an economy with a marginal insertion in the world market, such as that of Peru, concentrated all exports of a single species, so there seems to be no doubt that non-structural aspects such as timber consumption patterns within the different countries should certainly be considered when analysing the economic pressure on certain timber species. Now we will analyse the commercial situation of the *palo de mora* exported from Costa Rica.

The *palo de mora* business cycle from Costa Rica

According to the data available, the total value of timber exports from Costa Rica between 1883 and 1955 was US$4,899,386.91 and the total value of *palo de mora* exports, though concentrated in the period 1885–1940, represented 25.5 per cent of this total, making it a very representative export item.[50] In its commercialization, there were two clearly differentiated export cycles from Costa Rica: 1885–1911 and 1920–40 (see Figure 11.3). This is associated with a demand that is not located in the periods of the First World War or the Second World War, but rather seems to be due to a demand of the textile industry for non-war consumption, linked to fashion, contrary to what we had previously stated. The greatest demand was concentrated at the end of the nineteenth century, especially after 1887, where the year 1894 stood out, when palo de mora accounted for almost 75 per cent of the country's wood exports. This first export cycle was much more dynamic than the second, and 1921 was the most representative year, with 40 per cent of Costa Rica's total wood exports.

If we disaggregate the first export cycle, there is a decreasing trend due to the length of its commercial duration, as well as fluctuations, although there are very significant peak years in the total wood exports. It is worth mentioning that in the first part of this cycle, and even before then, the United States also imported *palo de mora* from Puerto Rico.[51]

In the second cycle, the proportion of *palo de mora* in the total wood exports was lower, with the exception of 1921, but it is clear that there was an important demand for natural dyes even after the invention/innovation of synthetic dyes, because some cotton textiles and other fabrics had problems to fix this type of dye. In addition, *khaki* (khakhi) later became a fashionable trouser style, which meant a growing demand, in spite of other producers of *palo de mora* in the world that supplied the European market.

The major markets for Costa Rican *palo de mora* in the period 1883–1955 were the United States (45.13 per cent) and France (40.18 per cent). While the European market also required this raw material, in the case of Germany (7 per cent) and the UK (5.4 per

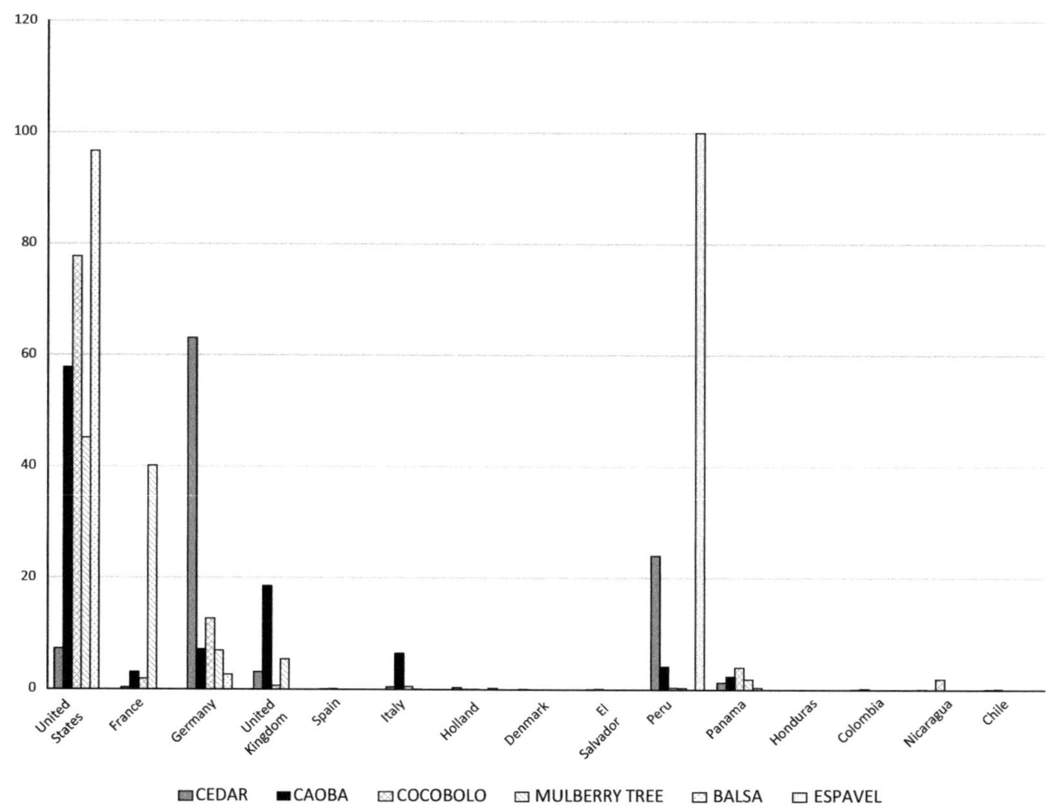

Figure 11.3 Costa Rica: Percentage share of *palo de mora* exports in total timber exports (main export cycles of 1885–1911 and 1920–40). *Source*: Dirección General de Estadística y Censos (DGEC). *Anuarios Estadísticos*, years 18831955 and *Resúmenes Estadísticos*, years 1883–1910. Authors' calculation.

cent) this product tended to be provided by suppliers from their colonies; while Italian demand was much lower (0.09 per cent). The European market imported 52.67 per cent of the Costa Rican *palo de mora* total exports, but if we add the other industrialized countries market, we see that together they imported 97.80 per cent. As for the Latin American market, it was undoubtedly limited. Panama imported 1.87 per cent of the total exports and Peru 0.33 per cent, for a total of 2.2 per cent.[52]

The economic importance and commercial dynamism of the mulberry tree – especially during the first cycle – is evidenced by the data of the newly found exports and by the sociopolitical dynamics of the time. The latter can be traced through the permits requested to the government for the exploitation of the tinctured wood in the national wastelands, as well as from the conflicts that arose as a result of the economic revenues from the activity. The conflicts occurred among diverse social and institutional actors, but especially among logging entrepreneurs and representatives of municipalities located in the exploitation centres.

For example, in 1889, when the first export cycle of mulberry was picking up, Juan José Matarrita y Gutiérrez, a resident of Nicoya, Guanacaste, addressed a letter to the Secretary of Public Works, in which he applied for a concession to exploit the timber on a vast stretch of state-owned land. The most remarkable thing about this communication is the justification. The applicant indicated that the land 'can produce more or less, a third or half of a shipment of mulberry wood, from whose exquisite dye (*sic*) is extracted abroad, a variety of colours, which as is well known have corresponded with good success in all circles of the commercial world'. The request also demonstrated that although the cost of developing such a company was high because of the distance and the difficulties of transporting the timber to the ports of embarkation, investors were in a position to take risks and develop such a business and make use of 'stagnated things in the center of our mountains', whose exploitation had been hindered either by lack of capital or 'arms' (the latter meaning 'human power').[53]

It is worth paying attention to the applicant's knowledge of the dynamics of the world market of the time and especially of the boom of the mulberry in the global market of natural dyes, which, as already mentioned, did not languish in its entirety after the appearance of synthetic dyes. However, it is also likely that the applicant's interest in the economic activity he proposed to undertake with high costs and difficulties presumably represented a way of circumventing the imposition of taxes or commissions by the government, aside from the ones already established. In the same sense, it affirms the statement that mulberry trees were 'stagnant in the center of our mountains', because exports of this product suggest that the mountains have lost much of their forest cover daily, being the mulberry one of the most exploited species.

The successful and lucrative business of the exploitation and export of this species, however, did not go unnoticed by local and national authorities. In 1893, for example, investor Félix Carazo sent a letter to the Ministry of Development of the Republic protesting the Executive's approval of a tax promulgated by the municipality of Nicoya, that consisted of a tax of fifty cents on each ton of mulberry and each 'construction tree' extracted from that canton in Guanacaste, which also confirms the preeminence of the Costa Rican North Pacific as a site of exploitation par excellence of fine woods extracted from the deciduous and semi-deciduous forest of that region, including the mulberry.

The plaintiff denounced, as in the previous case, the injurious nature of the tax due to the high costs of the wages, the price of each plant usually purchased from private individuals, and the low selling price of the good to exporting companies. To this he added the difficulties of transportation due to the poor condition of the roads, especially during the rainy season, combined with the unnecessary character of the tax in view of his generous contribution to the Municipal Treasury. The tax, the investor summarized, would discourage the logging activity, which he considered of great benefit 'because it circulates the monetary resources', to the point of making it disappear.[54]

The municipal body disagreed with the investor. In response to the communication sent to the Executive by Carazo and to the government request to submit a report regarding Carazo's demands, the Nicoya Municipal Council responded in very different terms to those put forward by the investor. The Council indicated that in the report, contrary to what he implied, the tax was established due to the lack of resources the municipality had to deal with, since the few rents it collected were insufficient to meet its most urgent needs, such as the construction of a municipal house, a prison, a trail, a cemetery, and the improvement of the roads of the canton among other works of 'recognized public utility'. It was implied that the tax did not harm the residents in any way, since the exploitation of the tree 'not just recently but for some time has been a business of certain entrepreneurs, to the detriment not only of the municipality but of the Nation', when exploited almost exclusively in public property (national forests and municipal lands) without paying any tax to the municipality. This would clearly be to the detriment of the community, according to the report, as logging entrepreneurs are the only ones to benefit from a lucrative activity from which they obtained 'huge profits'.[55]

The dispute helps to reaffirm the economic importance of the mulberry in the context of the boom of extractive exploitation of precious woods and the first cycle of the export of this dyewood. The fact that the tax in dispute was established specifically for such timber and those intended for construction seems to be a clear indicator that their export was an important business that did not go unnoticed by local and national authorities who did not hesitate to seek tax benefits from an activity that, at least according to their argument, was developed on public land. It should be noted that this was an eminently distributive conflict.[56] In the dispute there is no reference to the value of the timber and forest other than that of the strictly economic and specifically commercial dimensions. It was considered a resource to be exploited (whether for private gain or for the welfare of the community) without going into considerations about the need to conserve forest cover or regulate its exploitation, as it happened in other types of conflicts during the same period.[57] This utilitarian logic, present in the dispute, was an integral part of what in other places has been called the liberal environmental regime and would ultimately be detrimental to what both actors were ardently defending: the economy and society.[58]

Conclusion

Despite the diverse conditions that existed in the economic exploitation of forests in Costa Rica, deforestation emerged as a pervasive outcome of the broader process of territorial reorganization aimed at creating what O'Connor has called a 'new nature', specifically within a capitalist framework.[59] Although timber, as an exportable good, did not have the high profile of the 'star commodities' of the Costa Rican agro-export model, the intensive logging of the forests allowed for the expansion of these products, while the timber obtained from this process, although not exportable, was marketed internally to be used as firewood – being Costa Rica's energy base during much of our study period – or for other uses such as construction or the production of furniture and handicrafts, among many other possibilities.

The cycle of Costa Rican exports of the ten timber species in greatest demand during our study period illustrate the transition from a phase clearly dominated by the export of a few species of slow-growing precious woods, extractively harvested, especially in the North Pacific, where cedar and *palo de mora* paid the high ecological price of the world demand for Costa Rican timber, to a phase marked by the predominance of fast-growing 'cultivated' woods, especially balsa, and other extractive but mass-consumed woods such as cativo, harvested in new logging regions for export, such as the Costa Rican Atlantic/Caribbean, in the context of a diversified and increasingly mass-produced supply.

Regarding the specific case of *palo de mora*, the dyeing wood par excellence of Costa Rican exports of forest products during the period studied, it is noteworthy that its behaviour observed important economic cycles, even after the innovation of synthetic dyes, being the second most important export wood behind only cedar. We propose here that the massive industrial use of this species, in a period when artificial dyes driven by the development of the chemical industry displaced traditional dyeing, is related to the versatility of the yellow dye extracted from this wood, given its easy fixation on textiles to produce a range of colours ranging from yellowish to greenish. An example of this is the *khaki* colour (*kakhi*) typical of military uniforms, which can be obtained with *palo de mora*, despite the fact that, according to the analysis of the main economic cycles of this wood, and the scarce temporal relationship between the massification of *khaki* and the export boom of *palo de mora*, it was the non-warlike uses of the dye that boosted its demand. This leads us to conceive *palo de mora* as a specialty rather than a commodity, that is, a type of good with a selective demand for a particular market niche, especially as a natural dye for textiles.

It is worth noting that the *ecological debt* generated by the processes of forest exploitation analysed in this chapter is still with us today, which invites us to reflect on the enormous need to move towards more sustainable ways of relating to the biophysical environment in which ecological exchange (the guarantor of life on the planet) is placed above economic exchange. The latter only generates profits in a system that externalizes the past, present and future environmental impacts of the prodctive and extractive activities that take place within it, and excludes non-economic – or, more precisely, non-crematistic languages of the valuation of nature – and hence can contribute to the collapse of our societies.[60]

A final element that we do not wish to leave aside is the non-linear character of technonological and scientific change. Authors such as Samuel Lilley have elucidated that beneath the surface novelty of inventions – particularly within the context of industrial revolutions – lie a series of adaptations of pre-existing and well-known techniques, which in many cases are new ways of 'organizing' pre-existing technologies, rather than authentic innovations. We suggest that the replacement of natural dyes with synthetic dyestuffs occurred based on this logic. The diverse origin and costs of extraction and processing of natural dyes resulted in a gradual and differentiated substitution by synthetic dyes. In the specific case of *palo de mora*, we consider that the advantages of the dye extracted from this wood for industrial uses surpassed its artificial counterpart until well into the twentieth century, which could contribute to explain the notable weight of exports of this dyewood in our study period. This proposition, presented as a hypothesis, encourages the exploration of other natural raw materials that exhibited prolonged resistance to obsolescence despite the rapid pace of techno-scientific modernization.

Notes

1 The authors wish to thank Carlos Marichal Salinas and David Pretel for their valuable and generous observations to preliminary versions of this text. Gratitude is also due to Krista Beam of the University of Pittsburgh for her help in translating the text into English.

2 Michael McGerr, 'The Price of the "New Transnational History"', *The American Historical Review* 96, no. 4 (1991): 1056–67.

3 Leslie Bethell, *Historia de América Latina*, Vol. 1 (Barcelona: Cambridge University Press/Crítica, 1990); Victor Bulmer-Thomas, 'Centroamérica desde 1920: desarrollo económico en el largo plazo', *Anuario de Estudios Centroamericanos* 11, no. 1 (1985): 5–21; Victor Bulmer-Thomas, *La economía política de Centroamérica desde 1920* (San José, Costa Rica: Banco Centroamericano de Integración Económica (BCIE) EDUCA, 1989); Victor Bulmer-Thomas, 'La crisis de la economía de agroexportación (1930–1945)', in *Historia General de Centroamérica*, ed. Víctor Hugo Acuña, Vol. 4 (San José, Costa Rica: FLACSO, 1993), 325–97; Victor Bulmer-Thomas, *The Economic History of Latin America since Independence* (Mexico City: Fondo de Cultura Económica, 1998); Victor Bulmer-Thomas, *The Economic History of the Caribbean since Napoleonic Wars* (Cambridge: Cambridge University Press, 2012); Héctor Pérez Brignoli, 'The Economies of Central America. 1860–1940', in *An Economic History of Twentieth-Century Latin America*, ed. Enrique Cardenas, José Antonio Ocampo and Rosemary Thorp, Vol. 1 (Oxford: St. Antony's College/Palgrave, 2000), 85–118; Héctor Pérez Brignoli, 'Las ambigüedades del crecimiento: dos siglos de historia económica de Centroamérica, 1810–2010', in *Institucionalidad y desarrollo económico en América Latina*, comps. Luis Bértola and Pablo Gerchunoff (Santiago: ECLAC/OAICE, 2011), 93–134; Ronny Viales, 'La crisis de 1929 en América Latina: del viejo paradigma al nuevo paradigma explicativo. Alcances y limitaciones', *Revista de Historia de América* 126 (2000): 85–111; Rosemary Thorp

and Ame Bergés, 'Las raíces históricas de la exclusión en América Latina', in *La lucha contra la exclusión social en América Latina*, ed. Fernando Carrillo (La Paz, Bolivia: Plural Editores, 2009), 19–39; Rosemary Thorp, 'Economy, 1914–1929', in *Latin America. Economy and Society. 1870–1930*, ed. Leslie Bethell (Cambridge: Cambridge University Press, 1989), 57–81; Rosemary Thorp, *Progreso, pobreza y exclusión. Una historia económica de América Latina en el siglo XX* (Washington: IDB, 1998); Luis Bértola and Pablo Gerchunoff (Comps.), *Institucionalidad y desarrollo económico en América Latina* (Santiago: ECLAC/OAICE, 2011); Luis Bértola and José Antonio Ocampo, *Desarrollo, vaivenes y desigualdad. Una historia económica de América Latina desde la independencia* (Madrid: Secretaría General Iberoamericana, 2010); Luis Bértola, and José Antonio Ocampo, *The Economic Development of Latin America since Independence* (Oxford: Oxford University Press, 2012).

4 Iván Molina, *Costa Rica (1800–1850) El legado colonial y la génesis del capitalismo* (San José, Costa Rica: EUCR, 1991); Ronny Viales, 'La historia monetaria de Costa Rica en el período posterior a la independencia. Del caos monetario a los orígenes del patrón oro (1821–1896)', in *Nueva historia monetaria de Costa Rica. De la Colonia la década de 1930*, ed. Ronny Viales (San José, Costa Rica: Editorial de la Universidad de Costa Rica/Colección Historia de Costa Rica, 2012), 101–29; Jorge León, Justo Aguilar, Manuel Chacón, Gertrud Peters, Antonio Jara, and María Lourdes Villalobos, 'Las políticas económicas en Costa Rica en la primera mitad del siglo XX', in *Historia Económica de Costa Rica en el siglo XX: Crecimiento y las políticas económicas*, ed. Jorge León, Justo Aguilar, Manuel Chacón, Gertrud Peters, Antonio Jara, and María Lourdes Villalobos, Vol. 1 (San José, Costa Rica: EUCR/IICE/CIHAC/Escuela de Historia, 2014), 56–121; Eugenia Rodríguez Sáenz, *Campaña Nacional, crisis económica y capitalismo. Costa Rica en la época de Juan Rafael Mora. (1850–1860)* (San José, Costa Rica: ECR, 2014).

5 Emanuel Barrantes, Hilda Bonilla, and Olga Marta Ramírez, *Las subsistencias en una coyuntura de crisis. Producción, consumo y nivel de vida, Costa Rica* (San José, Costa Rica: EUCR, 2011); León, Aguilar, Chacón, Peters, Jara, and Villalobos, 'Las políticas económicas', 56–121; Frank Notten, *La influencia de la Primera Guerra Mundial sobre las economías centroamericanas. 1900–1929* (San José, Costa Rica: SIEDIN/CIHAC/School of History, 2012). Antonio Jara Vargas, 'Finanzas públicas y crecimiento del Estado: un siglo de ingresos y gastos estatales en Costa Rica. 1870–1980', in *Historia Económica de Costa Rica en el siglo XX: Crecimiento y las políticas económicas*, ed. Jorge León, Justo Aguilar, Manuel Chacón, Gertrud Peters, Antonio Jara, and María Lourdes Villalobos, Vol. 1 (San José, Costa Rica: EUCR/IICE/CIHAC/Escuela de Historia, 2014), 393–455.

6 Pérez Brignoli, 'The Economies of Central America', 85–118; Pérez Brignoli, 'Las ambigüedades del crecimiento', 93–134.

7 While Robert Allen makes a clear division of labour between economists, who seek the causes of development from timeless theories, and economic historians, who locate the causes in a dynamic process of historical change, we defend the idea that economic historians are located at the confluence of both perspectives. Robert Allen, *Global Economic History. A Very Short Introduction* (Oxford: Oxford University Press, 2011).

8 Andrea Montero and Ronny Viales, 'Agriculturización' y cambios en el paisaje. El banano en el Atlántico/Caribe de Costa Rica (1870–1930)', *Revista de Historia Ambiental Latinoamericana y Caribeña (HALAC)* 3, no. 2 (2014): 310–38. Also see Carlos Marichal, Steven Topik, and Zephyr Frank (coords.), *De la plata a la cocaína.*

Cinco siglos de historia económica de América Latina (México: Fondo de Cultura Económica, 2014).

9 David Díaz and Ronny Viales, *El impacto económico de la independencia en Centroamérica, (1760–1840). Una interpretación desde la historia global* (San José, Costa Rica: Editorial de la Universidad de Costa Rica, Serie Cuadernos de Historia de Centroamérica, 2016); José Antonio Fernández, *Pintando el mundo de azul. El auge añilero y el mercado centroamericano, 1750–1810* (San Salvador: CONCULTURA, 2003); David Mc Creery, 'Las cadenas de la materia prima índigo en los Imperios español y británico, de 1560 a 1850', in Marichal, Topik, and Frank, *De la plata a la cocaína*, 76–107.

10 Anthony Goebel, 'Land and Climate: Natural Constraints and Socio-Environmental Transformations', in *The Oxford Handbook of Central American History*, ed. Robert H. Holden (Oxford: Oxford University Press, 2021), 1–34 (Online Publication, Date: Feb 2021).

11 Miguel Herrera, 'Gentes de la mar. Marinos y comerciantes en el Pacífico Central mesoamericano, 1830–1860', *Revista de Historia* 43 (2001): 213–64.

12 David Pretel, 'Reacciones en cadena: cambio tecnológico global y frontera forestal en la península de Yucatán (ca. 1850–1950)', *Historia Mexicana* 70, no. 1 (2020): 259–311.

13 Carlos Marichal, 'Un capítulo olvidado del comercio internacional: la grana cochinilla mexicana y la demanda europea de tintes americanos, de 1550 a 1850', in Marichal, Topik, and Frank, *De la plata a la cocaína*, 108–31.

14 Anthony Goebel Mc Dermott, 'El mercado mundial de productos forestales y la presión económica sobre los bosques de la Costa Rica "liberal": hacia una tipología básica (1883–1955)', in *Costa Rica: Cuatro Ensayos de Historia Ambiental*, ed. Ronny Viales Hurtado and Anthony Goebel Mc Dermott (San José: Sociedad Editora Alquimia 2000, 2011), 125–96.

15 Quírico Jiménez Madrigal, *Árboles Maderables en peligro de extinción en Costa Rica* (Heredia, Costa Rica: Instituto Nacional de Biodiversidad, 1998), 128–36.

16 Carla Chízmar Fernández, *Plantas comestibles de Centroamérica* (Heredia, Costa Rica: Instituto Nacional de Biodiversidad, INBio, 2009), 241–3.

17 Anthony Goebel, *Los bosques del 'progreso'. Explotación forestal y régimen ambiental en Costa Rica: 1883–1955* (San José, Costa Rica: Editorial Nuevas Perspectivas, 2013), 143.

18 Chízmar, *Plantas comestibles.*

19 Tarek Ismail Kakhia, *Dyes, Colors and Pigments* (n.d.). https://tarek.kakhia.org/books_eng/Dyes_Colors_Pigments.Tarek_Kakhia.pdf.

20 Charles Mellor and Donald Stephen Lowell Cardwell, 'Dyes and Dyeing 1775–1860', *The British Journal for the History of Science* 1, no. 3 (1963): 265–79.

21 Ester Ferreira, Alison Hulme, Hamish McNab, and Anita Quye, 'The Natural Constituents of Historical Textile Dyes', *Chemical Society Reviews* 33 (2004): 329–36.

22 Editors of Encyclopaedia Britannica, 'Khaki Fabric' (Encyclopaedia Britannica). https://www.britannica.com/topic/khaki (accessed 7 July 2020).

23 Jane Tynan, 'Why First World War Soldiers Wore Khaki', *World War One Centenary: Continuations and Beginnings* (2014). http://ww1centenary.oucs.ox.ac.uk/?p=2969 (accessed 7 October 2018).

24 Mellor and Lowell Cardwell, 'Dyes and Dyeing'.

25 Alexander Engel, 'From Traditional Networks to Modern Companies. The Transition of the Markets for Dyestuffs in the 19th Century', in *International Business, International Organizations, and the Wealth of Nations*, Paper presented at

the 11th Annual Conference of the European Business History Association (Geneva, Switzerland, 2007), 17.

26 Clark S. Monson, 'Mulberry Trees: The Basis and Remnant of the Utah Silk Industry', *Economic Botany* 50, no. 1 (1996): 130–8.

27 For a recent work that analyses in depth the socio-environmental impacts of the modernization of the agrarian system in Guanacaste and the weight of extensive beef cattle ranching in this process, see: Anthony Goebel Mc Dermott, Andrea Montero Mora, Ronny Viales Hurtado, and Juan Infante Amate, *Las huellas de la ganadería vacuna en Costa Rica: transformaciones socioecológicas y cambios en el paisaje en Guanacaste (1890–2014)* (San José, Costa Rica: Centro de Investigaciones Históricas de América Central, 2023).

28 Luis Cayuela, *Deforestación y fragmentación de bosques tropicales montanos en los altos de Chiapas. Efectos sobre la diversidad de árboles* (Doctoral dissertation in Ecology) (Alcalá de Henares: University of Alcalá, 2006), 135–6.

29 Goebel, *Los bosques del 'progreso'*.

30 Joan Martínez Alier, *El ecologismo de los pobres. Conflictos ambientales y lenguajes de valoración* (Barcelona: Icaria Antrazyt-Flacso, 2002), 26–7; Joan Martínez Alier, 'Deuda Ecológica y Deuda Externa', *Ecología Política* 14 (1997): 157–73.

31 Jiménez Madrigal, *Árboles Maderables.*

32 Jiménez Madrigal, *Árboles Maderables*, 20.

33 Martínez Alier, *El ecologismo*, 150–2.

34 Cayuela, *Deforestación y fragmentación de bosques*, 61.

35 Martínez Alier, *El ecologismo*, 152.

36 Cayuela, *Deforestación y fragmentación de bosques*, 61.

37 Benjamin Orlove and Arnold J. Bauer, 'Giving Importance to Imports', in *The Allure of Foreign. Imported Goods in Postcolonial Latin America*, ed. Benjamin Orlove (Ann Arbor: University of Michigan Press, 1997), 1–29.

38 Dirección General de Estadística y Censos (DGEC). Anuarios Estadísticos, years 1883–1955 and Resúmenes Estadísticos, years 1883–1910. This takes into account that information is not available for all years on the destination of the exported species, since in some cases this information was not provided.

39 Jiménez Madrigal, *Árboles Maderables*, 137–43.

40 Jiménez Madrigal, *Árboles Maderables*, 78–81.

41 Michael Williams, *Deforesting the Earth: From Prehistory to Global Crisis: An Abridgment* (Chicago: The University of Chicago Press, 2006), 215.

42 Williams, *Deforesting the Earth*, 370.

43 The European deficit was, according to this study, 3.6 million cubic metres. What contributed to reduce this imbalance generated by the excessive US and European demand – although to a much lesser extent – in the developed world, was the inclusion of the Soviet Union, which had a favourable balance between production and consumption (+161.7) and the item 'others', which had the same characteristic (+13.6). Williams, *Deforesting the Earth*, 370.

44 Williams, *Deforesting the Earth*, 370.

45 Williams, *Deforesting the Earth.*

46 Michael Williams, 'Ecology, Imperialism and Deforestation', in *Ecology and Empire: Environmental History of Settler Societies*, ed. Tom Griffiths and Libby Robin (Edinburgh, Scotland: Keele University Press, 1997), 169–84, 175.

47 Williams, 'Ecology, Imperialism and Deforestation', 175.

48 Williams, 'Ecology, Imperialism and Deforestation'.

49 James Foreman-Peck, *Historia económica mundial. Relaciones económicas internacionales desde 1850* (Madrid: Prentice Hall), 118–55, 361–6.
50 Dirección General de Estadística y Censos (DGEC). Anuarios Estadísticos, years 1883–1955 and Resúmenes Estadísticos, years 1883–1910.
51 Carlos Domínguez, *Panorama histórico forestal de Puerto Rico* (San Juan, Puerto Rico: Editorial de la Universidad de Puerto Rico).
52 Dirección General de Estadística y Censos (DGEC). Anuarios Estadísticos, years 1883–1955 and Resúmenes Estadísticos, years 1883–1910.
53 Archivo Nacional de Costa Rica, 1889, f. 1.
54 Archivo Nacional de Costa Rica, 1893, f. 1 and 1.
55 Archivo Nacional de Costa Rica, 1893 f. 2-3 v.
56 Manuel González de Molina, Antonio Herrera, Antonio Ortega-Santos, and David Soto-Fernández, 'Peasant Protest as Environmental Protest: Some Cases from the 18th to the 20th Century', *Global Environment. A Journal of History and Natural and Social Sciences* 4 (2009): 48–77; Manuel González de Molina, 'Sociedad, naturaleza, metabolismo social. Sobre el estatus teórico de la historia ambiental', in *Agua, poder urbano y metabolismo social*, coords. Rosalva Loreto (México: Instituto de Ciencias Sociales y Humanidades, Benemérita Universidad Autónoma de Puebla, 2009), 238–43.
57 Anthony Goebel Mc Dermott, 'Este bosque es mío. Un acercamiento a la dinámica de los conflictos forestales en la Costa Rica liberal: concepciones, representaciones y acción de los sujetos (1880–1955)', *Memorias: Revista Digital de Arqueología e Historia desde el Caribe* 35 (2018): 107–32.
58 Goebel, *Los bosques del 'progreso'*.
59 James O'Connor, '¿Qué es la historia ecológica? ¿Por qué la historia ecológica?', *Ecología Política* 14 (1997): 114–31.
60 Martínez Alier, *El ecologismo de los pobres.*

12

Defying substitution

A Caribbean dyewood in the synthetic age

David Pretel

A lengthy article in the *American Dyestuff Reporter* – published in 1918 by the American Association of Textile Colourists and Chemists – noted that 'despite unwearying attempts by armies of chemists during the recent decades, no perfect substitute for logwood has yet been synthesized'.[1] The article praised 'the chemical superiority of logwood' as a 'true black'. Similar arguments praising the tinctorial attributes of logwood appeared in specialised journals and monographs during the 1910s and 1920s.[2] Evidence from production and trade data, as well as testimonies from the time, indicates that logwood continued to be utilized in large quantities during the first two decades of the twentieth century. In the context of increasing yet volatile international trade in colourants, the invention of aniline dyes in the second half of the nineteenth century did not immediately displace logwood – and other natural dyes such as madder and fustic – from global markets. On the contrary, the transition from natural colourants to synthetic dyestuffs was a long process that took over half a century.

This chapter concentrates on the transition period from natural to synthetic dyes. It argues that conventional histories of the rise of artificial dyestuffs suffer from an inaccurate evolutionary narrative of modernization in which chemical science enabled a rapid, almost abrupt, replacement of traditional materials and techniques with modern ones. This chapter, by contrast, shows that in the case of some Caribbean and Central American dyewoods, this substitution process was gradual and would last several decades, a long period during which tree extracts would continue to be of vital importance in the European and American textile industries, competing and coexisting with artificial dyes. Integrating natural colourants into the history of the so-called Second Industrial Revolution has the potential to challenge traditional distinctions between 'natural' and 'synthetic', disrupting standard narratives and periodization of the transition between the two.

The chapter concentrates on the case of logwood, a dyewood that managed to enjoy a production and commercial boom even as a global chemical industry emerged. It shows that the exports of Caribbean and Central American logwood remained strong into the early twentieth century, with a period of resurgence until the First World War, followed by a gradual decline between the 1920s and 1940s. The chapter also examines the factors explaining the persistence of the production, trade and use of logwood – from higher quality to a diversification of its uses – and thus offers a less linear history that questions the temporality of the diffusion of synthetic dyestuffs. More importantly, it shows the continuity of traditional dyewoods and practical colouring skills well into the synthetic age.

From logwood black to artificial black

In the global history of dyes, logwood occupies a prominent place. Not only was it an essential commodity in the Atlantic trade for more than three centuries, but its properties made it an indispensable part of textile production in both the New and Old Worlds. From pre-Hispanic times, Mayas had used this tropical hardwood for construction and its extract, obtained from the heartwood, as a dye and medicine. Beginning in the late sixteenth century, logwood was exported to Europe in large quantities along with other natural colourants produced in the Caribbean and Central America, such as indigo, madder and cochineal. The dye extracted from logwood, called haematoxylin, was used along with different colour-fixing mordants (iron, copper, alum or alumina) to obtain a wide range of rich colours such as blues, purples and yellows, with black being particularly valuable in the colouring and printing of textiles. It is noteworthy to mention that the chemical structure of hematoxylin resembles that of the dye brazilin, derived from brazilwood. The only difference is that hematoxylin possesses an additional oxygen atom in its molecular composition.

Until the mid-nineteenth century, the main area for the commercial exploitation of logwood trees worldwide was the Yucatan Peninsula. Specifically, much of the exploitation occurred along the banks of the Hondo River (on the southern border between Mexico and Belize) and the banks of the New River in Belize (including the Corozal Bay), as well as in the tropical forests of Laguna de Términos in the southern part of the present-day state of Campeche. Logwood trees occupied large swaths of the peninsula, where they crossed disputed political boundaries. Since the seventeenth century, the question of who controlled the regular supply of logwood dyes had sparked political clashes between the Spanish and British.[3] British industrialization since the late eighteenth century further stimulated the extraction of logwood in the Caribbean basin. Logwood production was, in fact, a central motivating factor behind the consolidation of British Honduras as a logging colony.[4] By 1851, the year of the Great Exhibition in London, logwood was the leading dye in terms of volume of imports into the UK, ahead of indigo and madder.[5]

The invention of synthetic dyes in the late 1850s marked a significant milestone in the textile and chemical industries. Developed in British and German laboratories, synthetic dyes increasingly posed a threat to natural dyes insofar as they yielded consistent colours and simplified production, among other advantages. The chemistry of synthetic anilines opened a new technological paradigm in the textile industry that would eventually lead to the substitution of artificial dyes for natural ones. The first artificial dye, mauve or purple aniline (obtained from coal tar), was accidentally invented in England in 1856 by the chemist and businessman William Henry Perkin. From then on, and throughout the later nineteenth century, artificial dyes would be isolated or synthesized in industrial research contexts, mostly German or English, thanks to the development of organic chemistry. In 1869 alizarin was synthesized, and in the 1880s, synthetic indigo.

The first black aniline was invented in 1862 in Manchester by the German chemist and textile colourist Heinrich Caro, who worked for the English manufacturer Roberts, Dale & Company.[6] Additional synthetic black dyes obtained from coal-tared hydrocarbons became commercially available after this invention. For instance, in 1863, the English chemist John Lightfoot patented a process for applying black aniline to cotton fabrics using copper chloride as a mordant. Black aniline, the artificial dye of choice for this colour until the 1930s, would not undergo significant advances for half a century.[7] Other synthetic substances, albeit to a lesser degree, were used as black dyes in textile printing since the 1880s, one example being 'black alizarin'. Similarly, artificial dyes derived from alkylated hydrocarbons and sulphur compounds were introduced during the following decades, such as the one invented by the Frenchman Henri Raymond Vidal in 1893.[8]

From the 1880s onwards, the German chemical industry became the undisputed international leader in the production of artificial dyes, accounting for 80 per cent of world production in 1914. Among the German companies born in the heat of the development of artificial dyes were such giants as BASF, Bayer and AGFA. These German manufacturing companies, along with several British and American firms, established industrial testing laboratories and employed many expert chemists and engineers. Chemical companies exhibited their new products and artificial dyeing processes at large industrial exhibitions and protected their inventions with patents and trade secrets.[9] The increased number of textile-related patents granted in Germany, the United States and Britain during the period 1850–1950 concentrated on coal-tar dyestuffs but also included inventions in natural dyes' extraction methods and colouring techniques (see Figure 12.1).[10]

The persistence of logwood

From the late nineteenth century until the First World War, logwood retained its popularity as a dye of choice, primarily because certain synthetic dyes encountered difficulties in fixing their colour on specific fabrics. Artificial blacks proved to be a more suitable choice for cellulosic materials, as they exhibited better resistance to fading.[11] This explains that the use of logwood to dye cotton was discontinued early on due to the complexities involved in applying it (along with the required mordants) onto vegetable fibres. However, logwood continued to find application in dyeing animal fibres such as wool, leather and, to a lesser extent, silk. This was especially favoured around the turn of the twentieth century, when there was a strong preference for black clothing.[12]

During the 1910s, logwood remained a prominent natural dyestuff worldwide, frequently employed in conjunction with synthetic dyes to enhance colour tones and

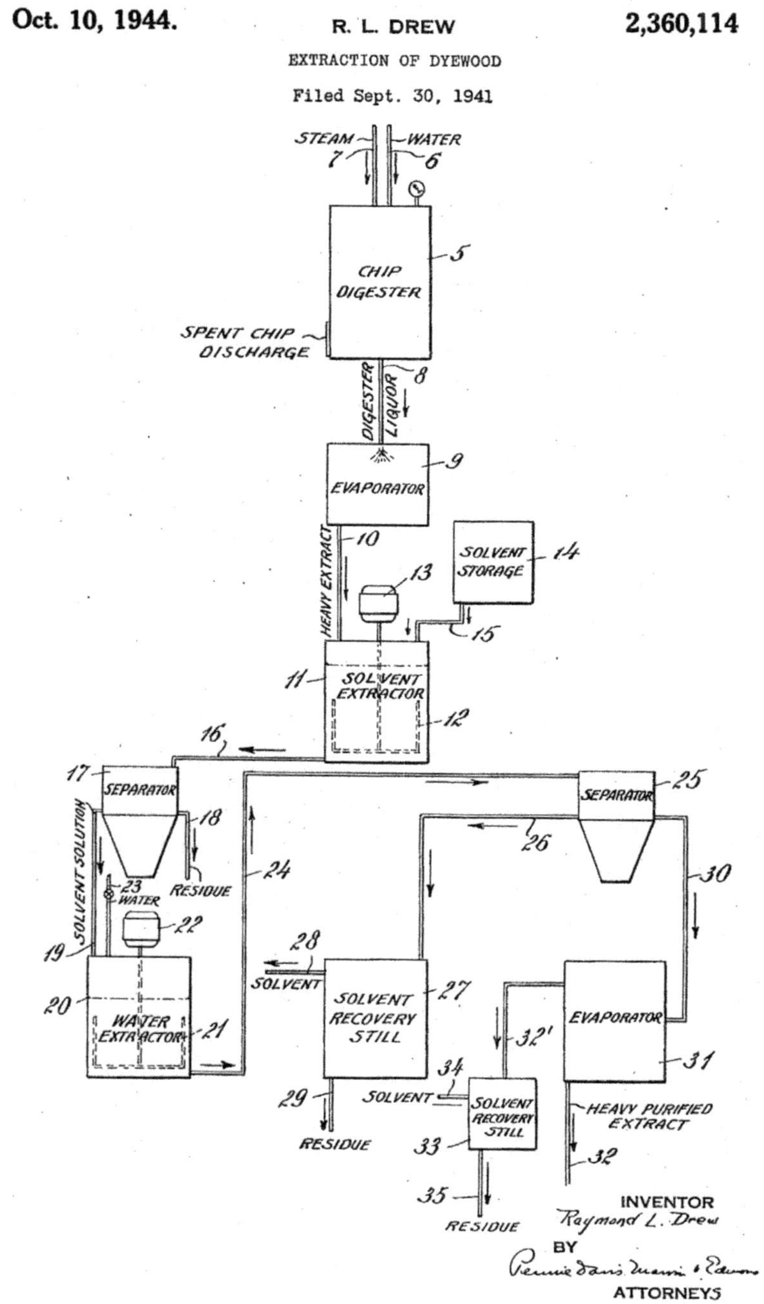

Figure 12.1 Extraction of Logwood, US patent granted to the American Dyewood Co. in 1941. *Source*: US Patent No. US2360114A.

prevent fading. The demand for logwood rose and fell with fashion trends, yet this dyewood continued to be selectively preferred within specific market segments. Some artisans and textile producers continued to value logwood for its unique properties in the black colouring of high-quality garments. An analogous example was the yellow fustic extract obtained from Central American mulberry trees (*palo de mora*).[13] Fustic dyes experienced a narrow, but significant, demand between the 1880s and the 1940s because khaki trousers, initially worn as part of military uniforms, became fashionable among civilians as well.

The textile industry during the late nineteenth and early twentieth centuries was characterized by a mature artisan culture that had become accustomed to using natural colourants, while the industrial use of artificial blacks in textile manufacturing remained challenging. The introduction of artificial dyes as substitutes for logwood in dyeing wool, linen and silk presented several challenges in their industrial implementation. Even after the initial corrosive effects of artificial chemicals on printing presses had been resolved, textiles coloured with synthetic dyes still encountered issues of discolouration and non-uniformity. Moreover, during the early years of the rise of a science-based chemical industry in Europe and the United States, the costs of manufacturing synthetic dyes remained relatively high. This can be attributed to the intricate process of synthesizing artificial substances, which frequently demanded costly raw materials.

Natural dyes were often used together with synthetic ones in the textile industries in the early twentieth century, with colouring processes often repeated to increase durability and improve appearance.[14] It was not only that consumers 'did not appreciate the flashy and unnatural colours produced by synthetic dyes' but that the mixing of the 'old' and 'new' dyes saved costs and allowed the colours to last longer.[15] Resistance to synthetic dyes came not only from the textile industry and dyewood manufacturers but also from craftsmen and artists. In the words of the celebrated British designer William Morris: 'Anyone wanting to produce dyed textile with any artistic quality in them must entirely forgo the modern and commercial methods'.[16] A significant problem was that artificial dyes did not allow for the same range of colours and shades as natural dyewoods. Nor did artificial blacks offer the same strength or intensity as logwood blacks. Artificial dyes were also controversial because of their presumed negative health effects, and initially they were even rejected by leading textile industrialists and food manufacturers.

It was not until later that synthetic anilines effectively surpassed the properties of logwood for black dyeing. The American monthly magazine *Textile Colorist* noted in several articles published in 1915 that logwood extract, used with appropriate mordants and modern mechanical and chemical techniques (such as large steam boilers), was the 'perfect dye' due to a wide range of favourable colouring properties such as penetration, brightness, versatility, richness and permanence.[17] As mentioned earlier, the advantages of logwood were far more considerable in the dyeing of wool and silk than cotton. In the case of wool, the use of chromium instead of ferrous sulphate as a mordant made the logwood extracts easier to apply, with better results and without damage to the yarn.

With the advent of the synthetic age, dying processes in textile manufacturing became even more complex, requiring the collaboration of experts in industrial chemistry and the continued use of natural colourants. Even when the development of organic chemistry had increased the knowledge of the chemical composition of natural dyes, professional colourists' practical experience was still required to obtain a wide range of shades. Moreover, innovations in the manufacture and testing of synthetic dyes had direct applications to natural dyes, reducing costs and improving their consistency and the precise use of mordants. Decades of scientific research had contributed to an understanding of natural dyestuffs chemistry and its industrial application to fibres and materials. As the American chemist and expert in dyewoods Edward Chapin put it in a 1918 article:

> The vegetable cell is a laboratory in which are carried out a most remarkable series of chemical reactions. . . . If we will thoroughly investigate these remarkable organic compounds and complexes, formed within the chemical laboratory of nature, we shall witness the dawning of a new era in pure and applied chemistry.[18]

During the 1920s, logwood extracts lost much of their significance in the textile industry. This decade witnessed a notable decline in the use of logwood as chromium-based dyes increasingly replaced it. The main disadvantage of logwood was its impurity and inconsistency relative to the newer iterations of artificial dyes. Furthermore, using logwood in dyeing required mordants, making the entire process expensive and intricate and necessitating expert knowledge. While the demand for logwood significantly declined with the rise of synthetic dyes during the interwar period, this natural colourant retained a small group of supporters who continued to use it for specific projects, particularly in wool dyeing and other special applications. For example, Ronald Horsfall, head of the dyestuff department of the British state-controlled British Dyestuff Corporation Ltd., noted in his book *The Dyeing of Textile Fibre* – co-authored with expert L. G. Lawrie and published in 1927 – that logwood was the only natural dye in widespread use and emphasized its low cost, strength, colour fastness and superior quality to synthetic dyes.[19] Indeed, as late as the 1950s, logwood continued to be used to dye wool and even new materials such as nylon, usually as a complement to artificial dyes. The occasional use of logwood as a textile dye, along with its adoption in various other sectors, was the reason that world production of logwood extracts reached 70,000 tons in 1943.[20]

Logwood production and trade

The so-called second industrialization witnessed a significant growth in mechanized textile production worldwide, leading to an increased demand for dyestuffs of different qualities and prices. As a result, the supply of chemical dyes and certain natural dyestuffs experienced a significant boost. Large quantities of Caribbean logwood continued to be produced and exported to Europe during the second half of the nineteenth century and the first two decades of the twentieth century.[21] British diplomatic reports showed that at the turn of the twentieth century Germany continued to import large quantities of logwood despite the considerable advance of its chemical industry.[22] For instance, in

1898 German imports of this dye amounted to 20,771 tons. In the United States, several companies, including the American Dyewood Company, Taylor-White Extracting Company, Boston Dyewood Company and the New York Color & Chemical Company, manufactured significant volumes of logwood extracts during the first three decades of the twentieth century.[23] These extracting companies expanded their production beyond logwood, manufacturing various dyewoods sourced from diverse geographical origins.

The surge in logwood exports in southeast Mexico took place in the 1880s and 1890s. Although logwood cutting and trade declined in the British colonial outpost of Belize during the late nineteenth century, it increased in the Mexican state of Campeche in the Yucatan Peninsula.[24] The ports of El Carmen and Campeche in the Gulf of Mexico would enjoy a golden period of logwood trade until the 1910s, during which time it was exported in large quantities to the United States, the United Kingdom, Germany and France. El Carmen even had a modern dyewood extraction factory inaugurated in 1881, featuring one of Mexico's earliest electric power plants and receiving support from German and Belgian investors. Nonetheless, it was ultimately abandoned during the early twentieth century.

In the last third of the nineteenth century, the Mexican federal government also granted several forest concessions for the exploitation of logwood in other parts of the Yucatan Peninsula.[25] A good example is the Compañía Agrícola del Cuyo & Anexas, established in 1876 on the northeastern coast of the Yucatan with the backing of German capital investment. This firm set up a large company town that exploited forestry resources for international markets. Its forest exploitation of logwood – and to a lesser extent chicle, cedar and wild rubber – entailed a narrow-gauge railway, a telephone network, large warehouses and a company pier. In 1895, German investors withdrew, probably because of overharvesting and competition from cheaper synthetic dyestuffs. Even so, 'El Cuyo' would continue to operate, diversifying its productive activity with a new focus on chicle resin production. Other examples of forest concessions in the area were the Mexican Exploration Company, created in 1892 and backed by British capital, and the Compañía Colonizadora de la Costa Oriental de Yucatan, established in 1896 with Mexican and British capital. Both firms produced logwood until at least the 1910s.

The price of logwood in international markets played a crucial role in the changing demand for this dyewood. Obtaining logwood had long been costly, involving the extraction, transportation and primary processing of raw materials in the inaccessible tropical forests of the Caribbean region and the Gulf of Mexico. Logwood shipment and trade costs had also been substantial across the Atlantic. However, the cost of logwood production fell considerably between the late 1870s and the First World War, especially compared to other natural dyes such as indigo (see Table 12.1).[26] In 1909, Emilio Noelting, director of the Mulhouse chemistry school and an inventor in the dye industry, noted that 'aniline black had to compete with logwood black, which could be dyed at a very low price'.[27] The still-high price of oxidizing agents required for the processing of artificial dyes dissuaded some manufacturers from using them. However, as chemical knowledge advanced, researchers and industrial chemists worked on improving synthesis methods and finding more cost-effective raw materials. As

Table 12.1 Prices of Natural and Artificial Dyes, 1872–1913 (£ / Ton of Textiles)

Years	Indigo	Madder	Cochineal	Fustic	Logwood	Artificial Dyes *
1872–1877	39.9	17.6	19.0	3.2	2.0	8.5
1878–1883	36.7	16.0	15.1	3.1	2.0	6.6
1884–1889	32.2	15.1	9.3	2.1	2.0	4.0
1890–1895	29.9	15.2	8.2	1.9	2.0	3.0
1896–1901	22.3	9.6	7.2	1.5	1.5	2.6
1902–1907	18.3	14.5	11.6	1.9	1.3	2.4
1908–1913	18.7	18.3	9.6	2.0	1.3	2.1

Source: Material cost for dyes according to the prices of the German trade statistics in Alexander 'Colouring Markets: The Industrial Transformation of the Dyestuff Business Revisited', *Business History*, 54:1 (2012): 10–29. For First World War prices, see P. W., Carleton, *Prices of Natural Dyestuffs and Tanning Chemicals* (Washington: G.P.O., 1919).
*Coal tar dyes except indigo.

synthetic dyes were produced on a large scale and industrial processes became more efficient, the costs of producing synthetic dyes gradually decreased.

From the mid-nineteenth century, logwood extracts were increasingly obtained and isolated by mechanical and chemical means, thereby decreasing shipping expenses. Logwood extraction involved chipping the wood and then digesting the chips using water and steam. Dyewood companies producing commercial extracts resorted to boiling and vacuum evaporation machines to increase the concentration of the dyestuffs.[28] This innovation enabled the logwood industry to cut costs by shifting its production towards liquid extracts, pastes, powders and crystals packed in barrels and casks rather than exporting raw trunks or wood chips.

During the first two decades of the twentieth century, new chemical tests and technical analyses allowed the industry to assess the purity of liquid and solid dyewood extracts, detecting adulterations such as glucose, molasses, dextrins, salts and other less valuable dye extracts.[29] As a result, logwood became more accessible and suitable for modern textile industries in Europe and the United States but less profitable for logwood producers in the Caribbean region. In this context, the biggest dyewood manufacturers, equipped with larger R&D capacities, were able to better confront the rise of synthetic dyes. For instance, the factories of the American Dyewood Company in the United States started to have teams of chemists working in industrial research laboratories to assist in producing logwood extracts, meet production standards and reduce the cost per colour unit.

Logging technologies, by contrast, did not undergo significant innovations during this period. Logwood cutting was a simple but strenuous task carried out by seasonal workers using hand tools such as axes and saws. Once the trees were cut, the logs were cleaned and squared into smaller pieces, leaving only the heartwood. The chunks of logwood were transported by road through the forest and down rivers, taking advantage of the wood's buoyancy. Logwood cutting and basic processing in the Caribbean was well documented by, among others, the British botanist Daniel Morris, who held relevant scientific and administrative positions in Jamaica, Ceylon, British Honduras and the Royal Botanic Gardens in Kew and was a Scientific Advisor in

Tropical Agriculture to the British Colonial Office until 1913.[30] In an 1883 publication, Morris described logwood production in Belize:[31]

> Logwood cutting appears to be a more simpler (*sic*) and much less laborious work. After the trees, which are seldom more than a foot in diameter, but often only half this size, are cut down, the outer or sap wood is removed, leaving nothing but the inner dark-coloured heart wood. When thus prepared, the logwood is carried on trucks or 'crooked' to the nearest bank, where to prevent it from sinking, it is packed in 'bark-logs' or light, buoyant cradles, capable of carrying a ton or two of logwood. An immense train of these heavily-freighted 'bark-logs' is often met with on its way down the river, or anchored at night in the middle of the stream.

Logwood production and trade were tied to external shocks and events, including the First World War. During the war, the supply of German synthetic dyes to the United States was cut off. In 1914, when the supply of low-priced aniline dyes was disrupted, US woollen manufacturers shifted their focus to logwood extract for producing blacks and blues and combining logwood with fustic for browns. During the war, the primary sources of supply for logwood to the United States were Jamaica and Haiti. Additionally, supplies were sourced from Mexico, Cuba, the Dominican Republic, Belize, the Bahamas and Trinidad (see Table 12.2). The downward trend in logwood prices had come to an end with the blockade of the supply of German artificial dyes during the First World War and the volatility of international markets. Faced with the shortage and scarcity of other synthetic and natural dyes, between 1913 and 1916, the incentives for using and consuming logwood increased, raising its price by 320 per cent and making it the most important natural dye in international trade.[32]

While British demand for Caribbean logwood significantly decreased in the first two decades of the twentieth century, American demand for this dyestuff remained high, even experiencing significant increases during the First World War.[33] After the logwood boom that occurred during the First World War, the demand for this natural colourant plummeted. By then, synthetic aniline dyes had become cheaper, readily available in large quantities and easy to use. This situation was apparent in the case of Belize, which, after the recovery of dyewood exports during the war, suffered a sharp

Table 12.2 Logwood Imports into the United States (1913–17) (Tons)

	1913	1915	1916	1917
British Honduras	2	-	1,486	1,477
Mexico	295	4,543	18,143	22,250
Jamaica	18,771	29,247	45,052	27,960
Dominican Republic	1,209	2,006	3,988	8,963
Haiti	15,607	18,495	64,419	58,722
Other Countries	1,083	768	1,541	3,422

Source: 'Vegetable Dyes from Latin America', Circular from Latin American Division, *Department of Commerce, Bureau of Foreign and Domestic Commerce, Washington*, No. 36 (3 July 1918).

decline in logwood production and trade during the 1920s, from which it would never recover.

Diversification and new producers

Despite infrastructural expansion and improvements in extraction processes, logwood producers in the Yucatan Peninsula entered a crisis during the 1910s and 1920s. For instance, Mexico's logwood production in the period 1910–11 fell to only 17,000 tons.[34] In Belize (then British Honduras) exports fell from 77,000 tons in 1901 to 4,500 tons in 1921.[35] For the years 1921-1925, a study by the governor general of British Honduras, John Burdon, clearly showed that the export of chicle, mahogany and cedar far outstripped that of logwood.[36] Easily accessible and older logwood trees – providing good-quality dyes – were in short supply. Overexploitation in coastal areas and the increasing inaccessibility of trees located deeper within the tropical forest may explain the crisis in logwood production in the Yucatan Peninsula.

Logwood producers in Mexico and Belize have experienced competition from other producers in the Caribbean since the late nineteenth century. Logwood was naturalized in other parts of the world, including Asia and the United States, but particularly in other places in the Greater Caribbean, such as Jamaica, Cuba and Haiti, where it was exploited on plantations. A 1902 British consular report showed that logwood exports from Jamaica had increased by 20.8 per cent in quantity and 5 per cent in value between 1897 and 1901, making it the island's main export item.[37] By the turn of the century, Germany and France surpassed Britain in their quantity of logwood imports, with Hamburg and Le Havre becoming the main ports of destination for Jamaican logwood extracts. In the late nineteenth century, Haiti had also become one of the world's leading exporters of logwood, boosted by foreign investment. Indeed, between the 1880s and 1920s, Haiti would be a significant player in the international market for logwood, supplying a large part of this dye to the American dyewood industry. In the 1880s, for instance, Haiti's annual total logwood exports reached a peak ranging from 100,000 to 145,000 tons.[38]

Several companies that processed and exported logwood extracts were active in Jamaica and Haiti between the late nineteenth and mid-twentieth centuries. In 1892, The West Indies Chemical Works Ltd., a manufacturing company that supplied logwood extracts and crystals to international markets for over half a century, was established in Spanish Town, Jamaica. The manufacturing plant boasted spacious warehouses and was well connected with modern roads and railways. Until 1942, the British Dyewood Company, headquartered in Glasgow, also operated a factory in Jamaica that produced logwood extracts. Similarly, the American Dyewood Company set up a subsidiary extract production plant in Port-au-Prince, Haiti, in 1895. Some of the shareholders of the American Dyewood Company established The Compagnie Haitienne, which operated its own pier in Port-au-Prince and installed a cable car whose cables stretched more than 20 kilometres to transport logwood from the forests to the coast. This company would eventually cease its activities due to protests from native communities and challenges posed by adverse environmental and climate conditions. Moreover, after a hundred years of exploitation, Haiti's reserves of high-quality logwood started to deplete. An official study published in 1927 reported that Port-au-Prince was no longer a heavy shipper of logwood, lamenting that 'logwood supplies have been largely exhausted, and compensating industries have not been developed'.[39]

As for dyewood production in Jamaica, it remained highly profitable throughout the First World War, evidenced by the presence of at least three modern factories producing logwood extracts.[40] By the early 1900s, the exports of logwood extracts increased in importance compared to raw logwood, a change that became more pronounced with the outbreak of the war. The war multiplied demand, which explains that associations of logwood growers in Jamaica and Haiti promoted higher prices for the raw materials of its members.

From the late nineteenth century, improvements in botanical techniques enabled logwood to be cultivated on commercial plantations. Logwood trees required little care and were adapted to various soils, although cutting and primary processing remained a labour-intensive activity. Nonetheless, producers continued cutting self-sown logwood trees found in groves and unproductive lands unfit for cultivation. After all, the production of logwood in commercial plantations implied a high risk for foreign investors and companies. In this regard, forestry expert Tom Gill – who would go on to become president of the International Society of Tropical Foresters – stated in his 1931 study *Tropical Forests of the Caribbean* (a publication of the Tropical Plant Research Foundation) that logwood's unique status among natural dyes was coming to an end:[41]

> As a dyewood this species is the sole survivor, and the chances are that in the not distant future logwood, too, must succumb to the advances of modern chemistry and relegated to the comparatively useless woods of the world.

In his writings, Tom Gill issued a warning regarding the expansion of dyewood plantations in the Caribbean due to the slow growth of these trees and the potential emergence of new chemical substitutes that would inevitably lead to the decline in the importance of this tropical forest resource. Gill's concerns were well founded, as the advancement of chemistry would eventually bring about a definitive improvement and cost-effectiveness in synthetic dyes, finally bringing to a close the long cycle of this tree's exploitation. Despite all the advancements in botany and transportation, logwood could no longer withstand the relentless progress made by the chemical industry.

While logwood's use as a dye underwent a gradual decline, its applications diversified. Although it was still used primarily for textile dyeing, it began to be used additionally in the dyeing of paper and glass, as well as in the industrial manufacture of black writing ink.[42] The dark red extract obtained from logwood had already been used as an astringent since the end of the eighteenth century. All these varied new uses required a significantly lower logwood production volume compared to industrial textile dyeing. More important was the late nineteenth-century discovery of its attributes for biological staining in histology and pathology, in conjunction with a mordant like alum.[43] Other natural dyes from the Americas, such as cochineal, brazilwood, madder and indigo, were also tried as biological stains but were of little significance. By the turn of the century, purified hematein – which results from the oxidation of logwood extract by natural or artificial means – became the standard histological stain for microscopy in laboratories. Although the extract of logwood was first successfully used as a biological staining formulation in 1865, it was not until the 1920s that researchers began to realize its full value. Since then, logwood extracts have been employed to stain tissue samples for observation under a microscope – often in conjunction with artificial eosin. As one medical researcher put it, logwood 'is, to a very large extent, the foundation upon which modern histopathological diagnosis is built'.[44]

Conclusion

The fluctuating cycles of logwood exploitation were largely determined by the technical innovations – especially in industrial chemistry and textile techniques – that occurred between the second half of the nineteenth century and the first half of the twentieth century. However, the rise of artificial dyes, contrary to what is often highlighted, was not a story of abrupt change but of continuity.[45] The conventional histories of the inexorable rise of artificial dyes suffer from an erroneous linear narrative in which chemical science enabled a rapid replacement of a traditional technological system with a modern one. In reality, the transition was gradual. For decades, logwood was used alongside artificial dyes. There was continuity not only in the use of this natural dyewood, especially in extract form, but also in traditional textile printing techniques. Notwithstanding a proliferation of new machines and methods, the textile industry continued to depend on the know-how of expert colourists trained in a culture that was experimental as well as practical.

The science-based innovations that were introduced to replicate natural dyes did eventually cause a sharp decline in logwood production in the Caribbean and Central America during the interwar period, but this decline can only be understood in terms of a long-term cycle. Not only did logwood blacks and synthetic blacks coexist for more than half a century, but there were rearrangements and relocations in logwood production in diverse regions of the world in response to external geopolitical shocks and the development of plantations. Paradoxically, between the 1850s and at least the 1920s, the technological progress made in the textile and chemical industries played a pivotal role in maintaining the use of logwood. Simultaneously, advancements in extraction and logistical processes also contributed to its continued utilization. As a curator at the Division of Textiles of the Smithsonian Museum of American History eloquently put it in a 1968 publication: 'Logwood has certainly proved to be history's most tenacious natural dyestuff, defying substitution almost until the beginning of the Second World War'.[46] Although logwood is no longer commercially valuable today, its appeal in specific niche markets enables it to enjoy a modicum of demand. Artisans and traditional craftspeople have continued to use logwood dye for special applications, particularly handmade textiles and high-quality crafts. Additionally, in certain sustainable fashion movements, there has been a resurgence of interest in natural dyes like logwood due to their environmentally friendly characteristics.

Notes

1 'Logwood, The Historic and Standard Black', *American Dyestuff Reporter*, 2:20 (17 June 1918): 1–6.
2 See, for instance, Matthews J. Merritt, *Application of Dyestuffs to Textiles, Paper, Leather and other Materials* (New York: Wiley, 1920); Edward S. Chapin, 'National Dyestuffs: A Revival of Long Abandoned Methods', *Scientific American* 2132 (11 November 1916): 318–19.
3 Joseph M. Gilbert, 'British Loggers and Spanish Governors: The Logwood Trade and Its Settlements in the Yucatan Peninsula: Part II', *Caribbean Studies* 15, no. 4 (1976): 43–52.
4 Alan K. Craig, 'Logwood as a Factor in the Settlement of British Honduras', *Caribbean Studies* 9, no. 1 (1969): 53–62.
5 *Reports by the Juries. Exhibition of the Works of Industry of all Nations* (London: Spicer Bros., 1852): 85–6 (logwood imports for the years 1848–1850).
6 Carsten Reinhardt and Anthony S. Travis, *Heinrich Caro and the Creation of Modern Chemical Industry* (New York: Springer, 2000), 57–9.
7 Anthony S. Travis, 'From Manchester to Massachusetts via Mulhouse: The Transatlantic voyage of Aniline Black', *Technology and Culture* 35, no. 1 (1994): 70–99.
8 K. G. Ponting, 'Logwood: An Interesting Dye', *Journal of European Economic History* 2, no. 1 (1973): 109–19.
9 Joris Mercelis, 'Corporate Secrecy and Intellectual Property in the Chemical Industry through a Transatlantic Lens, *c.*1860–1930', *Entreprises et Histoire* 82 (2016): 32–46.
10 For research and patent activity in the chemical industry, including natural and artificial dyes, between 1820 and 1975, see Vivien Walsh, 'Invention and Innovation in the Chemical Industry: Demand-Pull or Discovery-Push?', *Research Policy* 13 (1984): 211–34.
11 Ronald S. Horsfall and L. G. Lawrie, *The Dyeing of Textile Fibres* (New York: D. van Nostrand Company Inc., 1927). See also 'American Dyes: Can they compete with Foreign Products?', *New York Times* (15 September 1918).
12 Alexander Engel, 'Colouring Markets: The Industrial Transformation of the Dyestuff Business Revisited', *Business History* 54, no. 1 (2012): 10–29.
13 See the chapter by Anthony Goebel McDermott and Ronny J. Viales Hurtado in the present volume.
14 Berthold Wuth, 'Substitutes for Indigo, Aniline Black, Logwood & c.', *Journal of the Society of Dyers and Colourists* 25, no. 4 (1909). See also, Reinhart and Travis, *Heinrich Caro*, 59–61, 96; Travis, 'From Manchester', 74.
15 Diego Tamburini et al., 'Exploring the Transition from Natural to Synthetic Dyes in the Production of 19th-century Central Asian Ikat Textiles', *Heritage Science* 8, no. 114 (2020): 24.
16 William Morris, 'The Art of Dyeing', *The Decorator and Furnisher* 19, no. 6 (March 1892): 217–18.
17 Edward S. Chapin, 'The Uses of Natural Dyestuffs', *Textile Colorist* 37, no. 433–437 (1915): 97–101 and 'The True Value of the Natural Dyes', *Textile Colorist* 37, no. 442 (1915): 273–5.
18 Edward S. Chapin, 'Natural Dyestuffs: An Important Factor in the Dyestuff Situation', *The Journal of Industrial and Engineering Chemistry* 10 (1918): 795–8.
19 Horsfall and Lawrie, 'The Dyeing', 218–19.

20 M. Titford, 'The Long History of Hematoxylin', *Biotechnic & Histochemistry* 80, no. 2 (2005): 3–78.
21 David Pretel, 'Reacciones en cadena: cambio tecnológico global y frontera forestal en la península de Yucatán (*c.* 1850 - 1950)', *Historia Mexicana* 70, no. 1 (2020): 259–311.
22 British Parliamentary Papers, 'Diplomatic and Consular Reports. Germany. Report on Chemical Instruction in Germany and the Growth and Present Condition of the German Chemical Industry', Microfiche no. 107.719-731 (1901): 60.
23 'Development of the Dyewood Industry in America', *American Dyestuff Reporter* 2, no. 13 (29 April 1918): 21–3.
24 Pascale Villegas and Rosa Torras, 'La extracción y exportación del palo de tinte a manos de colonos extranjeros: El caso de la B. Anizan y Cía', *Secuencia* 90 (2014): 79–93.
25 Martha Villalobos, 'Las concesiones forestales en Quintana Roo a fines del Porfiriato', *Relaciones* XIX, no. 53 (1993): 87–112.
26 *Transactions of the National Association of Cotton Manufacturers* (Boston, 28–29 April 1915): 410–12. See also Engel, 'Colouring Markets', 20–1, and Travis, 'From Manchester', 70–99.
27 Emilio Noelting, *Aniline Black and its Applications in Dyeing and Printing* (London: Heywood, 1909), 150.
28 Von Georgievics, *The Chemical Technology of Textile Fibres* (London: Scott, Greenwood, 1902).
29 Samuel P. Sadtler, *A Handbook of Industrial Organic Chemistry* (Philadelphia: J. B. Lippincott Company, 1906), 461–8.
30 Daniel Morris, *The Colony of British Honduras: Its Resources and Prospects with Particular Reference to its Indigenous Plants and Economic Productions* (London: E. Stanford, 1883), 48–9, 62–3, 90–2 and 117–18.
31 Morris, *The Colony*, 49.
32 P. W. Carleton, *Prices of Natural Dyestuffs and Tanning Chemicals* (Washington: G.P.O., 1919).
33 Melchor Campos García and Carlos Leyva Morales, 'Talar y exportar. La depredación del palo de tinte en la península de Yucatán, 1845–1917', *Revista De Economía* 40, no. 101 (2023): 59–91.
34 Pretel, *Reacciones en cadena*, 280.
35 British Parliamentary Papers, 'Report of the Committee on agricultural research and administration in the non-self-governing dependencies', Microfiche no. 135.73 (1926): 79.
36 John A. Burdon, *Brief Sketch of British Honduras: Past, Present and Future* (London: The West India Committee, 1928), 40–1.
37 British Parliamentary Papers, 'Colonial Reports-Annual. No. 373. Jamaica', Microfiche no. 108.565 (1902): 34.
38 A. Bellande, 'Haïti dans le marché mondial du bois aux 19ème et 20ème siècles', *Journal of Haitian Studies* 22, no. 1 (2016): 130–46.
39 W. W. Cumberland, E. A Colson and J. S. Stanley, *Annual Report of the Financial Adviser-General Receiver for the Fiscal Year October 1926–September 1927* (New York: Evening Post Job Printing Office, Inc., 1928).
40 'Jamaica's Dyewood Industry: The Island's Valuable Exports of Logwood and Fustic Products', *Fibre & Fabric*, 46 (1907). See also 'Logwood Industry in Jamaica', *Journal of the Royal Society of Arts* 70, no. 3632 (30 June 1922): 584.

41 Tom Gill, *Tropical Forests of the Caribbean* (Baltimore: Read-Taylor, 1931), 259.

42 Han Neevel, 'Logwood Writing Inks: History, Production, Forensics, and Use', *Restaurator. International Journal for the Preservation of Library and Archival Material* 42, no. 4 (2021):169–91.

43 H. J. Conn, 'Standardized Nomenclature of Biological Stains', *Science* 54, no. 1487 (1929): 743–6; H. J. Conn, 'The History of Staining Logwood Dyes', *Stain Technology* 4, no. 2 (1929): 37–48; C. Ortiz-Hidalgo and S. Pina-Oviedo, 'Hematoxylin: Mesoamerica's Gift to Histopathology. Palo de Campeche (Logwood Tree), Pirates' Most Desired Treasure, and Irreplaceable Tissue Stain', *International Journal of Surgical Pathology* 27, no. 1 (2019): 4–14.

44 R. T. Allison, 'Haematoxylin from the Wood', *Journal of Clinical Pathology* 52, no. 7 (1999): 527–8.

45 Agustí Nieto-Galán, *Colouring Textiles: A History of Natural Dyestuffs in Industrial Europe* (Boston: Kluwer Academic Publishers, 2001), chapter 6.

46 Rita J. Adrosko, *Natural Dyes in the United States* (Washington: Smithsonian Institution Press, 1968), 47.

13

Epilogue

Dominique Cardon

Scientific research being a never-ending story, this epilogue cannot be only a conclusion. It certainly is the place to highlight how the diversity of approaches and case studies offered in this book contributes to a wider understanding of the huge importance of colour, of the discovery of colourant sources, and of the development of colouring technology in humankind's history. Such major issues are illustrated here from a viewpoint that had not yet received all the attention it deserves, that is, the production and trade of natural dyes from the Americas. Vast zones of the subject still remain to be explored to expand the progress made and to further highlight the 'complex interplay between local practices and global exchanges', characteristic of the dye sources of prime importance examined in this book. Additionally, since recurring analyses in this book reveal the disastrous impact that the massive exploitation and production of some dye plants have had on both the natural and the social environments of too many parts of the Americas, it is indispensable to discuss these issues in the present context of global revival of the production and uses of natural colourants. There fortunately exist examples today to demonstrate that natural dyes from the Americas, and not only those discussed in this book, present great potentials for a cultural revival and for economic and social developments, in an alternative approach of regenerative economy.

An interesting approach to further delve into the new insights provided by this book regarding the exploitation of several major American dye plants would be to try and follow them to their final destination; that is, to look at them from the point of view of their most important consumers, the European professional dyers. More specifically, as a first case study, from the point of view of the dyers of a type of fine wool broadcloths massively exported in the late seventeenth and eighteenth centuries from the province of Languedoc, in the south of France, to the Eastern Mediterranean and beyond, to all parts of the Ottoman Empire. Preceding the Industrial Revolution of the nineteenth century linked with the development of a massive cotton-weaving industry in Europe, wool broadcloth production was the most important textile industry at the time when the first European explorers arrived to different parts of the continent that became known as America. The wool broadcloth production of Languedoc, for which an exceptional wealth of precise quantitative data is preserved, provides a typical example of industrial-scale textile production specially developed for exportation in an already global economy. The region produced a median number of 60,000 cloth pieces per year during the last three quarters of the eighteenth century, with peaks of 100,000 up to 130,000 pieces in the 1760–70s.[1] Each piece of cloth measured around 19 metres in length, 1.40 metres in width, for a weight of a little more than 10 kilos.[2] They were all dyed in a vast array of colours directly on the cloth, answering the tastes of their Oriental customers. Dyeing 60,000 pieces, amounting to 11,400 kilometres and 600 metric tons of fabric, obviously required a huge and regular supply of dyestuffs. By an extraordinary luck, several dyers' books have also been preserved, with precise recipes illustrated by dyed samples of the very type of broadcloth that was exported.[3] Additionally, a number of preserved sample sheets represent the orders sent from the Levant for various assortments of colours to be arranged in the cloth bales, following the evolution of colour fashions in the customers' countries. Combined together, all these documents allow to figure out plausible orders of magnitude for the consumption of the main dyestuffs then imported from the Americas.

To examine them, I shall follow the same order as the European wool dyers of the past. They recognized five 'primary' or 'matrix' colours: blue, red, yellow, fawn and black.[4] This classification reflects their empirical understanding of the chemical similarity of the colourants obtained from various groups of plants and invertebrate animals, and their impressive practical knowledge of the extraction and dyeing techniques best suited to each category.

Blue. In this book, it is represented by the indigo pigment obtained from the cultivation of several species of indigo bushes (*Indigofera* spp.): at least two indigenous ones, *I. suffruticosa* Mill. and *I. micheliana* Rose, one or both of which were called *xiuhquilitl* in Nahuatl, the language of the Aztecs; and at least one introduced by European colonizers: *I.tinctoria L*. From this book, one learns a lot on the chronology and organization of the development at large scale of these crops and of the indigo factories in which the leaves were transformed through a complex biochemical process into dark blue, dry light blocks of compacted indigo pigment, easily commercialized at a good price, making it the ideal commodity for transatlantic travels. It now emerges from this book that from the sixteenth century onwards, attempts at producing indigo commercially were made in many regions of the Caribbean and of Central and South America which had not – or still have not – been documented exhaustively enough: Yucatan, as mentioned in Chapter 3, Venezuela, as discussed in Chapter 8, Guatemala and, of course, El Salvador, where excellent indigo is still produced commercially nowadays. In the Caribbean, Tristan Yvon's extensive archaeological and documentary research has so far resulted in the discovery of as many as 125 indigo factories active on the islands of Guadeloupe at the end of the seventeenth century, before the bulk of indigo production for exportation to France moved to the island of Saint-Domingue.[5] Thanks to the typical architecture of these factories, now well known, and mentioned in Chapter 9 as having been adopted in the indigo plantations of South Carolina, future archaeological research should help identify remains of many more indigo factories, in more regions of former production, besides the sites where they are still being used, as is the case on some estates in El Salvador.[6] This new vision of a much wider diffusion of indigo production in the Americas than previously realized, has an interesting parallel in the results of recent historical research into woad production. It now also emerges that woad was being produced at different scales in many more regions of Europe than the main, routinely mentioned ones.[7] What still needs to also be better understood generally, is that the production of woad did not at all disappear when the use of exotic indigo was allowed for wool dyers, at different dates in the different European regions of industrial-scale textile production. On the contrary, the production and uses of these two plant sources of indigo developed together for nearly three more centuries. This could happen because the standard process for indigo dyeing of wool and woollen textiles was the woad-and-indigo vat, a bacterial reduction process in which couched (twice fermented) woad acts not only as a comparatively weak source of blue indigo colourant, but more importantly as fermenting and reducing agent, very efficient in reducing indigo to the soluble form in which it can impregnate the wool fibres to be dyed. Indeed, woad is such a rich source of bacterial flora with a high reduction potential, that the process went on being practised at industrial scale in the main European centres of wool broadcloth production until the beginning of the twentieth century, its final demise only being due to the general adoption of synthetic indigo.[8] The process evolved in time, to allow more and more successive additions of indigo imported from tropical countries, in ever bigger proportions, to the vats initially set with couched woad. In eighteenth-century Languedoc, the vats, of a traditional volume of 4,500 litres, were set with 298 kilos of couched woad; 6.624 kilos of indigo pigment were added some hours later, as soon as the reducing power of the vat liquor permitted. According to Antoine Janot, one of the master dyers who described the process from first-hand experience and illustrated it with samples of dyed cloth, one vat set in this way allowed to dye forty cloth pieces blue, of all the gradations from the darkest, nearly black Persian blue to a range of very light off-white blues, as the colouring power of the vat was getting exhausted.[9] The substantial series of sample sheets and shipping lists preserved for the years 1748 to 1775 show that an average of 60 per cent of the pieces were dyed blue of various degrees, either to remain blue or to be dyed one of the colours based on a ground of blue, namely various degrees of the gradations of purples and greens, a range of blueish greys, as well as the fastest blacks.[10] This implies that, out of a median production of 60,000 pieces, 36,000 would have been dipped into a woad-and-indigo vat. At the rate of forty pieces dyed blue per vat requiring a first addition of 6.624 kilos of indigo, the average yearly consumption of indigo pigment by the dyers of wool broadcloth of Languedoc may have been in the order of at least 5,962 kilos, or approximately 6 metric tons. From the experience of current producers of organic indigo pigment today, about 60 kilos of indigo (at a concentration of around 30 per cent of indigotin) can be produced per year, per hectare planted in *Indigofera tinctoria*.[11] The yearly consumption of the dyers of wool broadcloth of Languedoc in the eighteenth century may therefore have corresponded to the production obtained from 100 hectares cultivated in indigo plants.

Red. The most important source of fast red exported to Europe from the Americas is discussed in Chapter 5 of this book: it was the domesticated scale insect cochineal, *Dactylopius coccus* (Costa, 1829), bred on the prickly pear, exported from Mexico since the beginning of the sixteenthth century. However, it was only thanks to the invention of the tin mordant around 1630 by Cornelis Drebbel, a Dutch engineer and inventor emigrated to England, that European dyers could begin to imitate the vermilion red of scarlet from kermes (a parasite of the kermes oak) using the newly available red insect dye, which had the huge advantages of being comparatively less costly than kermes and above all, easier to obtain regularly in sufficient quantities. The new mordant consisted in dissolving tin in nitric acid, with a little sal-ammoniac added. What was revolutionary in the process was that it did not only induce a strong chemical bonding between the colourants of cochineal and the wool fibres of the broadcloths, but it also changed the natural purplish red colour of cochineal, producing a range of more vivid, orange-red shades.[12] From then on, depending on the mordanting process used, cochineal could be used to produce the whole range of prestigious reds called the 'strong colours' by dyers: not only crimson and winesoup (or 'wine colours'), intense reds with purple hues fixed onto the fibres by alum mordants, but also the bright scarlet hues obtained by using the tin mordant. This was the beginning of a massive use of cochineal, not only for silk textiles but, even more importantly, for the higher qualities of wool broadcloths. The three nuances of red of the 'strong colours' were the colours most constantly favoured by the Oriental elites who bought the wool broadcloths from Europe. With an aim to win and to go on satisfying this market, the mercantilist policy developed by the king of France Louis XIV and his minister Jean-Baptiste Colbert was implemented by a wealth of incredibly detailed regulations, including the minimum quantity of cochineal to be put into the dyeing of scarlet, crimson or winesoup colours: by a decision of the Council of State in 1713, at least 1 ¾ *poids de marc* pound (0.855 kilos) had to be used per piece of fine broadcloth produced in Languedoc.[13] Compliance with this regulation was ensured by quality controls combining visual comparisons with standard samples of each colour and chemical tests in dubious cases.[14] This allows to evaluate the minimum quantity of cochineal consumed in the centres of broadcloth production of Languedoc with some accuracy. To meet the constant demand from the customers in the Levant, at least four cloth pieces dyed scarlet – or occasionally one of the two other 'strong colours' – were included in each bale of twenty pieces shipped from Marseille. On the basis of the median production of 60,000 pieces per year already proposed, this meant that 3,000 bales were exported, each including four pieces of cloth dyed with 0.855 kilos of cochineal. The minimum amount of cochineal needed by the dyers of Languedoc was therefore in the order of 10,260 kilos per year, only for the 'strong colours'. Many other shades of pink and orange, and the whole gradation of purples dyed on an indigo ground, also used cochineal in lesser quantities, but these are difficult to evaluate because the dyers had devised ways to systematically recycle the cochineal dyebaths until their complete exhaustion. The minimum amount of 10,260 kilos proposed here represented the yield of approximately 30 hectares of cochineal-infested cactus fields. At the rate of more or less 116 dry insects in 1 gram of commercial cochineal, one billion and 190,160 millions of insects had to be collected and dried to meet the demand of the Languedocian dyers of fine wool broadcloths.[15]

Yellow, fawn and black. For the purpose of evaluating the impact of colourants early exported from the Americas on the practice of the eighteenth- century dyers of wool broadcloths of Languedoc, the remaining three 'matrix' colours can be examined together. One reason is that the only American dyestuffs mentioned in the dye processes for these three colour groups are all dyewoods. Another common point is that at the time, they are still used as minor ingredients. Moreover, contrary to what could be done concerning indigo and cochineal, it is not possible to provide plausible evaluations of how many cloth pieces were dyed in the various colours requiring their use, because such colours only appear sporadically in the shipping lists and sample sheets reflecting the evolution of colour tastes in the Levant. The only American source of yellow mentioned in the recipes is the heartwood of *Maclura tinctoria* (L.) D. Don ex Steud., the tree called *palo de mora* (mulberry wood) in Spanish, old fustic in English, *bois jaune* (yellow wood) or *mûrier des teinturiers* (dyers' mulberry) in French. It does not appear in recipes before the middle of the eighteenth century. The dyers of Languedoc did not need it for their ranges of fast yellows and greens because these were traditionally based on locally available herbs or bushes, cultivated or collected in their natural environment. Consequently, old fustic only appears in connection with a new range of colours, the Saxon greens (*verts de Saxe*), in which it is combined in different proportions with Saxon blue, a new dye invented in 1748, consisting in a dissolution of indigo in sulphuric acid. The new greens, in which old fustic was used in quantities of up to 7.5 kilo per piece of cloth, were experimented by dyers in Languedoc around the years 1751–2 and one or two pieces of Saxon green or Saxon celadon were included in some bales and exported to the Levant in the following years. However, their poor lightfastness caused the fashion for these greens to be short-lived. During the second half of the eighteenth century, old fustic is used by the dyers of fine broadcloths either in very small quantities as a component in some complex recipes of fawn colours, or in bigger proportions in some recipes of fast blacks. The matrix colour called *fauve* (fawn) by French dyers included a wide range of beiges and browns mainly obtained from tannin-rich plants locally available, or from the Aleppo galls imported from Syria. But their richly varied shades and nuances were achieved by additions of subtle combinations of yellowish, reddish and purplish colourants: for these particular uses, exotic dyewoods were adopted as soon as regulations permitted. Old fustic figured in recipes for some nuances of *noisette* (hazelnut) and *tabac* (tobacco). The more reddish nuances of these same colours and some purplish-brown reds required the use of the red dyewoods commonly referred to as 'brazilwoods', mentioned in different chapters of the present book: not only the heartwood of *Paubrasilia echinata* (Lam.) Gagnon, H.C. Lima & G.P. Lewis, massively exported from Brazil; but also those of two smaller trees, containing similar colourants in different proportions: brasiletto, called *brésillet* in French, *Caesalpinia brasiliensis* Sw., exported from the Caribbean, and peachwood, *Haematoxylum brasiletto* H. Karst., called *bois de Sainte Marthe* in French, exported from Colombia around Santa Marta and from Venezuela.[16] In the recipes, however, they only figure in proportions of around 1–4 per cent of the weight of the cloth pieces – that is in quantities of 100 grams to 0.414 kilos, and the colours for which they are used are seldom included in the assortments ordered from the Levant. Logwood also was used as a minor component in recipes for various nuances of fawns, greys and dark greens. It only figured in more substantial amounts in various recipes of fast blacks, including the prestigious Sedan black, first elaborated for high qualities of dense broadcloth produced at the royal Manufacture of Sedan, in the north of France, since 1646. These new blacks were based, like most fast blacks since the medieval ages, on grounds of dark indigo blue completed by 'galling baths' of tannin-rich plants and an addition of ferrous sulphate – copperas – to produce a black ferro-gallic precipitate on the textile. The new additional use of logwood, in quantities of up to 2.9

kilosg per piece of cloth, with also 1.24 kilos of old fustic added, brought so much colour that it became possible to substantially save on the degree of blue of the indigo ground, while achieving an impressive depth of black.[17] Documents in the archives of Languedoc show that black cloths dyed in this way were appreciated in the Levant.[18] Unfortunately, since they seem to have been shipped separately, it is not possible to evaluate how many were dyed each year and, hence, the quantities of logwood and old fustic consumed.

Of course, what precedes only gives a glimpse of the dyestuffs consumption corresponding to one category of textile production. Even so, this case study comes as a necessary complement to the previous chapters of this book, to provide a useful ground for reflection, in the present global context of renewed interest in natural sources of colourants and at the same time, of questioning about the sustainability of their uses.

This history of the exploitation of some major American plant sources of colourants tragically abounds in descriptions of destructions of natural environments and of the indigenous societies who inhabited them and had discovered the colouring power of the plants. But another approach is now being explored, in different regions of the Americas. It consists in valorizing the rich diversity of colouring resources discovered by indigenous civilizations, focusing on those that can be produced in a sustainable way. This is the case of annatto, the seeds of the bush or small tree *Bixa orellana* L., whose Tupi name, *uruku* (red colour), well proves its core importance in the civilizations of many Amazonian people who use it ritually as body paint. Now largely cultivated and used internationally as food and cosmetic colourant, annatto was used as a textile dye in Europe since the sixteenth century, giving beautiful shades of fiery oranges to silk and printed cotton.[19] Another interesting example is the liana *chica* or *carajuru*, *Fridericia chica* (Bonpl.) L.G. Lohmann, which has been identified as the source of deep reds in pre-Columbian archaeological textiles from Peru and more recently from Muisca, Guane and U'wa sites in the Eastern Ranges of Colombia.[20] It is now cultivated sustainably in Brazil as a source of red colourants for the Natura Company, a leading brand in the cosmetics market and the first listed B Corp company in Latin America. It is also cultivated in Costa Rica at the female-led agro-ecological farming cooperative Coopecuna to produce colourant extracts, developed through the expertise of the Cenibiot Laboratory of biotechnologies in San José.[21] In Guadeloupe, indigo production is being revived in a vision of regenerative economics. The blue indigo pigment from *Indigofera suffruticosa* is produced by the local company Phytobôkaz from organic and symbiotic crops combining plants with medicinal, cosmetic and colouring properties. Its concentration in indigotin is exceptional: 50 per cent minimum, up to 80 per cent – the highest ever recorded in a natural indigo pigment.[22]

Many other similar projects are currently emerging in various parts of North, Central and South America: the history of the natural dyes of the Americas is indeed a never-ending story.

Notes

1 Jean-Michel Minovez, *La Puissance du Midi* (Rennes: Presses Universitaires, 2012), 123.

2 Dominique Cardon, *The Dyer's Handbook – Memoirs of an 18th century Master Colourist* (Oxford and Philadelphia: Oxbow Books, 2016), 15 note 14.

3 Cardon, *The Dyer's Handbook*; Dominique Cardon, *Des Couleurs pour les Lumières – Vie et Oeuvres d'Antoine Janot (1700–1778)* (Paris: CNRS Editions, 2019); Dominique Cardon and Iris Brémaud, *Le Cahier de Couleurs d'Antoine Janot / Workbook, Antoine Janot's Colours*, bilingual edition (Paris: CNRS Editions, 2020); Dominique Cardon and Iris Brémaud, *Les 157 Couleurs de Paul Gout/Paul Gout's 157 Colours*, bilingual edition (Mercin: Vieilles Racines et Jeunes Pousses, 2022).
4 Cardon and Brémaud, *Paul Gout's 157 Colours*, 32.
5 Tristan Yvon, *La production d'indigo en Guadeloupe et Martinique (XVIIe – XIXe siècles), Histoire et archéologie* (Paris: Éditions Karthala, 2015).
6 Illustrated in Dominique Cardon, *Natural Dyes - Sources, Tradition, Technology and Science* (London: Archetype Books, 2007), 366.
7 Dominique Cardon, *Le Monde des teintures naturelles*, 2nd augmented ed. (Paris: Belin, 2014), 366–8; Dominique Cardon, 'Le pastel en Languedoc: une histoire qui ne finit jamais', in *Retour au Pays de cocagne – Nouvelles perspectives sur l'histoire du pastel languedocien (XIIIe-XVIIIe siècle)*, ed. Judicaël Petrowiste, Annales du Midi, t. 132 (Toulouse: Privat, 2020), 69–88.
8 Cardon and Brémaud, *Paul Gout's 157 Colour*, 37–44; Dominique Cardon, Zvi Koren, and Hisako Sumi, 'Woaded Blue: A Colorful Approach to the Dialectic between Written Historical Sources, Experimental Archaeology, Chromatographic Analyses, and Biochemical Research', Heritage 6, no. 1 (2023): 681–704. https://doi.org/10.3390/heritage6010037.
9 Cardon, *Des couleurs pour les Lumières*, 100.
10 Shipping lists and sample sheets preserved in the Archives départementales de l'Hérault in Montpellier, refs. C/2160, C/2241, C/2514, C/5550; Archives départementales de l'Aude in Carcassonne, ref. 3 J 342.
11 Data obtained from NCC Agro Industries, Villupuram, Tamil Nadu, https://www.nccagroindustries.com/, communicated by Hisako Sumi, https://north-indigo.com.
12 Cardon and Brémaud, *Paul Gout's 157 Colours*, 52–65.
13 *Arrêt du Conseil d'Etat* of 18 April 1713, Archives départementales de l'Aude in Carcassonne, ref. 9 C 11.
14 A 'matrix' or standard sample for scarlet is illustrated in Cardon and Brémaud, *Paul Gout's 157 Colours*, 57.
15 Calculations based on the data kindly communicated by Lorenzo Perez, third-generation producer of cochineal in Gran Canaria, director of Canaturex, https://www.canaturex.com/: 3,000 cactus plants can be cultivated on 1 hectare of land. Each cactus must be two and a half to three years old before it can be infested and can produce from 350 to 400 grams of live cochineals. Hence from 1 hectare of nopaleria, a maximum of 1,200 kilos of live cochineals can be collected. From 3.5 kilos of live cochineals, 1 kilo of dry cochineal can be obtained.
16 Cardon, *The Dyer's Handbook*, 47.
17 The process was reproduced in June and August 2023, during workshops aimed at testing the recipes published in Cardon and Brémaud, *Paul Gout's 157 Colours*. The samples of black wool broadcloth produced during these workshops were analysed with a spectrophotometer and were found identical to the original sample of black in Paul Gout's manuscript, with a CIELab Delta E colour difference of less than 1, which means that they cannot be distinguished by a human eye. Furthermore, to our delight, lightfastness tests performed according to the standard suntest methodology gave a fastness of 8 on the Blue Wool Scale, which is the highest possible rating.

18 Cardon, *The Dyer's Handbook*, 138–41.
19 Illustrated in Cardon, *Natural Dyes*, 315–16.
20 Illustrated in Cardon, *Natural Dyes*, 262.
21 Rafael Ocampo, TRAMIL programme, personal communication.
22 https://www.green-ingredients.com/produit/indigo-suffruticosa-de-guadeloupe/; https://www.phytobokaz.fr/indigo-de-guadeloupe-1.

Selected Bibliography

Alden, Dauril. 'The Growth and Decline of Indigo Production in Colonial Brazil: A Study in Comparative Economic History'. *The Journal of Economic History* 25, no. 1 (1965): 35–60.

Appadurai, Arjun, ed. *The Social Life of Things: Commodities in Cultural Perspective*, Cambridge: Cambridge University Press, 1986.

Avíla Blomberg, Alejandro and Carlos Sánchez Silva, eds. *La grana y el nopal en los textos de Alzate*, Mexico: Archivo General de la Nación, Conaculta, 2005.

Balfour-Paul, Jenny. *Indigo*, London: Archetype, 2006.

Balfour-Paul, Jenny. *Indigo in the Arab World*, London and New York: Routledge, 2016.

Baskes, Jeremy. *Indians, Merchants and Markets: A Reinterpretation of the Repartimiento and Spanish Indian Economic Relations in Colonial Oaxaca, 1750–1821*, Stanford: Stanford University Press, 2001.

Bechtold, Thomas and Rita Mussak. *Handbook of Natural Colorants*, Chichester: Wiley, 2009.

Beckert, Sven, Ulbe Bosma, Mindi Schneider, and Eric Vanhaute. 'Commodity Frontiers and the Transformation of the Global Countryside: A Research Agenda'. *Journal of Global History* 16, no. 3 (2021): 435–50.

Berthe, Jean-Pierre. 'El cultivo del pastel en Nueva España'. *Historia Mexicana* 9, no. 3 (1960): 1–28.

Biggam, Carole P. and Kirsten Wolf, eds. *A Cultural History of Color*, Volumes 1–6, London: Bloomsbury, 2021.

Blaszczyk, Regina L. and Uwe Spiekermann, eds. *Bright Modernity. Color, Commerce, and Consumer Culture*, New York: Palgrave Macmillan, 2017.

Bristol, Joan. 'A Trail of Precious Goods: Colonial Latin American Commodity History'. *History Compass* 11, no. 11 (2013): 948–56.

Brunello, Franco. *The Art of Dyeing in the History of Mankind*, Vicenza: Neri Pozza, 1973.

Brusatin, Manlio. *Histoire des Couleurs*, Paris: Flammarion, 1986.

Butler, Amy. *A Perfect Red: Empire, Espionage and the Quest for the Color of Desire*, New York: Harper Collins, 2006.

Campos García, Melchor and Carlos Leyva Morales. 'Talar y exportar. La depredación del palo de tinte en la península de Yucatán, 1845–1917'. *Revista de Economía* 40, no. 101 (2023): 59–91.

Cardon, Dominique. *Le monde des teintures naturelles*, Paris: Belin, 2003.

Cardon, Dominique. *Mémoires de teinture. Voyage dans le temps chez un maître des couleurs*, Paris: CNRS Éditions, 2013.

Cardon, Dominique. 'Natural Dyes, Our Global Heritage of Colors'. *Textile Society of America Symposium Proceedings*, 2010.

Cardon, Dominique. *Natural Dyes: Sources, Tradition, Technology and Science*, London: Archetype, 2007.

Cardon, Dominique and Gaëtan du Chatenet. *Guide des teintures naturelles: Plantes, lichens, champignons, mollusques et insectes*, Paris: Delachaux et Niestlé, 1990.

Contreras, Alicia. *Capital comercial y colorantes en la Nueva España, segunda mitad del siglo xviii*, Mérida: Universidad Autónoma de Yucatán and El Colegio de Michoacán, 1996.

Contreras, Alicia. 'El palo de tinte, motivo de un conflicto entre dos naciones, 1670–1802'. *Historia Mexicana* 37, no. 1 (1987): 49–74.
Contreras, Alicia. *Historia de una tinórea olvidada: El proceso de explotación y circulación del palo de tinte, 1750–1807*, Mérida: Universidad Autónoma de Yucatán, 1990.
Coon, David L. 'Eliza Lucas Pinckney and the Reintroduction of Indigo Culture in South Carolina'. *Journal of Southern History* XLII, no. 1 (1976): 69–75.
Craig, Alan K. 'Logwood as a Factor in the Settlement of British Honduras'. *Caribbean Studies* 9, no. 1 (1969): 53–62.
Cromwell, Jesse. 'Life on the Margins: (Ex) Buccaneers and Spanish Subjects on the Campeche Logwood Periphery, 1660–1716'. *Itinerario* 33, no. 3 (2009): 43–71.
Dean, Warren. *With Broadax and Firebrand. The Destruction of the Brazilian Atlantic Forest*, Berkeley: University of California Press, 1995.
Delamare, François and Bernard Guineau. *Colour: Making and Using Dyes and Pigments*, London: Thames and Hudson, 2000.
Eastsugh, Nicholas, et al. *The Pigment Compendium. A Dictionary of Historical Pigments*, Amsterdam: Elsevier, 2004.
Edelson, Max S. *Plantation Enterprise in Colonial South Carolina*, Cambridge: MA and London: Harvard University Press, 2006.
Engel, Alexander. 'Colouring Markets: The Industrial Transformation of the Dyestuff Business Revisited'. *Business History* 54, no. 1 (2012): 10–29.
Fairlie, Susan. 'Dyestuffs of the Eighteenth Century'. *The Economic Historical Review* 17 (1965): 490–2.
Feeser, Andrea. *Red, White, and Black Make Blue*, Athens: University of Georgia Press, 2013.
Feeser, Andrea, Maureen D. Goggin, and Beth K. Tobin, eds. *The Materiality of Colour: The Production, Circulation, and Application of Dyes and Pigments, 1400–1800*, Surrey: Ashgate, 2012.
Fernández Molina, Jose Antonio. *Pintando el mundo de azul: El auge añilero y el mercado centroamericano, 1750–1810*, El Salvador: Biblioteca de Historia Salvadoreña, 2003.
Fox, Robert and Agustí Nieto-Galan, eds. *Natural Dyestuffs and Industrial Culture in Europe, 1750–1880*, Canton, MA: Science History Publications, 1999.
Garfield, Simon. *Mauve. How One Man Invented a Colour that Changed the World*, New York: Canongate Books, 2016.
Gilbert, Joseph M. 'British Loggers and Spanish Governors: The Logwood Trade and its Settlements in the Yucatan Peninsula: Part II'. *Caribbean Studies* 15, no. 4 (1976): 43–52.
Goebel McDermott, Anthony. *Los bosques del progreso. Explotación forestal y régimen ambiental en Costa Rica: 1883–1955*, San José, Costa Rica: Editorial Nuevas Perspectivas, 2013.
Heers, Jacques. 'La búsqueda de colorantes'. *Historia Mexicana* XI, no. 1 (1961): 1–27.
Hopkins, A.G., ed. *Globalization in World History*, London: Pimlico, 2002.
Hopkins, Terence and Immanuel Wallerstein. 'Commodity Chains in the World Economy Prior to 1800'. *Review* 10, no. 1 (1986): 57–70.
Kirby, Jo, et al. *Natural Colorants for Dyeing and Lake Pigments. Practical Recipes and their Historical Sources*, London: Archetype - CHARISMA, 2014.
Kumar, Prakash. *Indigo Plantations and Science in Colonial India*, New York: Cambridge University Press, 2012.
Langue, Frédérique. 'El añil en la Venezuela ilustrada. una historia inconclusa'. *Revista de Indias* 58, no. 214 (1998): 637–53.

Marichal, Carlos. 'Mexican Cochineal and the European Demand for American Dyes, 1550–1850'. In *From Silver to Cocaine: Latin American Commodity Chains and the Buiding of the World Economy, 1500–2000*, edited by Steven Topik, Carlos Marichal, and Zephyr Frank, 76–92. Durham and London: Duke University Press, 2006.

McCreery, David. 'Indigo Commodity Chains in Spanish and British Empires, 1560–1860'. In *From Silver to Cocaine: Latin American Commodity Chains and the Building of a World Economy, 1500–2000*, edited by Zephyr Topik, Carlos Marichal, and Steven Topik, 62–3. Durham and London: Duke University Press, 2006.

McKinley, Catherine. *Indigo: In Search of the Colour that Seduced the World*, London: Bloomsbury, 2011.

Miller, Shawn William. *Fruitless Trees. Portuguese Conservation and Brazil's Colonial Timber*, Stanford: Stanford University Press, 2000.

Nadri, Ghulam. *The Political Economy of Indigo in India, 1580–1930: A Global Perspective*, Leiden: Brill, 2016.

Newton, Ross A. '"Good and Kind Benefactors": British Logwood Merchants and Boston's Christ Church'. *Early American Studies* 11, no. 1 (2013): 15–36.

Nieto-Galan, Agustí. *Colouring Textiles. A History of Natural Dyestuffs in Industrial Europe*, Dordrecht, Boston and London: Boston Studies in the Philosophy of Science, Kluwer Academic Publishers, 2001.

Offen, Karl H. 'British Logwood Extraction from the Mosquitia: The Origin of a Myth'. *Hispanic American Historical Review* 80, no. 1 (2000): 113–35.

Offen, Karl H. 'English Designs on Central America: Geographic Knowledge and Imaginative Geographies in the Seventeenth Century'. *Early American Studies* 18, no. 4 (2020): 399–460.

Pacheco Troconis, Germán. 'El comercio colonial americano del añil: Guatemala y Venezuela, 1774–1810'. *Boletín Americanista* 61 (2010): 151–75.

Pacheco Troconis, Germán. 'Las haciendas de añil en los valles de Aragua en las últimas décadas del período colonial (1767–1830)'. *Estudios de Historia Social y Ecónomica de América* 11 (1994): 349–64.

Padilla, Carmela and Barbara C. Anderson, eds. *A Red Like No Other One: How Cochineal Colored the World. An Epic Story of Art, Culture, Science and Trade*, New York: Skira Rizzoli, 2015.

Pastoureau, Michel. *Bleu. Histoire d'une couleur*, Paris: Le Seuil, 2000.

Peck, Amelia, et al., eds. *Interwoven Globe: The Worldwide Textile Trade, 1500–1800*, New York: Metropolitan Museum of Art, 2013.

Phipps, Elena. *Cochineal Red: The Art History of a Color*, New York: Metropolitan Museum of Art, 2010.

Phipps, Elena. 'Global Colors: Dyes and the Dye Trade from the Sixteenth to the Eighteenth Century'. In *The Interwoven Globe: Worldwide Textile Trade 1500–1800*, edited by Amelia Peck, 28–45. New York: Metropolitan Museum, 2013.

Phipps, Elena. 'Textile Colors and Colorants in the Andes'. In *Colors between Two Worlds: The Florentine Codex of Bernardino de Sahagún*, edited by Gerhard Wolf, Joseph Connors, and Louis A. Waldman, 257–80. Villa I Tatti, Florence: Harvard University Press, 2012.

Pretel, David. 'Reacciones en cadena: Cambio tecnológico global y frontera forestal en la península de Yucatán (ca. 1850–1950)'. *Historia Mexicana* 70, no. 1 (2020): 259–311.

Roque, Georges. *La cochenille, de la teinture à la peinture. Une histoire matérielle de la couleur*, Paris: Gallimard, 2021.

Roque, Georges, ed. *Rojo mexicano. Coloquio internacional sobre la grana cochinilla en el arte*, Mexico: Museo del Palacio de Bellas Artes, 2017.

Roquero, Ana. 'Materias tintóreas en la época de Felipe II: Introducción en Europa de los Colorantes Indianos'. In *Los Ingenios y las máquinas en la época de Felipe II*, edited by Ignacio González, 262–75. Madrid: Real Jardín Botánico, 1998.

Rubio Sánchez, Manuel. *Historia del cultivo de la grana o cochinilla en Guatemala*, Ciudad de Guatemala: Tipografía Nacional, 1994.

Sánchez Diaz, Gerardo. 'Cultivo, producción y mercado del añil en Michoacán en el siglo XIX'. *Nuestra Historia* 1 (1991): 35–44.

Sarabia Viejo, María Justina. *La grana y el añil. Técnicas tintóreas en México y América Central*, Sevilla: EEHA, 1994.

Siracusano, Gabriela. *Pigments and Power in the Andes: From the Material to the Symbolic in Andean Cultural Practices, 1500–1800*, London: Archetype, 2011.

Smith, Robert S. 'Forced Labor in the Guatemalan Indigo Works'. *The Hispanic American Historical Review* 36, no. 3 (1956): 319–28.

Stubbs, Jean, William G. Clarence-Smith, Jonathan Curry-Machado, and Jelmer Vos, eds. *Oxford Handbook of Commodity History*, Oxford: Oxford University Press, 2024.

Topik, Steven, Carlos Marichal, and Zephyr Frank, eds. *From Silver to Cocaine: Latin American Commodity Chains and the Buiding of the World Economy, 1500–2000*, Durham and London: Duke University Press, 2006.

Torras, Rosa and Pascale Villegas. 'La extracción y exportación de palo de tinte a manos de colonos extranjeros. El caso de la B. Anizan y Cía'. *Secuencia* 90 (2014): 79–93.

Torras, Rosa. 'Palo de tinte y territorialidad en la península de Yucatán, en las postrimerías del periodo colonial novohispano'. *Temas Americanistas* 43 (2019): 231–48.

Trichaud-Buti, Danielle and Gilbert Buti. *Rouge Cochenille, Histoire d'un insecte qui colora le monde: xvie-xxie siècle*, Paris: CNRS Éditions, 2021.

Vadillo López, Claudio. *La región del palo de tinte: El partido del Carmen, Campeche: 1821–1857*, Campeche: Fondo Estatal para la Cultura y Las Artes, 1994.

Viales Hurtado, Ronny and Anthony Goebel McDermott, eds. *Costa Rica: Cuatro Ensayos de Historia Ambiental*, San José, Costa Rica: Sociedad Editora Alquimia, 2011.

Williams, Michael. *Deforesting the Earth: From Prehistory to Global Crisis: An Abridgment*, Chicago: The University of Chicago Press, 2006.

Wilson, Arthur M. 'The Logwood Trade in the Seventeenth and Eighteenth Centuries'. In *Essays in the History of Modern Europe*, edited by Donald C. McKay, 1–15. New York: Harper & Brothers, 1936.

Zabalza Seguín, Ana. 'Añil, grana y vainillas. Los comerciantes navarros en América Central (XVII-XVIII)'. In *Identidad y estructura de la emigración vasca y navarra hacia Iberoamérica siglos (XVI-XXI). Redes sociales y desarrollo socioeconómico*, edited by José Manuel Azcona Pastor, 86–109. España: Thomson Reuters Aranzadi, 2015.

Index

Index

www.ingramcontent.com/pod-product-compliance
Ingram Content Group UK Ltd.
Pitfield, Milton Keynes, MK11 3LW, UK
UKHW020806290926
15470UKWH00003B/976